POLLUTION SCIENCE, TECHNOLOGY AND ABATEMENT

CHEMICAL POLLUTION CONTROL WITH MICROORGANISMS

POLLUTION SCIENCE, TECHNOLOGY AND ABATEMENT

Additional books in this series can be found on Nova's website under the Series tab.

Additional e-books in this series can be found on Nova's website under the eBooks tab.

POLLUTION SCIENCE, TECHNOLOGY AND ABATEMENT

CHEMICAL POLLUTION CONTROL WITH MICROORGANISMS

NASER A. ANJUM
EDITOR

nova
science publishers
New York

Library of Congress Cataloging-in-Publication Data

ISBN: 978-1-53611-034-0

Published by Nova Science Publishers, Inc. † New York

CONTENTS

PREFACE

The desire for a more efficient life coupled with the methods of production and pollution brought about by the Industrial Revolution have degraded the environment. Reports concerning sustainable strategies for the control of pollutants released into the environment are meager at best. Notably, the significance of sustainable/bio-remediation energy using either plants or bacteria has been elucidated recently as a primary method to decontaminate such polluted environments. Through different scholarly manuscripts contributed by eminent researchers and scientists from all over the globe, this edited volume aims to discuss insights into the control of pollutants in environmental sectors with microorganisms. The designing and execution of innovative studies encompassing microorganisms and their role in making our planet free of chemical pollutants can be provoked by the outcomes of the deliberations of scientists and researchers.

Chapter 1 - Unchecked and uncontrolled release of toxic chemicals into varied environmental compartments has become inevitable. That, in turn, has been severely impacting every walk of an organism's life. Hence, exploring sustainable strategies for minimizing the level of toxic chemicals in environmental compartments including soil and water has become one of the major aims of current environmental research. This chapter briefly introduces the volume entitled '*Chemical Pollution Control with Microorganisms*' and also presents an overview of major points discussed in different scholarly manuscripts contributed to the book by eminent researchers and scientists from over the globe.

Chapter 2 - Microbes play a vital role in bioremediation technology because of their potential to degrade, transform or accumulate contaminants. Bacteria and fungi possess good capacity for removing various inorganic and organic contaminants including metals, radionuclides, polychlorinated biphenyls, polyaromatic hydrocarbons, pesticides, dyes from various components of the environment. Enzymes and proteins assist transformation and degradation of contaminants in microbes. The use of genetic engineering technology has improved the potential of microbes for removing inorganic and organic contaminants by overexpressing genes responsible for synthesis of enzymes/proteins involved in removal/transformation/degradation of contaminants. The potential of microbes can thus be exploited for developing large-scale bioremediation technologies for cleaning the environment.

Chapter 3 - Phytoremediation is the use of plants to recover contaminated soils; is a nondestructive and cost-effective *in situ* technology to clean up contaminated soils. In the case of contamination with petroleum hydrocarbons, plants enhance microbial degradation of the contaminant in the rhizosphere. Venezuela is one of the largest oil producers in the world. For the rehabilitation of oil-contaminated sites, phytoremediation, using arbuscular mycorrhizal fungi to remediate soils contaminated with hydrocarbons have high potential and represents a promising technology whereby plants are used to enhance biodegradation processes in soil. The feasibility of using plant-Glomeromycota fungi association in the biorestoration of contaminated soil was studied in an oil pit, located in Anzoategui state, Venezuela. The results and discussion presented in this chapter aim to highlight the role of mycorrhizal symbiosis as a promising tool for phytoremediation of soils contaminated with hydrocarbons; rhizodegradation seemed to be the most feasible mechanism.

Chapter 4 - Covering 75% of the earth surface, water is the most important and vital molecule of our planet. However, the pollution of water (aquatic) system with varied pollutants including trace metals is inevitable due to high industrial growth. Trace metals entering through industrial sources to the aquatic system can be fatal for both aquatic organisms and major organisms (animals and humans) closely associated with on this system. Microorganisms have immense inherent capability to control the trace metal levels in aquatic system. This chapter aims to overview trace metal pollution in aquatic system, and to discuss the role and underlying mechanisms of varied microorganisms-mediated control (sorption, degradation, transformation and metabolism) of major trace metals in aquatic system.

Chapter 5 - The contamination of different environmental compartments due to anthropogenic and natural sources is increasing day by day because of increase in population, industrialization and urbanization. It is imperative and crying need of the time to explore sustainable measures to control the levels and impacts of environmental contaminants. This chapter aims to overview the belowground processes associated with the plant rhizosphere covering the interactions between plants and microbes, their chemical communications, the biological processes they sustain, and the costs and benefits to plants associated with these interactions particularly in phytoremediation.

Chapter 6 - The boom in the development of industries and modernization as a whole has no doubt made our life easier. On the contrary, the environment has been hit hard due to unchecked and uncontrolled release of toxic wastes into the environment. This leads to the accumulation of detrimental wastes to the soil, thereby affecting not only agronomically but also the health of the animals and human beings. Contamination of soil is one of the major environmental issues which need to be addressed by the environmentalists and scientists at the foremost level. The issue can be solved by employing physicochemical remediation treatment methods and the bioremediation technology. Physicochemical remediation of the contaminated soil adopts one of these methods *via* physical dumping, incineration, *in situ* acid treatment, electrokinetic reclamation and soil washing. However, being expensive and disruptive in nature, their effectiveness is questioned economically and impact wise. Bioremediation in contrast, is a low cost, environmental friendly and practical technology utilized in order to reclaim contaminated soil. Rhizospheric dwelling Plant Growth Promoting Rhizobacteria (PGPR) play significant role in providing a base for contaminants in soil and plant to interact during the process of phytoremediation. PGPR releases chelating and bioleaching agents in the rizospheric region during phytoremediation process. They cause acidification, enzymatic reduction and redox changes enhancing the mobility and availability of contaminants to the plant. There arises a necessity to understand and study potential mechanisms underlying the removal and mobilization of contaminants by microbes especially PGPR in order to sustainably restore and reclaim environmental compartments with chemical pollution.

Chapter 7 - Environmental pollution resulting from organic and inorganic pollutants is a widespread problem for soil and water resources. Several of these pollutants are highly persistent and accumulated in soil beyond the permissible limits. They are extremely toxic to soil biology and may be accumulated to higher trophic levels through food chain contamination. Thus, removal of such contaminants from the soil environment is not only mandatory but also a topic of interest nowadays. During the recent years, important role of bioremediation using either plants or bacteria has been elucidated as a primary method to decontaminate such polluted soils.

However, a number of studies also indicate that phytoremediation (bioremediation by plants) can be partnered by rhizospheric or endophytic bacteria. The partnership of bacteria with plants not only improves plant health but also accelerates the remediation process of environmental pollutants. Sometimes bacteria perform this function by improving the bioavailability of pollutants to plants (hyper accumulation of metals) and sometimes by degrading the pollutants into plant nutrients or useable forms. The present book chapter presents an overview of the studies, which have been conducted to evaluate the potential of synergistic plant-bacteria partnerships for remediation of different organic and inorganic pollutants including polycyclic aromatic hydrocarbons, polychlorinated biphenyls, pesticides, explosives, heavy metals, metalloids and radionuclides from the soil environment. Furthermore, this book chapter also reports different factors, which may influence the performance of bacteria-plant partnership in remediation of contaminated sites.

Chapter 8 - In recent years, the ability of hairy root cultures (HRCs) to simulate enzyme based degradation, tolerance and hyperaccumulation of pollutant molecules has been scientifically focused. The versatility and close resemblance with native plants have made HRCs a preferred system for studying various *in planta* physiological and biochemical events during pollutant stress. Additionally, use of HRCs also facilitated the study of catabolic pathways and identification of related enzymes involved in pollutant metabolism. This chapter presents a comprehensive account on various aspects of remediation of environmental pollutants utilizing HRCs of different plant species. In addition, the enzymatic degradation of phenol and its derivatives, polyaromatic hydrocarbons, polychlorinated biphenyl ethers, azodyes, antibiotics etc. have also been focused. Besides, development of up-scale culture system for large scale and incessant production of effective enzymes, genetic engineering for expression of candidate genes for the enhancement of remedial potential and microbe assisted hairy root remediation system has also been discussed as potential prospects for exploring HRCs based phytoremediation.

Chapter 9 - Air, water and soil can be impacted by a myriad of chemical pollutants. Notably, the health of animal and human significantly depends on the quality of air, water and soil. Therefore, the level of chemical pollutants in these compartments must be kept at minimum. The volume entitled '*Chemical Pollution Control with Microorganisms*' made an effort to discuss various aspecpts related with the theme.

In: Chemical Pollution Control … ISBN: 978-1-53611-034-0
Editor: Naser A. Anjum

Chapter 1

CHEMICAL POLLUTION CONTROL WITH MICROORGANISMS: AN INTRODUCTION

Naser A. Anjum*
CESAM-Centre for Environmental and Marine Studies,
Department of Chemistry, University of Aveiro, Aveiro, Portugal

ABSTRACT

Unchecked and uncontrolled release of toxic chemicals into varied environmental compartments has become inevitable. That, in turn, has been severely impacting every walk of an organism's life. Hence, exploring sustainable strategies for minimizing the level of toxic chemicals in environmental compartments including soil and water has become one of the major aims of current environmental research. This chapter briefly introduces the volume entitled '*Chemical Pollution Control with Microorganisms*' and also presents an overview of major points discussed in different scholarly manuscripts contributed to the book by eminent researchers and scientists from over the globe.

Keywords: chemical pollution, microorganisms, introduction

---

* Corresponding Author: Naser A. Anjum, CESAM-Centre for Environmental and Marine Studies, Department of Chemistry, University of Aveiro, 3810-193 Aveiro, Portugal; E-mail: anjum@ua.pt

1. BACKGROUND

A variety of chemical pollutants from a wide range of sources are released into the natural environment, and are negatively affecting the air, water and soil, and also dwelling organisms. Notably, air and water is becoming the house of more than one billion pounds of toxins released therein. According to an estimate, about 1,000 new products are synthesized annually and between 60,000 and 95,000 chemicals are commercially used (Shukla et al., 2010). Immediate, acute effects, as well as chronic effects, may result from long-term exposures to chemicals. Additionally, such chemical poisoning has caused an annual death of about 47000 persons (mostly children and adolescents) (http://www.who.int/ceh/risks/cehchemicals/en/). Therefore, it is imperative to explore sustainable ways for promoting global health through chemical pollution control and making environment safe for life.

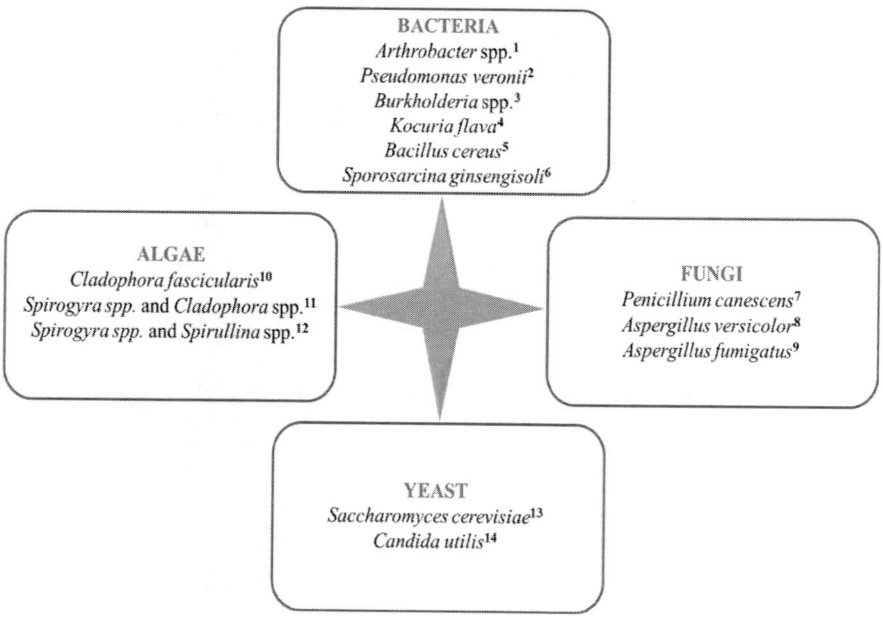

Figure 1.1. Examples of microorganisms studied and strategically used in bioremediation treatments for trace metals ([1]Roane et al., 2001; [2]Vullo et al., 2008; [3]Jiang et al., 2008; [4]Achal et al., 2011; [5]Kanmani et al., 2012; [6]Achal et al., 2012; [7]Say et al., 2003; [8]Tastan et al., 2010; [9]Ramasamy et al., 2011; [10]Deng et al., 2007; [11]Lee and Chang, 2011; [12]Mane and Bhosle, 2012; [13]Machado et al., 2010; [14]Kujan et al., 2006).

Notably, minimizing the loads of chemical pollutants in environmental compartments has been the primary focus of current environmental research in order finally to protect environmental and organism′s health (Anjum et al., 2012). As a sustainable effort the use of microbial community has been considered as a representative huge biological resource with potential for removing chemical pollutants of various types from soil, water, sediments (Gao et al., 2010; Shukla et al., 2010; Fuchs et al., 2011; Durve et al., 2013; Rachitra et al., 2013). For instance, soil bacterial isolates were evidenced to perform the degradation of the organophosphorus insecticide diazinon (Mahiudddin et al., 2014). Bacteria and fungi-mediated degradation/ transformation toxicants is possible through biological and chemical modifications (Rachitra et al., 2013). Microbes can degrade aromatic compounds (Fuchs et al., 2011). Arbuscular mycorrhizal fungal hyphae have been reported to contribute to the uptake of polycyclic aromatic hydrocarbons by plant roots (Gao et al., 2010). Major examples of microorganisms studied and strategically used in bioremediation treatments for trace metals have been summarized in Figure 1.1.

2. Chapter Highlights

Bhupinder Dhir (*Chapter 2*) overviews major chemical pollutants removed by bacteria, and fungi. Mechanisms underlying bacteria and fungi-mediated degradation/transformation chemical pollutant types were also well discussed. The chapter also throws lights on major molecular tools for enhancing remediation capacity. Taking into account a research study, Marcia Toro et al. (*Chapter 3*) discuss the role of arbuscular mycorrhizal fungi in the remediation of soils contaminated with hydrocarbons. The authors studied the feasibility of using plant-Glomeromycota fungi association in the biorestoration of contaminated soil was studied in an oil pit, located in Anzoategui state, Venezuela. Microorganisms have immense inherent capability to control the trace metal levels also in aquatic system. To this end, Singh and Singh (*Chapter 4*) overview the trace metal pollution in aquatic system, and discuss the role and underlying mechanisms of varied microorganisms-mediated control (sorption, degradation, transformation and metabolism) of major trace metals in aquatic system. Under the title ′Biochemical talk in the rhizospheric microbial community for phytoremediation′, Mazhar Rafique et al. (*Chapter 5*) overview the belowground processes associated with the plant rhizosphere covering the interactions between plants and microbes, their chemical communications, and the biological processes they sustain. Additionally, the authors also highlight the costs and benefits to plants associated with the mentioned interactions in phytoremediation. Considering 'Plant-bacteria partnerships' as a major focus, C. Karthik et al. (*Chapter 6*) and Iftikhar Ahmad et al. (*Chapter 7*) discuss in details various aspects of 'Plant-bacteria partnerships' in the remediation of major chemical pollutants (including organic and inorganic pollutants). In the last invited contribution, Mehrotra and Srivastava (*Chapter 8*) summarize the significance of hairy root *in-vitro* systems and consider this system as a suitable biological matrix for plant based remediation of environmental pollutants.

Conclusion

Overall, a conceptual overview of various aspects of chemical pollution control with microorganisms was provided by the chapters set out in the current volume. The outcomes of the deliberations are expected to provoke the designing and execution of innovative studies encompassing microorganisms and their role in making our planet free of chemical pollutants.

References

Achal, V., Pan, X., Fu, Q., Zhang, D. (2012). Biomineralization based remediation of As (III) contaminated soil by *Sporosarcina ginsengisoli. Journal of Hazardous Materials* 201-202, 178-184.

Achal, V., Pan, X., Zhang, D. (2011). Remediation of copper-contaminated soil by *Kocuria flava* CR1, based on microbially induced calcite precipitation. *Ecological Engineering* 37 (10), 1601-1605.

Anjum, N. A., Pereira, M. E., Ahmad, I., Duarte, A. C., Umar, S., Khan, N. A. (2012). *Phytotechnologies: Remediation of Environmental Contaminants*. Boca Raton, FL (USA). CRC Press/Taylor and Francis Group, Boca Raton, FL (USA).

Deng, L., Su, Y., Su, H., Wang, X., Zhu, X. (2007). Sorption and desorption of lead (II) from wastewater by green algae *Cladophora fascicularis*. *Journal of Hazardous Materials* 143 (1-2), 220-225.

Durve, A., Naphade, S., Bhot, M., Varghese, J., Chandra, N. (2013). Quantitative evaluation of heavy metal bioaccumulation by microbes. *Journal of Microbiology and Biotechnological Research* 3 (6), 21-32.

Fuchs, G., Boll, M., Heider, J. (2011). Microbial degradation of aromatic compounds - from one strategy to four. *Nature Reviews Microbiology* 9, 803-816.

Gao, Y. Z., Cheng, Z. X., Ling, W. T., Huang, (2010). Arbuscular mycorrhizal fungal hyphae contribute to the uptake of polycyclic aromatic hydrocarbons by plant roots. *Bioresource Technology* 101, 6895-6901.

Jiang, C. Y., Sheng, X. F., Qian, M., Wang, Q. Y. (2008). Isolation and characterization of heavy metal resistant Burkholderia species from heavy metal contaminated paddy field soil and its potential in promoting plant growth and heavy metal accumulation in metal polluted soil. *Chemosphere* 72, 157-164.

Kanmani, P., Aravind, J., Preston, D. (2012). Remediation of chromium contaminants using bacteria. *International Journal of Environmental Science and Technology* 9, 183-193.

Kujan, P., Prell, A., Safár, H., Sobotka, M., Rezanka, T., Holler, P. (2006). Use of the industrial yeast *Candida utilis* for cadmium sorption. *Folia Microbiologica* 51 (4), 257-260.

Lee, Y. C., Chang, S. P. (2011). The biosorption of heavy metals from aqueous solution by Spirogyra and *Cladophora filamentous* macroalgae. *Bioresource Technology* 102 (9), 5297-5304.

Machado, M. D., Soares, E. V., Soares, H. M. (2010). Removal of heavy metals using a brewer's yeast strain of *Saccharomyces cerevisiae*: chemical speciation as a tool in the prediction and improving of treatment efficiency of real electroplating effluents. *Journal of Hazardous Materials* 180(1-3), 347-353.

Mahiudddin, M., Fakhruddin, N. M., Abdullah-Al-Mahin, M. A. Z., Chowdhury, M. A., Rahman, M. K. (2014). Degradation of the organophosphorus insecticide diazinon by soil bacterial isolate. *The International Journal of Biotechnology* 3(1), 12-23.

Mane, P. C., Bhosle, A. B. (2012). Bioremoval of some metals by living *Algae Spirogyra* sp. and *Spirullina* sp. from aqueous solution. *International Journal of Environmental Research* 6(2), 571-576.

Rachitra, H. R., Sowmya, H. S., Jyothi, V. A. (2013). Biosorption of heavy metals by fungi and bacteria. *Journal of Biotechnology and Biosafety* 1, 37-43.

Ramasamy, R. K., Congeevaram, S., Thamaraiselvi, K. (2011). Evaluation of isolated fungal strain from e-waste recycling facility for effective sorption of toxic heavy metal Pb (II) ions and fungal protein molecular characterization: A Mycoremediation approach. *Asian Journal of Experimental Biological Sciences* 2(2), 342-347.

Roane, T. M., Josephson, K. L., Pepper, I. L. (2001). Dual-bioaugmentation strategy to enhance remediation of cocontaminated soil. *Applied and Environmental Microbiology* 67 (7), 3208-3215.

Say, R., Yimaz, N., Denizli, A. (2003). Removal of heavy metal ions using the fungus *Penicillium canescens*. *Adsorption Science and Technology* 21 (7), 643-650.

Shukla, K. P., Singh, N. K., Sharma, S. (2010). Bioremediation: developments, current practices and perspectives. *Genetic Engineering and Biotechnology Journal* 3, 1-20.

Tastan, B. E., Ertugrul, S., Donmez, G. (2010). Effective bioremoval of reactive dye and heavy metals by *Aspergillus versicolor*. *Bioresource Technology* 101(3), 870-876.

Vullo, D. L., Ceretti, H. M., Daniel, M. A., Ramírez, S. A., Zalts, A. (2008). Cadmium, zinc and copper biosorption mediated by *Pseudomonas veronii* 2E. *Bioresource Technology* 99 (13), 5574-5581.

In: Chemical Pollution Control …
Editor: Naser A. Anjum

ISBN: 978-1-53611-034-0

Chapter 2

BIOREMEDIATION OF ORGANIC/INORGANIC CONTAMINANTS: ROLE OF MICROBIAL GENERA

Bhupinder Dhir*
Department of Genetics, University of Delhi South Campus,
New Delhi, India

ABSTRACT

Microbes play a vital role in bioremediation technology because of their potential to degrade, transform or accumulate contaminants. Bacteria and fungi possess good capacity for removing various inorganic and organic contaminants including metals, radionuclides, polychlorinated biphenyls, polyaromatic hydrocarbons, pesticides, dyes from various components of the environment. Enzymes and proteins assist the transformation and degradation of contaminants in microbes. Use of genetic engineering technology has improved the potential of microbes for removing inorganic and organic contaminants by overexpressing genes responsible for synthesis of enzymes/proteins involved in removal/transformation/degradation of contaminants. The potential of microbes can thus be exploited for developing large-scalebioremediation technologies for cleaning the environment.

---

*Corresponding author: Bhupinder Dhir, Department of Genetics, University of Delhi South Campus, New Delhi-110021, India; E-mail: bhupdhir@yahoo.co.in.

Keywords: organic/inorganic contaminants, bioremediation, microbial genera

1. Introduction

Microbes represent a huge biological resource with potential for removing contaminants from soil, water, and sediments. Bacteria and fungi assist in removing various inorganic and organic contaminants (Tables 2.1 and 2.2). Inorganic compounds include metals, radionuclides, while the organic compounds include hydrocarbons, explosives, pesticides, dyes. Both living and non-living biomass possess possess sites for binding contaminants and hence can be equally exploited for use in bioremediation technology (Rhodes, 2013; Sharma et al., 2013).

Table 2.1. List of major contaminants removed by bacteria

Contaminants	
Inorganic	**Organic**
• Radionuclides • Heavy metals • Nutrients	• Explosives • Monoaromatic hydrocarbons • Polycyclic aromatic hydrocarbons (PAHs) • Polychlorinated biphenyls (PCBs) • Herbicides • Pesticides • Dyes

Table 2.2. List of major contaminants removed by fungi

Contaminants	
Inorganic	**Organic**
• Radionuclides • Heavy metals • Nutrients	• Explosives • Polycyclic aromatic hydrocarbons • Polychlorinated biphenyls (PCBs) • Herbicides • Dyes • Organic solvents • Endocrine disrupting chemicals (EDCs) • Dioxins • Polychlorinated dibenzofurans

2. Mechanism of Removal

Bacteria and fungi degrade/transform toxicants through biological and/or chemical modifications (Rachitra et al., 2013). The mechanism of removal/ degradation of contaminants includes bioaccumulation, biosorption or adsorption (Figure 2.1). Biosorption involves the use of living cells through a metabolism dependent process. The rate of degradation or transformation of contaminant depends upon the species, type of adsorbate, and physical conditions (Gelagutashvili, 2013). Several enzymes degrade contaminants aerobically or anaerobically (Table 2.3). Aerobic degradation involves introduction of oxygen atoms and is mediated by enzymes such as monooxygenases, dioxygenases, hydroxylases and oxidative dehalogenases. Anaerobic degradation involves activation reactions followed by oxidative catabolism mediated by electron acceptors. Degradation of most of the organic contaminants (such as pesticides, hydrocarbons and aromatic compounds) occurs effectively under aerobic conditions.

2.1. Remediation of Different Contaminants

2.1.1. Inorganic

Metals, nutrients and radionuclides are the major inorganic contaminants removed by microbes.

Table 2.3. Enzymes involved in degradation of various contaminants

Name of the enzyme	Mechanism	Compounds degraded
Monooxygenases	Catalyses of oxidative reactions of substrates	Alkanes, steroids and fatty acids
Dioxygenases	Introduction of molecular oxygen into their substrate	Metals and aliphatic compounds
Oxidoreductases	Assistance in the transfer of electrons from a reduced organic substrate (donor) to another chemical compound (acceptor)	Phenolic or anilinic compounds
Laccases	Oxidation	Phenolic and aromatic substrates
Peroxidases	Oxidation	Lignin
Lignin Peroxidases	Oxidation	Lignin, phenolic compounds

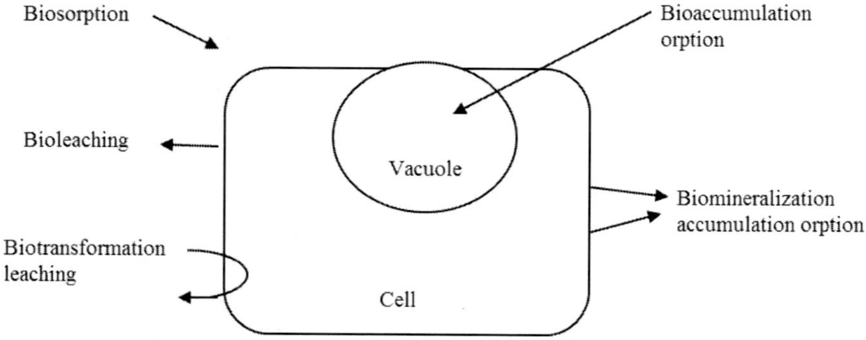

Figure 2.1. Microbial processes involved in remediation of contaminants.

2.1.1a. Metals

Non-living and living biomass of bacteria and fungi equally possess good potential to remove heavy metals (Table 2.4). Metals are either immobilized or mobilized by microbes. Immobilization of heavy metals changes the physical or chemical state of the toxic metals i.e., reduces elements to a lower redox state making them less bioactive, while mobilization involves leaching, chelation, methylation and redox transformation (Wang and Chen, 2009; Jha et al., 2014).

Biosorption mechanism in bacteria involves the sequestration of positively charged metal ions to the negatively charged cell membranes and polysaccharides or extracellular polymeric substances (EPS) secreted on the outer surfaces. Carboxyl groups aid in the binding of metal cations to O-side-chains of lipopolysaccharide (LPS). The anionic functional groups such as uronic acids, phosphoryl groups, carboxylic groups, teichoic acids and teichuronic acids associated with membrane components, peptidoglycan, phospholipids and lipopolysaccharides of Gram-positive and Gram-negative bacteria are primarily responsible for binding of metals (Gutnick and Bach, 2000; Durve et al., 2013). The ionizable functional groups such as carboxyl, phosphoric, amine, and hydroxyl groups enable EPS to sequester heavy metals. Other metal-binding components in bacterial cells include proteinaceous S-layers, and sheaths composed of polymeric materials including proteins and polysaccharides (Liu et al., 2001). The adsorption of heavy metals by EPS is non-metabolic energy independent process. Extracellular polymeric substance produced by bacterial species such as *Pseudomonas* sp has shown capacity to remove Cd, Zn, and Cu from an aqueous system (Guo et al., 2010; Nagashetti et al., 2013). *Pseudomonas* sp. is considered one of the most efficient species because of its high metal adsorption capacity (Hussein et al., 2004; Dixit et al., 2015). Binding involves chelation by metabolites, biosynthesis of metallothionins and other proteins.

Subsequent to binding, heavy metal ions pass into the cell across the cell membrane through the cell metabolic cycle. This requires the input of energy; hence, the mode of metal uptake is referred as active uptake. Thereafter, bacteria transform the contaminants through oxidation, reduction and alkylation reactions. Transformation changes the ionic state of metal. Some bacteria utilize metals as terminal electron acceptors for anaerobic respiration.

The bacterial transformation of metals involves summarized below different reactions:

a. *Methylation*: This reaction removes metals like Hg and produces gaseous compounds (methyl Hg). The bacterial species showing these reactions include *Alcaligenes faecalis*, *Bacillus pumilus*, *Pseudomonas aeruginosa* and *Brevibacterium iodinium*.
b. *Precipitation*: In this reaction, metals get precipitated as insoluble sulfides by the metabolic activity of sulphate reducing bacteria (SRB). These bacteria utilize organic substrates with SO_4^{2-} as the terminal electron acceptor. Hydrogen sulfide (H_2S) is produced as the end product (Asha et al., 2013). SRB play a crucial role in metal sulfide immobilisation in anaerobic sediments. This process is catalysed by the crystallization of inorganic phosphate liberated via the activity of enzyme periplasmic acid phosphatase (named PhoN). The removal of metals such as Cr, As has been reported by SRB. The bacterial species showing metal precipitation include *Desulfovibrio, Desulfomicrobium, Desulfobacter, Desulfosarcina, Desulfotomaculum, Archaeoglobus*.
c. *Redox reactions*: Microorganisms act as oxidizing agent for heavy metals and cause them to lose electrons, which are accepted by alternative electron acceptors (nitrate, sulphate and ferric oxides). The process occurs under anaerobic conditions. Acidophilic bacteria like *Acidithiobacillus ferrooxidans* leach high concentrations of metals such as As, Cd, Cu, Co and Zn through this process. Reduction of Cr(VI) to Cr(III), Se(VI) to elemental Se, U(VI) to U(IV) and Hg(II) to Hg(0) is also reported via this process (Pathak and Dikshit, 2012).
d. *Dissimilatory sulfate reduction pathway*: In this process, the transformation of the target metal takes place and the reaction is catalysed by membrane-bound c-type cytochromes or

enzymes associated with cytochromes.Removal of metals such as Se, Cr, Au has been reported via this mechanism in *Desulfovibrio, Desulfomicrobium, Desulfobacter, Desulfosarcina, Desulfotomaculum, Archaeoglobus*.

Table 2.4. Toxic heavy metals removed by bacteria and fungi

Organism	Element
BACTERIA	
Citrobacter sp.	Pb, Cd, Ni, Zr
Thiobacillus ferrooxidans	Ag
Bacillus cereus	Cd
Bacillus subtilis	Cr, Cu, Zn
Pseudomonas aeruginosa	Cr, Cu, Zn
Micrococcus luteus	Sr, Pb
Saccharomyces cerevisiae	Cd, Zn
Alcaligene	Cr, Cu, Zn
Rhodococcus	Cr, Cu, Zn
Mycobacterium.	Cr, Cu, Zn
Streptomyces spp.	Au, Cu, Cr, Zn, Co
Spirulina platensis	Cd, Ag, Au
Arthrobacter species.	Cr
Klebsiella aerogenes	Hg
Deinococcus radiodurans	Cr
FUNGI	
Rhizopus arrhizus	Hg
Aspergillus niger	Th
Aspergillus lentulus	Cu
Aspergillus flavus	Pb, Cu
Neurospora crassa	Pb, Cu
Lentinus edodes	Cd
Hymenoscyphus ericae	Hg
Neocosmospora vasinfecta	Hg
Verticillium terrestre	Hg
Phanerochaete chrysosporium	Pb
Aspergillus parasiticus	Zn, Pb
Ganoderma carnosum	Zn, Pb

Some bacteria develop resistance against metals. This occurs by two mechanisms- i) transformation, and ii) active efflux. Efflux pumps remove toxic metal out from the cells, while transformation changes their redox state or results in formation of complexes. Heavy metal resistant bacteria namely *Klebsiella oxytoca* BAS-10, *Allescheriella* sp. DABAC 1, *Stachybotrys* sp. DABAC 3, *Phlebia* sp. DABAC 9, *Pleurotus pulmonarius* CBS 664.97, and *Botryosphaeria rhodina* DABAC P82. *B. rhodina* has been isolated from different sources such as industrial effluents, soil, and sediments (Hadis et al., 2011). The isolates were identified using 16S rRNA sequencing method.

In fungi, cell walls bind metals through chitins, glucans, mannans, proteins, polysaccharides, and lipids (Dursun et al., 2003). Fungal phenolic polymers and melanins possess metal-binding sites with oxygen-containing groups including carboxyl, phenolic, carbonyl and methoxyl. Fungi mobilize, incorporate and/or transform metals or immobilize them in the mycorrhizosphere. Organic acids (citrate and oxalate) increase metal solubility through acidification and metal-complexing structures. Subsequent to binding, metal uptake occurs through specific transporters. Cytoplasmic vesicles, vacuoles and fungal hyphae help in translocation of metals to other parts of the mycelium and plant symbionts. After incorporation into the fungus, metals are chemically transformed, stored in different parts of the cell or translocated along fungal hyphae and plant symbionts. The transformation involves methylation and redox reactions, which change the transition state of the metal making it non-toxic. Intracellular metal immobilization involves storage in vacuoles and complexation by cytoplasmic metallothioneins and phytochelatins.

The white-rot basidiomycetes including *Phanerochaete chrysosporium* and filamentous fungi such as *Aspergillus* spp., *Rhizopus* spp, *Penicillium* sp. have been found very effective in the removal of metals (Kapoor et al., 1999; Akponah, 2013; Chaturvedi et al., 2015). Some fungi such as *Cunninghamella elegans* have shown potential for removing heavy metals from textile wastewater.

2.1.1b. Radionuclides

Bacteria and fungi remove radionuclides from the environment (Table 2.5). Cell surface, structure and composition of bacterial species play a major role in the uptake of radionuclides (Lloyd, 2003; Sasaki et al., 2013). *Pseudomonas, Citrobacter freudii, Firmicutes, Neurospora crassa* strains have shown potential for removing radionuclides such as U and Th from wastewater (Haferburg *et al.,*2007; N'Guessan *et al.,*2008; Xie *et al.,*2008; Prakash et al.,

2013).Basidiocarps have shown accumulation of Cs to the levels as high as 50 kBq/kg dry weight. Radionuclides are biosorbed and degraded (Law *et al.*, 2010; Hegazy and Emam, 2011). The translocation of radionuclides (^{137}Cs, ^{86}Rb) occurs through intact mycelia via diffusion. Radionuclides are solubilized by direct and indirect enzymatic reactions through oxidation–reduction. They transform the radioactive metals from an oxidized soluble form to a reduced insoluble form (Tawfik et al., 2005). They also use radioactive metal as the final electron acceptor. Biotransformation of radionuclides is greatly influenced by electron donors, acceptors, nutrients, environmental factors and metabolic activities of microorganisms.

The radionuclides are transformed via:

a. Enzymatic bioreduction triggered through indirect reduction of soluble contaminants by metal-reducing or sulfate-reducing microorganisms.
b. Reductive precipitation mediated by enzymes followed by precipitation within the cell. Wildung et al. (2000) reported enzymatic reduction of U(VI) and Tc(VII) by *Shewanella putrefaciens, Geobacter metallireducens, G. sulfurreducens* and *Desulfovibrio desulfuricans*. Highly soluble Np(V) is reduced to insoluble Np(IV) by *S. putrefaciens. E.coli* reduced Tc(VII) with the help of enzyme formate hydrogenase, while *Desulfovibrio fructosovorans* utilized enzyme periplasmic Ni-Fe hydrogenase (Lloyd *et al.*,1997,2001; De Luca *et al.,*2001). Iron-reducing bacteria such as *Geobacter*sp., *Citrobacter* sp and *Rhodoferax ferrireducens* also showed metabolic potential to reduce actinides enzymatically (Lloyd *et al.*,2000; Kim *et al.,*2012). In sulfate-reducing bacteria, *Desulfovibrio,* the enzymatic reduction of U(VI) is mediated by cytochrome c3 and U(VI) reductase in combination with hydrogenase as its physiological electron donor (Lovley, 2003).A homologous cytochrome (PpcA), a trihaem periplasmic cytochrome *c*7 assist in reduction of U(VI) in the Fe(III)-reducing bacterium *Geobacter sulfurreducens* (Lloyd et al., 2003).
c. Chelation and immobilization by compounds such as organic acids, siderophores and extracellular metabolites. *Pseudomonas aeruginosa* have shown extracellular chelating agents in bioaccumulation of polynuclear radionuclide complexes formed with citrate (John et al., 2001). *Clostridium sphenoides* (ATCC 19403) also reported accumulation of bidentate Fe(III)–citrate complex.

d. Change of ionic state. *Geobaccter* species reduce the U soluble state (U^{6+}) to insoluble state (U^{4+}). This bacterium reduced Fe(III) to Fe(II) (Macaskie *et al.*,2000).

Table 2.5. Bacteria involved in the removal of major radionuclides

Bacterial species	Radionuclides
Bacillus pantothenticus *Bacillus megaterium* *Pseudomonas putida* *Pseudomonas chlororaphis*	U
Ochrobactrum sp. MH1000 *Cupriavidus metallidurans* CH34	Am
Rhodobacter sphaer- oides SSI	Cs, Sr
Geobacter metallireducens	U

2.1.1c. Nutrients

Nitrogen fixing and phosphate solubilizing bacteria were reported to remove nitrogen and phosphorus (Akpor et al., 2013; Wu et al., 2014; Jena et al., 2015). Bacteria involved in nitrification are heterotrophic and cause direct oxidation of ammonium to nitrates. *Staphylococcus, Micrococcus* spp., *Streptococcus, Pseudomonas, Bacillus, Serratia* spp., *Vibrio* spp. and *Pasteurella* spp. show involvement in nitrification. The depletion of nutrient removal rate is directly correlated with availability of electron acceptor i.e., nitrate-N. *Acinetobacter calcoaceticus, Aeromonas hydrophila, Azotobacter vinelandii, Klebsiella spp., Moraxella spp., Alcaligenes spp., Enterobacter sp.* and *Pseudomonas spp.* have been identified as predominant phosphate removers. Some bacteria such as *Corynebacterium, Rhodocyclus and Paraccocus* have shown capacity to remove both nitrate-N and phosphate. It is proposed that exhaustion in nitrate-N concentration generates stress that probably leads to breakdown of Poly P and release of phosphate to the medium. Among fungal species *Neurospora crassa, Fusarium sp, Absidia sp, Aspergillus niger* and *Aspergillus flavus* show phosphate and nitrate removal capacity (Cramer et al., 1980). The nitrate is supposed to be removed by nitrification and denitrification pathways reported in fungi (Guest and Smith, 2002).

2.2. Organic Contaminants

Microorganisms degrade, mineralize or transform organic compounds such as hydrocarbons, pesticides, solvents, explosives and dioxins (Becker and Seagren, 2010; Megahraj et al., 2011; Rivera et al., 2015). Microbes utilize the contaminant/target compound as the carbon source (mineralization). The target compound is either enzymaticaly attacked (metabolized) or it is taken up and concentrated within the organism (bioaccumulation).

Bacterial strains degrade compounds by two ways:

- *Catabolism*: In this process organic chemical or a portion is completely degraded (e.g., mineralized) and the energy or nutrient gained contributes to cell growth.
- *Metabolism or co-metabolism*: In this process, partial degradation of organic chemical takes place with the help of normal metabolic activities of the microorganisms. Metabolic activities cause modifications in the structure and properties of the contaminants converting them into less toxic/non-toxic products.

Plant associated bacteria such as endophytes and rhizophytes contribute to biodegradation of toxic organic compounds. Endophytic bacterial species *Acetobacter, Arthrobacter, Bacillus, Burkholderia, Enterobacter, Herbaspirillum and Pseudomonas possess the* capacity to remove contaminants including polychlorinated biphenyls (PCBs), pesticides, industrial solvents, petroleum products, dioxins and explosives. Microbes use energy generated via biochemical reactions mediated by enzymes to cleave chemical bonds or to assist the transfer of electrons from a reduced organic substrate (donor) to another chemical compound (acceptor). Various enzymes including oxidoreductases detoxify xenobiotics such as phenolic or anilinic compounds, through polymerization, copolymerization with other substrates, or binding to humic substances.

Fungi degrade contaminants via metabolism, mineralization and accumulation mechanism but are more proficient at cometabolism and bioaccumulation. Fungi primarily co-metabolize organic pollutants, aliphatic or aromatic compounds, including volatile organics. Cytochrome P450 (CYP)-mediated oxygenation reactions play an important role during fungal metabolism of recalcitrant xenobiotics (Yadav et al., 2000). Free-living and mycorrhizal fungi transport hydrophobic organic chemicals inside their hyphae. Thereafter, they transform organic contaminants metabolically or co-metabolically. The enzymes involved in cleavage of compounds include lignin-peroxidase (LiP), manganese peroxidase (MnP), and laccase. These enzymes cause cleavage of carbon-carbon and carbon-oxygen bonds resulting in the depolymerisation of lignin and the subsequent degradation of aromatic and aliphatic fragments (Bending et al., 2002; Adenipekun and Lawal, 2012). Fungi attack organic compounds using a range of extracellular oxidoreductases using lignocellulose as substrates. Laccases (*p*-diphenol: dioxygen oxidoreductase) catalyze the oxidation of d phenolic and aromatic substrates and decarboxylate them by attacking their methoxy groups (demethylation). Haem–thiolate peroxygenases produced by certain basidiomycetes catalyze H_2O_2-dependent hydroxylation of various pollutants, including PAHs and dibenzofuran. They cleave ether bonds between aromatic and aliphatic parts of molecules as well as in alicyclic and aliphatic ethers. Extracellular peroxidases degrade chlorophenols, EDCs, PAHs, pesticides, and synthetic dyes.

Among fungi, white-rot fungi have shown immense potential to remove/degrade various xenobiotic compounds including polychlorinated biphenyls, polycyclic aromatics, chlorinated aromatics, dioxins, pesticides, solvents, and dyes effectively (Cameron et al., 2000; Singh, 2006). Major species involved in degradation include *Pleurotus ostreatus, Pleurotus tuber-regium, Pleurotus pulmonarius, Trametes versicolor, Bjerkandera adusta, Lentinula edodes, Irpex lacteus,* and *Agaricus bisporus* (Singh, 2006; Adenipekun and Lawal, 2012). *P. chrysosporium* is the most ideal candidate for xenobiotic metabolism. It is effective in degrading dibenzo-*p*-dioxins (Kamei et al., 2005), chlorophenols (Ehlers and Rose, 2005), polycyclic aromatic hydrocarbons (Cambria et al., 2008), and dyes (Asgher et al., 2008).The mineralization of these pollutants is correlated with their ligninolytic activities. *Phanerochaete chrysosporium* has been used in bioremediation of textile effluents, paper effluents, polycyclic aromatic hydrocarbons and anthracene-contaminated waters (Freitas et al., 2009; Mohammadia and Nasernejad, 2009).

2.2.1. Hydrocarbons

Microbes possess the ability to completely mineralize aliphatic, aromatic, and heterocyclic hydrocarbons (Finley et al., 2014). Degradation of hydrocarbons occurs under aerobic and anaerobic conditions. Bacterial species such as *Bacillus, Corynebacterium, Staphylococcus, Streptococcus, Shigella, Alcaligenes, Acinetobacter, Arthrobacter, Escherichia, Klebsiella, Aeromonas, Rhodococcus, Mycobacterium, Burkholderia, Sphingomonas,* and

Enterobacter assist in hydrocarbon degradation (Kafilzadeh et al., 2012). Bacterial species such as *Pseudomonas fluorescens, P. aeruginosa, Bacillus subtilis, Gordonia, Alcaligenes sp., Acinetobacter, Brevibacterium, Aeromicrobium, Dietzia, Burkholderia, Mycobacterium, Sphingomonas, Flavobacterium sp., Micrococcus roseus, Corynebacterium sp* and *Enterobacter* isolated from petroleum contaminated soil degrade/remove hydrocarbons (Throne-Holst et al., 2007; Pathak et al., 2008; Marco-Urrea et al., 2009a; Igwo-Ezikpe et al., 2010; Jain et al., 2010a, b). The microbial preference for degradation of hydrocarbons includes linear alkanes, branched alkanes, and cyclic alkanes. These species utilize n-alkanes (chain length C10–C40) as a sole source of carbon. The degradation reactions convert toxic compound into less toxic compound. The first step of degradation is oxidation catalyzed by oxygenases and peroxidases (Ulfig et al., 2006).Cytochrome P450 alkane hydroxylases, ubiquitous heme-thiolate monooxygenases also play an important role in degradation of oil, chlorinated hydrocarbons, and fuel additives.

Among the filamentous fungi *Cladosporium* and *Aspergillus* showed capacity for aliphatic hydrocarbon biodegradation. *Aspergillus, Cephalosporium* and *Pencillium* was also found to be potential degrader of crude oil hydrocarbons (Jahangeer and Vikram, 2013). *Candida* species namely *C. maltosa*, *C. tropicalis*, and *C.apicola* use n-alkanes, aliphatic hydrocarbons as a source of carbon and energy. Fungal species such as *Amorphoteca, Neosartorya, Talaromyces, Graphium* and yeast genera namely *Candida, Yarrowia,* and *Pichia*, isolated from petroleum-contaminated soil possess capacity for degradation of hydrocarbons. The initial intracellular attack of oil, chlorinated hydrocarbons, fuel additives is an oxidative process and involves the role of enzymes such as oxygenases, laccase, manganese peroxidase, and lignin peroxidase released from fungal mycelium. Peripheral degradation pathways convert organic pollutants into intermediates. The biodegradation of *n*-alkanes such as tridecane, tetradecane, pentadecane, hexadecane, heptadecane, octadecane (C13-C18) has been reported by the fungi *Aspergillus niger*, *A. ochraceus, Trichoderma asperellum* strain TUB F-1067 (SA4), *T. asperellum* strain Tr48 (SA5), *T. asperellum* strain TUB F-756 (SA6). The degradation of aliphatic hydrocarbon *n*-eicosane by *Trichoderma* sp. S019 produced nonadecanoic acid, *n*-octadecane, hexadecanoic acid, oleic acid and stearic acid as reaction products.

2.2.1a. Polyaromatic Hydrocarbons

The transformation of aromatic hydrocarbons occurs via different peripheral or upper and central or lower pathways. Central/lower pathways convert the intermediates into metabolites such as acetyl-CoA, succinyl-CoA and pyruvate. The intermediates obtained from central pathway are activated for oxidative ring cleavage, owing to the electron-rich substituents in their*ortho*or*para* positions. The calevage of the aromatic ring relies on the use of O_2 to form a non-aromatic epoxide. Aerobic peripheral pathways provide O_2as a powerful means to oxidize and cleave the aromatic ring. The reaction is catalyzed by oxygenases and further hydroxylated to produce catechol, protocatechuate, gentisate (2,5-dihydroxybenzoate) or homogentisate (2,5-dihydroxyphenylacetate) as the end products (Fuchs et al., 2011).

Fungal species from Ascomycota, Basidiomycota and Mucoromycotina hydroxylate PAHs intracellularly and convert them to water-soluble products non-toxic products (Hammel, 1995). Species such as *Aspergillus, Penicillium, Paecilomyces, Coriolus, Pycnoporus, Pleurotus, Fomitopsis,* C*ladosporium, Phanerochaete chrysosporium, Trametes* spp., *Cunninghamella*, *Mucor, Rhizopus, Nematoloma* spp., *Pleurotus* spp. and *Daedalea* degrade PAHs. Among bacteria species such as *Escherichia coli, Pseudomonas putida, Azoarcus* spp.,*Cupriavidus* spp. and *Arthrobacter* have shown capacity for degradation of aromatic hydrocarbons (Fuchs et al., 2011).

Mineralisation of benzopyrene (BaP) has been reported by range of bacteria including *Pseudomonas*, A*grobacterium, Bacillus, Burkholderia, Sphingomonas* and fungal species including *P. chrysosporium, Pleurotus Pycnoponus cinnabarinus, Trametes versicolor, Bjerkandera* sp. strain BOS55, *Trichoderma* sp., *Aspergillus niger*, *Penicillium chrysogenum, Mucor* sp., *Fusarium* sp. E033and *Nematoloma frowardii* (Aitken et al., 1998; Chen and Aitken, 1999; Bhattacharya et al., 2014). Bacteria cause hydroxylation and ring cleavage of the benzo ring of BaP (Ye et al., 1996). The oxidation of BaP is catalysed by extra-cellular peroxidases and laccases (Majcherczyk et al., 1998). BaP mineralization potential has been reported in *P. chrysosporium, P. cinnabarinus, Trametes versicolor, Pleurotus* strains, *Bjerkandera* sp. strain BOS55 and *Nematoloma frowardii.* BaP degradation involve a dioxygenase-catalyzed oxidation to anthracene cis-1,2-dihydrodiol, further oxidation to 1,2-dihydroxyanthracene, and subsequent extradiol (meta) ring-fission. Oxidation of BaP in *Cunninghamella elegans* and *Aspergillus ocharaceus* showed production of trans-dihydrodiols, trans-7,8-dihydroxy-7,8-dihydroBaP and trans-9,10-dihydroxy-9,10-dihydroBaP. The reaction is mediated by cytochrome P-450. In some species including *A. ochraceus, Penicillium* *janthinellum* and *Syncephalastrum racemosum*, further oxidation of BaP dihydrodiols to tetraols has also been demonstrated. Fungi such as *Cunninghamella elegans* showed the role of monooxygenases to activate the aromatic ring and produce anthracene-1,2-oxide which is further isomerized non-enzymatically to yield 1-or 2-anthrol or undergo enzymatic hydration to give anthracene trans-1,2-dihydrodiol. *Cunninghamella elegans* IM 1785/21Gp gave the metabolites *trans*-1,2,3,4- and 9,10-dihydrodiols, phenols, diphenols (diols), and glycoside conjugates of 1-,2-,3-,4-, and 9-phenanthrols.

Phenanthrene degradation in white-rot fungi *P. chrysosporium* produces diphenic acid as a major oxidation product via synthesis of 9,1 0-quinone. Biodegradation of phenanthrene by *Thrichoderma* sp. S019 produced 1-hydroxy-2-naphthoic acid, salicylaldehyde, salicylic acid, and catechol as intermediates. Ligninolytic fungi secrete peroxidases, which oxidize Mn^{2+} to Mn. Manganese peroxidases (MnPs) support Mn-dependent lipid peroxidation, and MnP oxidizes phenanthrene to diphenic acid. *Bjerkandera adusta* degrade hexachlorocyclohexane (HCH) isomers giving 1-(3-chloro-4-methoxyphenyl) ethanone and (2,4-dichloro-3-methoxy)-1-benzenecarbonyl chloride. Dermatophytes such as *Microsporum gypseum, M. canis, Trichophyton terrestre, T. ajelloi, T. mentagrophytes var. mentagrophytes, T. rubrum* and keratinolytic fungi, *Aphanoascus reticulisporus* and *Scopulariopsis brevicaulis* showed capacity to remove hexane and toluene.

Biodegradation of styrene produce metabolites including 2-phenyl ethanol, benzoic acid, cyclohexadiene-1,4-dione, butanol and succinic acid in *Phanerochaete chrysosporium* KFRI 20742, *Trametes versicolor* KFRI 20251, and *Daldinia concentric* KFRI 40-1 (Wick et al., 2007). The effective biodegradation of pyrene has also been reported in *Coriolus versicolor, Trichoderma* sp., *Aspergillus niger*, and *Fusarium* sp. Lignolytic fungi, *Allescheriella* sp. strain DABAC 1, *Stachybotrys* sp. strain DABAC 3, and *Phlebia* sp. strain DABAC 9 degrade naphthalene, dichloroaniline isomers, *o* hydroxybiphenyl and 1,1'-binaphthalene. *Fusarium flocciferum*, *Trichoderma* species, *Tremetes versicolor*, and *Pleurotus ostreatus* showed capacity for degradation of benzoanthracene, benzofluoranthene, benzopyrene, and chrysene.

Pseudomons species are regarded as one of the most common species of bacteria degrading phenolic compounds (Collins et al., 2005; Lin et al., 2008; Sinha et al., 2011). The cleavage of the extradiol ring of phenol resulting in the formation of 2-hydroxymuconate semialdehyde (2-HMS) (El-Sayed et al., 2003). White-rot fungi exhibit biodegradation capabilities for phenols. *Fusarium* sp. HJ01, *Trametes versicolor* was able to grow using phenol as the only carbon resource and producing catechol as intermediate biotransformation product. *Aspergillus awamori* NRRL 3112, *Peniophora cinerea, Psilocybe castanella, Trametes villosa, Agrocybe perfecta, Trichaptum bisogenum* degraded phenol, catechol, 2,4-dichlorophenol and 2,6-dimethoxyphenol. *P. involutus* 211804 was reported to perform the degradation of ferulic, *o*-coumaric, and *o*-hydroxyphenylacetic acids. Metabolism of phenol degradation under aerobic condition may involve two different pathways namely the *meta*- or the *ortho* pathways. Peroxidases such as *Caldariomyces fumago* use haem–thiolate chloroperoxidase (CPO) and *Coprinopsis cinerea* use peroxidase (CiP) for phenol degradation. Peroxidases such as *Caldariomyces fumago* haem–thiolate chloroperoxidase (CPO) and *Coprinopsis cinerea* peroxidase (CiP) act on phenols.

2.2.1b. Heterocyclic Aromatic Hydrocarbons: Dioxins and Furans

Polychlorinated dibenzo-*p*-dioxins (PCDDs) and polychlorinated dibenzofurans (PCDFs), dioxins (PCDD/Fs) are removed by microbes via oxidation owing to the electronegativity of the chlorine atoms withdrawing electrons from the aromatic rings (Chan, 2008; Kasai et al., 2010). Aerobic and anaerobic bacterial biotransformation of PCDD 2,3,7, 8-tetrachlorodibenzo-*p*-dioxin (2,3,7,8-TCDD) is slow and produces chlorinated aromatic products. *Phlebia radiata* I-5-6, *P. acerina, P. lindtneri* and *P. brevispora, Cordyceps sinensis, Phanerochaete chrysosporium* DSM 6909, *P. chrysosporium* DSM 1556, *Irpex* sp. W3, *Trametes* sp. CH2, *Fusarium* sp. VSO7, and *Pleurotus pulmonarius* hydroxylate and methoxylate PCDDs including 2,8-dichlorodibenzo-*p*-dioxin (2,8-DCDD), 2,3,7,8-tetrachlorodibenzo-*p*-dioxin (2,3,7,8-TCDD), 2,7-dichlorodibenzo-*p*-dioxin (2,7-DCDD), and 2,4,8-trichlorodibenzofuran (2,4,8-TCDF)and for the intermediate compounds such as 4-chlorocatechol and 4,5-dichlorocatechol. Transformation of 2,3,7, 8-TCDD into chlorocatechols has been noted in *Phanerochaete sordid*, while complete mineralization has been reported by *P. chrysosporium* (Nakamiya et al., 2005).

2.2.1c. Volatile Compounds

Owing to their proteolytic or keratinolytic activities, several volatile organic compounds can be removed by bacteria and fungi (Huang et al., 2012). Proteins such as keratin are the main source of nitrogen, carbon and sulfur for the growth of fungi. The aerobic degradation of trichlorothylene (TCE), perchloroethylene (PCE) has been reported by *Trametes versicolor, Ganoderma lucidum* and *Irpex lacteus*. Pentachloroethylene ((PCE) degradation to produce 2,2,2-trichloroethanol and CO_2 as the main byproducts and trichloroacetic acid (TCA) as the intermediate has been noted in several fungi including *Bjerkandera adusta*, *Anthracophyllum discolour*, *Phanerochaete chrysosporium*, *Trametes versicolor* isolate HR131, *Boletus edulis*, *Suillus luteus*, *Cortinarius russus*, *Suillus grevillei*, *Gomphidius viscidus*, *Laccaria bicolor*, *Leccinum scabrum*, *Xerocomus chrysenteron*, *Heboloma crustuliniforme*, *H. longicaudum* and *Trametes* sp. isolate HR577 (Zheng et al., 2007). Benzene, toluene, ethylbenzene and xylene (BTEX) compounds found in petroleum products and methyl tertiary butyl ether (MTBE), a fuel oxygenate has been known to be degraded by microbes (Weelink et al., 2010).

2.2.1d. Polychlorinated biphenyls (PCBs)

Anaerobic and aerobic bacteria are capable of biotransforming PCBs. Higher chlorinated PCBs are subjected to reductive dehalogenation by anaerobic microorganisms. Aerobic bacteria belonging to the genera *Pseudomonas, Burkholderia, Ralstonia, Achromobacter, Sphingomonas Comamonas* and some Gram-positive strains such as *Rhodococcus, Janibacter, Bacillus, Paenibacillus* and *Microbacterium* oxidize lower chlorinated biphenyls.

The catabolic pathway for PCB degradation comprises of steps catalysed by the enzymes, biphenyl dioxygenase (BphA), dihydrodiol dehydrogenase (BphB), 2, 3-dihydroxybihenyl dioxygenase (DHBD) (BphC) and hydrolase (BphD). Aerobic bacteria such as *Burkholderia xenovorans* LB400 and *Rhodococcus* sp. strain RHA1, cleave chlorinated biphenyl rings to yield chlorinated benzoates and pentanoic acid derivatives, while anaerobic bacteria convert highly chlorinated PCB congeners into less chlorinated biphenyls by reductive dechlorination.

The white-rot fungus *Phlebia brevispora* degrade PCBs producing *m*-methoxylated, *p*-dechlorinated and *p* methoxylated metabolites. *Phlebia brevispora* was shown to be able to degrade PCBs to produce *m*-methoxylated, *p*-dechlorinated and *p* methoxylated metabolites. *Phanerochaete* sp. degrade 4,4'-dichlorobiphenyl (4,4'-DCB) and metabolites namely 2-hydroxy-4,4'-DCB and 3-methoxy-4,4'-DCB, 4-chlorobenzoic acid, 4- chlorobenzaldehyde, 4-chlorobenzyl alcohol, and 4-hydroxy-3,4'-DCB. Fungi *Phlebia* sp., *Phanerochaete chrysosporium*, and *Mortierella* sp. were also able to degrade different chlorobenzoic acids (CBA) by a hydroxylation and dechlorination pathway. Chlorinated aliphatic compounds are degraded by oxygenases. Fungi *Basidioradulum molare* and *Schizopora paradoxa* degrade phenolic compounds such as 4-*tert*-octylphenol and dehydroabietic acid (DHA) by *Trametes versicolor* and *Phlebiopsis gigantea* (van Beek et al., 2007).

2.2.2. Organic Solvents

Gram positive bacteria such as *Rhodococcus, Bacillus* and some Gram negative *Pseudomonas aeruginosa, Mycobacterium tuberculosis* and *Salmonella enteric* bacteria possess capacity to tolerance solvents. Bacteria transform organic solvents into nontoxic byproducts by membrane rigidification, formation of compartmentalized vesicles to remove solvents and the extrusion via energy dependent process. Bacteria prefer efficient and active efflux systems to remove toxic solvents out of the cell. These include RND (Resistance-Nodulation-Cell division) efflux pumps MexAB-OprM, MexCD-OprJ, MexEF-OprN efflux system and AcrAB-TolC efflux pump. Efflux pumps are the proteins, which extrude solvents out of cells. *Mycobacterium tuberculosis* and *Salmonella enteric* extensively use efflux pumps for tolerating solvents.

P. fluorescens, Fusarium sp. HJ01, *P. aeruginosa* AT2, *Aspergillus awamori* NRRL 3112, *Peniophora cinerea*, *Psilocybe castanella*, *Trametes villosa*, *Agrocybe perfecta*, *Trichaptum bisogenum* and *B. cepacia* were capable of utilizing a wide range of phenolic compounds with different levels of halo-substitution as the carbon sources. *Phanerochaete chrysosporium*, *Pleurotus ostreatus, Trametes versicolor* and *Bjerkandera* sp. BOL13 degrade nonylphenol (Soares et al., 2003; Melchias et al., 2012). Bisphenol A was biodegraded by white rot fungi *Irpex lacteus*, *T. versicolor*, *Ganoderma lucidum*, *Polyporellus brumalis*, *Pleurotus eryngii*, *Schizophyllum commune* and *T. versicolor* strains MrP 1 and MrP 13. *Cladophialophora*, *Exophiala*, *Leptodontium* and *Pseudeurotium zonatum* have been found as toluene-degrading fungi. They use toluene as sole carbon and energy source (Edward et al., 2012).

2.2.3. Pesticides

Microorganisms degrade chlorinated and phosphate pesticides (Table 2.6). The metabolism of pesticides involves a three-phase process. Phase I metabolism involves the transformation of the parent compound through oxidation, reduction or hydrolysis to produce water-soluble less toxic product. Enzymes such as esterases, glutathione S-transferases (GSTs) and cytochrome P450 catalyze the previous step (Hussain et al., 2009; Bass and Field, 2011) (Table 2.7). Phase II conjugation involves the binding of pesticide to a sugar or amino acid, which increases the water solubility and reduces. The Phase III involves the transformation of Phase II metabolites into secondary conjugates, which are non-toxic. Notably, the biodegradation of pesticides is greatly influenced by both biotic and abiotic factors including moisture, temperature, pH, organic matter and presence of microorganisms (Aislabie and Lloyd-Jones, 1995; Scott et al., 2008; Alves et al., 2010). Fungi and bacteria produce intracellular or extracellular enzymes including hydrolytic enzymes, peroxidases, oxygenases that help in the degradation of pesticides (Ortiz-Hernández et al., 2011).

2.2.4. Organochlorine

Bacteria degrade organochlorine pesticides and the degradation involves co-metabolism under anaerobic conditions (Singh et al., 1999). Reductive dechlorination is the major mechanism for the microbial conversion of dichlorodiphenyltrichloroethane (DDT) to DDD (Aislabie et al., 1997; Kwon et al., 2005; Fang et al., 2010). Reductive degradation of DDT has been reported by bacterial species such as *Enterobacter aerogenes*, *Pseudomonas fluorescens*, *Escherichia coli*, and *Kleibsiella pneumaniae*. The reaction involves the substitution of aliphatic chlorine for a hydrogen atom. The dechlorination reactions of DDT yield 2,2-bis(*p*-chlorophenyl)ethylene (DDNU), which is further oxidized to 2,2- bis(*p*-chlorophenyl)ethanol (DDOH). Oxidation of DDOH yields bis(*p*chlorophenyl) acetic acid (DDA) which is decarboxylated to bis(*p*-chlorophenyl)methane (DDM). DDM is metabolized to 4,4'dichlorobenzophenone (DBP) or cleaved in one of the aromatic rings to form *p*-chlorophenylacetic acid (PCPA) (Barraga´n-Huertaa et al., 2007; Fang et al., 2010).

Bacterial strains showed potential to degrade endosulfan (6,7,8,9,10,10-hexachloro-1,5,5a,6,9,9a-hexahydro-6,9-methano-2,4,3-enzodioxathiepine3-xide) (Awasthi et al., 2003). Endosulfan gets transformed to endosulfan diol and endosulfan sulfate through oxidation to produce endosulfan hydroxyether and endosulfan lactone. The reaction is catalyzed by the enzyme mono-oxygenase. The compound is completely mineralized to carbon dioxide after metabolism. Mixed bacterial cultures (*Stenotrophomonas maltophilia, Rhodococcus erythropolis* and *Achromobacter xylosoxidans* CS5) showed capacity to degrade endosulfan a and b respectively with endodiol as an intermediate (Kumar et al., 2007; Li et al., 2009).Bacteria also showed capacity to degrade hexachlorocyclohexane (HCH) under aerobic conditions (Gupta et al., 2000; Camacho-Pérez et al., 2012; Ramadevi et al., 2012).

Table 2.6. Microorganisms involved in degradation of pesticides

Pesticides	Microorganisms	
	Bacteria	**Fungi**
Organochlorine		
PCP	*Arthrobacter* sp. *Flavobacterium* sp.	
1,4-Dichlorobenzene	*Pseudomonas sp.*	
DDT	*Aerobacter aerogenes* *Trichoderma viridae* *Pseudomonas sp.* *Micrococcus sp.* *Arthrobacter sp.* *Bacillus sp.* *Sphingobacterium* *sp.Staphylococcus* *Stenotrophomonas*	*Pseudomonas aeruginosa* *P. putida* *Stenotrophomonas* *maltophilia Flavimonas* *oryzihabitans* *Morganella morganii* *P. aeruginosa* *F. oryzihabitans* *Penicillium miczynskii* *Aspergillus sydowii* *Trichoderma* sp. *Penicillium raistrickii* *Aspergillus sydowii*
Lindane	*Basea thiooxidans* *Sphingomonas paucimobilis* *Streptomyces sp.* *Pleurotus ostreatus*	*Bjerkandera* sp. strain BOL13 *Phanerochaete* *chrysosporium* *Trametes hirsutus* *Bjerkandera adusta* *Pleurotus* sp.
DDE	*Phanerochaete chrysosporium*	
DDD	*Trichoderma sp.*	
Heptachlor epoxide	*Phlebia sp.* *Phanerochaete chrysosporium*	
Toxaphene	*Bjerkandera sp*	*Bjerkandera* sp. strain BOL13 *Phanerochaete* *chrysosporium* *Trametes hirsutus* *Bjerkandera adusta* *Pleurotus* sp.
Aldrin	*Trichoderma viridae* *Pseudomonas sp.* *Micrococcus sp.* *Bacillus sp.*	
Endrin	*Trichoderma viridae* *Pseudomonas sp.* *Micrococcus sp.* *Arthrobacter sp.* *Bacillus sp*	

Table 2.6. (Continued)

Pesticides	Microorganisms	
	Bacteria	**Fungi**
Endosulfan	*Staphylococcus* sp. *Pseudomonas aeruginosa* *Klebsiella oxytoca* *Klebsiella pneumonia* *Bacillus circulans* I, II	*Pseudomonas aeruginosa* *Pseudomonas putida* *Pseudomonas fluorescens* *Stenotrophomonas maltophilia* *Flavimonas oryzihabitans* *Morganella morganii* *F. oryzihabitans* *Chaetosartorya stromaloides Aspergillus terricola* *Aspergillus niger* *Aspergillus terreus* *Trichoderma harzianum* *Phanerochaete chrysosporium*
Hexachlorocyclohexane (HCH)	***Bacillus*** *circulans* *Bacillus brevis* ***Sphingomonas*** *paucimobilis*	
Profenofos		*Pseudomonas raistrickii* CBMAI 931 *Aspergillus sydowii* CBMAI 935 *Trichoderma* sp. CBMAI 932
Orthophosphate		
Chlorpyrifos	*Bacillus* *Pseudomonas* *Arthrobacter* *Micrococcus* *Flavobacterium sp.* ATCC 27551 *Enterobacter* strain B-14	*Rhizopus* *Trichoderma harznaium*
Parathion	*Bacillus subtilis* *Klebsiella sp.* *Pseudomonas stutzeri* *P. aeruginosa* *P. fluorescens*	

Pesticides	Microorganisms	
	Bacteria	**Fungi**
Diazinon	*Pseudomonas* *Burkholderia* *Brevundimonas* *Flavobacterium sp.* *Agrobacterium sp.* *Serratia marcescens* D1101 *Stenotrophomonas malthophilia* *Proteus vulgaris* *Vibrio metschinkouii* *Serratia ficaria* *Serratia* spp. *Yersinia enterocolitica*	
Fenitrothion	*Burkholderia* sp. strain NF100	
Malathion	*Bacillus pseudomycoides* S4 *Bacillus licheniformis* S5 *Klebsiella sp.* *Flavobacterium sp.* *Arthrobacter*	*Pseudomonas sp* *Aspergillus oryzae* *Pseudomonas aeruginosa* S1 *Aspergillus flavus* *Aspergillus sydowii* *Aspergillus niger*
Diuron	*Mortierella*	
Fenamiphos	*Pseudomonas putida* *Acinetobacter rhizosphaerae*	
Ethion		*Rhizopus nodosus* *Trichoderma harznaium*
Carbamates		
Carbofuran	*Pseudomonas* *Flavobacterium* *Achromobacterium* *Sphingomonas* *Arthrobacter*	*Gliocladium* *Mucor ramannianus*
Aldicarb	*Stenotrophomonas maltophilia*	
Other		
*N,N*diethyl-*m*-toluamide (DEET)		*Cunninghamella elegans* ATCC 9245 *Mucor ramannianus* R-56 *Aspergillus niger* VKMF-1119 *Phanerochaete chrysosporium* BKMF-1767
Pyrethroids		
Deltamethrin	*Serratia marcescens*	

The degradation in fungi occurs via extracellular enzymes excreted in *Phanerochaete chrysosporium, Gloeophyllum, Daedalea*, and *Fomitopsis.* DDT is metabolized to formation of 1,1-dichloro-2,2-bis (4-chlorophenyl) ethane (DDD), 1,1-dichloro-2,2-bis(4-chlorophenyl)ethylene (DDE), and 4,4-dichlorobenzophenone (DBP) (Pinedo-Rivilla et al., 2009; Premjet et al., 2009; Ortega et al., 2011; Xiao et al., 2011). Fungal species degrade α-and β-endosulfan, toxaphene and lindane (Romero et al., 2006; Pinedo-Rivilla et al., 2009; Reddy et al., 2012). Enzymes such as manganese peroxidase, laccase convert endosulfan-to-endosulfan hydroxyether and endosulfan lactone in *Aspergillus niger* ARIFCC 1053, *Chaetosartorya stromatoides*, *Aspergillus terricola*, and *A. terreus* (Bhalerao and Puranik, 2007; Pinedo-Rivilla et al., 2009; Bhalerao, 2012). The profenofos degradation by fungi involved the cleavage of the phosphorothioate ester bond to yield 4-bromo-2-chlorophenol followed by conjugation with glucose. Malathion degradation involves esterases and oxidoreductase enzymes (Zheng et al., 2007; da Silva et al., 2012). *Bjerkandera* sp. strain BOL13 degrades toxaphene (Romero et al., 2006).

Pentachlorophenols (PCP), chlorinated organic pesticide are removed by *Streptomyces sp, Arthrobacter, Pseudomonas, Sphingobium chlorophenolicum* strain ATCC 39723, and *Serratia marcescens*. Fungi namely *Bjerkandera adusta*, *Anthracophyllum discolour*, *Phanerochaete chrysosporium*, *Trametes versicolor* isolate HR131, *Trametes* sp. isolate HR577, *Boletus edulis*, *Suillus luteus*, *Cortinarius russus*, *Suillus grevillei, Gomphidius viscidus*, *Laccaria bicolor*, *Leccinum scabrum*, *Xerocomus chrysenteron*, *Heboloma crustuliniforme*, and *H. longicaudum* degrade pentachlorophenol. The degradation of pentachlorophenol is related to the oxidoreductase activity. Fungi namely *Bjerkandera adusta*, *Anthracophyllum discolour*, *Phanerochaete chrysosporium*, *Trametes versicolor* isolate HR131, *Trametes* sp. isolate HR577, *Boletus edulis*, *Suillus luteus*, *Cortinarius russus*, *Suillus grevillei, Gomphidius viscidus*, *Laccaria bicolor*, *Leccinum scabrum*, *Xerocomus chrysenteron*, *Heboloma crustuliniforme*, and *H. longicaudum* degrade pentachlorophenol. The degradation of pentachlorophenol is related to the oxidoreductase activity. Fungi namely *Bjerkandera adusta*, *Anthracophyllum discolour*, *Phanerochaete chrysosporium*, *Trametes versicolor* isolate HR131, *Trametes* sp. isolate HR577, *Boletus edulis*, *Suillus luteus*, *Cortinarius russus*, *Suillus grevillei, Gomphidius viscidus*, *Laccaria bicolor*, *Leccinum scabrum*, *Xerocomus chrysenteron*, *Heboloma crustuliniforme*, and *H. longicaudum* degrade pentachlorophenol. The degradation of pentachlorophenol is related to the oxidoreductase activity

(Wolski et al., 2006). Bacterial strains *Herbaspirullum* sp K1, *Sphingomonas*strains K74 and MT1, *Nocardioides* sp K44 degrade TCP (Mannisto *et al., 2001*). Among fungi, *Phanerochaete chrysosporium, Gloeophyllum striatum* and *G. trabeum* mineralizes 2,4,6-trichlorophenol. The degradation of 2,4,6-trichlorophenol is mediated by lignin peroxidase (LiP) and manganese peroxidase (MnP). The multistep dechlorination reaction produce 2,6-dichloro-1,4-benzoquinone. The quinone is reduced to 2,6-dichloro-1,4-dihydroxybenzene, which is reductively dechlorinated to yield 2-chloro-1,4-dihydroxybenzene. The dechlorinating bacteria *Dehalococcoides ethenogenes, Dehalobacter restrictus, Desulfitobacterium dehalogenans, Dehalospirillum multivorans, Desulfuromonas chloroethenica,* and *Desulfomonile tiedjei* dehalogenate PCE.

Table 2.7. Major enzymes involved in the degradation of the selected pesticides*

Enzyme	Microbe	Pestcide type
Oxidoreductases	*Pseudomonas* sp. *Agrobacterium*	Glyphosate
Monooxygenases	*Mycobacterium* sp. *Arthrobacter* sp. *Pseudomonas putida*	Endosulphan Endosulfan Aldrin Malation DDDT Hexachlorobenzene and pentachlorobenzene
Dioxygenases	*Pseudomonas putida*	Trifluralin
Phosphodiesterases	*Delftia acidovorans* *Agrobacterium radiobacter* *Pseudomonas diminuta* *Flavobacterium* sp.	Parathion Methyl parathion Malathion Coumaphos
Phosphonatase	*Escherichia coli* *Sinorhizobium meliloti*	Organophosphorus compounds
Haloalkane Dehalogenases	*Sphingobium* sp. *Shingomonas* sp. *Pseudomonas* sp. *Nocardioides* sp.	Hexachlorocyclohexane Chloro-s-trazine Hexachlorocyclohexane 2,4-dichlorophenoxyacetic acid\ Pyridyl-oxyacetic

*Modified from Ortiz-Hernandez et al. (2013)

2.2.5. Organophosphate

Microbes use organophosphates as a source of phosphorus and/or carbon (Karpouzas and Singh, 2006; Ghosh et al., 2010). *Pseudomonas* species, *Acinobacter* and *Acetobacter faecalis* species have showed capacity to degrade organophosphorus pesticides (Liu et al., 2007; Lakshmi et al., 2008; Ortiz-Hernández and Sánchez-Salinas, 2010; Olawale et al., 2011). The degradation is catalyzed by enzyme phosphotriesterase. In bacterial degradation of chlorpyrifos(O, O-diethyl O-(3, 5, 6-trichloro-2-pyridyl phosphorothioate), hydrolyzation to TCP produces 2-pyridinol i.e., 3,5,6-trichloro-2-pyridinol (TCP) and diethylthiophosphoric acid (DETP) via dechlorination and further cleavage of the pyridine ring (Singh et al., 2004; Bhagobaty et al., 2006; Xu et al., 2007; Chen et al., 2012; Agarry et al., 2013). Aerobic bacteria *Flavobacterium sp.* ATCC 27551, *Arthrobacter sp.*, *Enterobacter* strain B-14, *Stenotrophomonas* strain YC-1, and *Cupriavidus* sp. DT-1 transform chlorpyrifos (Yang et al., 2005, Lu et al., 2013). Transformation of TCP by ring breakage results in complete detoxification in species such as *Cladosporium cladosporioides* (Chen et al.,2012; Gao et al.,2012).

Ethion (O,O,O,O-tetraethyl S,S-methylene diphosphorodithioate) removal has been reported by *Streptomyces sp.* Bacteria have shown the ability to degrade methyl parathion (*O,O*-dimethyl-*O*-(*p*-nitro-phenylphosphorothioate) and its conversion to p-nitrophenol, 4-nitrocatechol by parathion hydrolase (phosphotriesterase) and p-nitrophenol hydroxylase (Hong et al., 2005;Singh and Walker, 2006; Karunya and Reetha, 2012).Parathion is hydrolyzed to p-nitrophenol and diethyl thiophosphate with the help of enzyme monooxygenase (Singh and Walker, 2006). *Serratia* sp. strain DS001 and *Nocardiopsis* species degrade methyl parathion to produce p-nitrophenol (PNP) with the help of the enzyme parathion hydrolase (Dudhagara et al., 2012). Bacterial strains degrade diazinon, fenitrothion and malathion (Hayatsu et al., 2000; Drufovka et al., 2008; Cycon et al., 2009; Ortiz-Hernández and Sánchez-Salinas, 2010; Abo-Amer and Aly, 2011; Karunya and Reetha, 2012; Karunya and Reetha, 2012; Thabit et al., 2013; Mahiuddin et al., 2014). Bacteriadegrade fenamiphos (FEN) and hydrolyze it tophenol and ethyl hydrogen isopropylphosphoramidate (IPEPAA) (Chanika et al., 2011).

Fungus including *Aspergillus* degrades OPs such as diuron (Malghani et al., 2009; Ellegaard-Jensena et al., 2013). Fungal isolates have shown potential for removing ethion and chloropyrifos. The hydrolytic degradation of ethion produce ethion monoxon, ethion dioxon, O,O-diethylthiophosphate and thio-formaldehyde. The enzyme hydrolase was involved in the degradation (Jilani and Khan, 2004; Yang et. al., 2006; Harish et al., 2013). Diuron, metalaxyl, atrazine, terbuthylazine, lindane, and malathion degradation has been noted in *Coriolus versicolor, Hypholoma fasciculare, Stereum hirsutum* and *Phenerochaete chryosporium* (Ningfeng et. al., 2004; Goda et. al.,2010; Ramadevi et al., 2012). *Phanerochaete chrysosporium*, *Trametes hirsutus, Bjerkandera adusta,* and *Pleurotus* sp. showed lindane degradation.

2.2.6. Carbamate

The carbamates are transformed via hydrolysis, biodegradation, oxidation, photolysis, and the biotransformation reactions in microbes (Soriano et al., 2001; Mi et al., 2007; Chanika et al., 2011). Metabolization of carbamates is initiated by hydrolysis at the ester linkage (De Schrijver and De Mot, 1999). Hydrolyzation resulted in formation of carbofuran phenol and intermediates such as 2-hydroxy-3-(3-methylpropan-2-ol) phenol or 7a-(hydroxymethyl)-2,2-dimethylhexahydro-6H-furo[2,3-b]pyran-6-one (Kim et al., 2004; Mi et al., 2007; Seo et al., 2007). Fungi *Gliocladium, Mucor ramannianus* and bacteria *Pseudomonas putida* are some microbial species capable of degradingcarbofuran (Kim et al., 2004).Aldicarb [2-methyl-2-(methylthio) propionaldehyde) O-(methylcarbamoysl)-oxime] is oxidized to aldicarb sulfoxide [2-methyl-2-(methylsulfininyl) propionaldehyde O-(methylcarbamoyl)oxime], or aldicarb sulfone [2-methyl-2-(methylsulfonyl)-propionaldehydeO-(methylcarbamoyl)oxime]. The oxidation of aldicarb is carried out under aerobic conditions, while under unaerobic conditions aldicarb sulfone and sulfoxide may revert to the parent compound (Caracciolo et al., 2002). The enzymatic hydrolysis is catalyzed by enzymes esterase and amidase (Karayilanoglu et al., 2008; Rocha et al., 2009; Martins et al., 2011; da Silva et al., 2013).

2.2.7. Pyrethroids

Complete mineralization of pyrethroid to CO_2 and H_2O has been reported in microbial species (Lone and Wani, 2012). Degradation of cypermethrin [(+/-)-a-cyano-3-phenoxybenzyl (+/-)-cis, trans-3 (2,2-dichlorovinyl)-2,2-dimethylcyclopropanecarboxy-late] has been reported in bacteria isolated from soil and water (Nirmali et al., 2005; Tallur et al.,2008; Arulazhagan et al.,2010; Murugesan et al., 2010; Bhonsle and Nasreen, 2012; Lone and Rafiq, 2012; Gur et al., 2014). *Stenotrophomonas maltophilia* MHF ENV b20, *Stenotrophomonas maltophilia* MHF ENV 22 and *Sphingobacterium thalpophilum* MHF ENV 23 have been identified as potential degraders of cypermethrin and fenvalerate respectively. Cyfluthrin transformation to 4-fluoro-3-phenoxybenzaldehyde and 3-(2,2-dichlorovinyl)-2,2-dimethyl-cyclopropane carboxylic acid has been reported in *Aspergillus nidulans* and *Sepedonium aheswarium* (Pinedo-Rivilla et al., 2009). Cyfluthrin was biotransformed by *Aspergillus nidulans* and *Sepedonium maheswarium* to 4-fluoro-3-phenoxybenzaldehyde and 3-(2,2-dichlorovinyl)-2,2-dimethyl-cyclopropane carboxylic acid.

2.2.8. Herbicides

Fungi *Chrysosporium* sp., *Eupenicillium* sp., *Paecilomyces* sp., *Trichoderma viride*, *Metarhizium anisopliae*, *Phoma eupyrena*, *Mucor hiemalis*, *Mortierella* sp. and *Beauveria bassiana* are capable of degrading herbicides such as sulfentrazone, fluometuron, isoproturon (Pinedo-Rivilla et al., 2009). The compounds are hydroxylated or methylated to form metabolites such as N-(4-(2-hydroxy-1-methylethyl) phenyl)-N',N'-dimethylurea, N-(4-isopropylphenyl)-N'-methylurea (MDIPU). Bacteria species such as *Phlebia brevispora* TMIC33929 degrade chlornitrofen to produce metabolites such as monomethoxylated compounds and 2,4,6-trichlorophenol.Biotransformation of N, N diethyl-m-toluamide (DEET) to N,N-diethyl-m-toluamide- N-oxide, N-ethyl-m-toluamide-N-oxide, and N-ethyl-m-toluamide has also been reported in fungus *Cunninghamella elegans* ATCC 9245, *Mucor ramannianus* R-56, *Aspergillus niger* VKMF-1119, and *Phanerochaete chrysosporium* BKMF-1767 (Seo et al., 2007; Pinedo-Rivilla et al., 2009). *Coriolus versicolor, Salmonella typhimurium* metabolize chloronitrofen (2, 4, 6-trichloro-4′-nitrodiphenyl ether, CNP) and nitrofen (2, 4-dichloro-4′-nitrodiphenyl ether, NIP), diphenyl ether herbicides. The degradation involves reductive dechlorination and nitro-reduction reactions followed by hydroxylation and ring-cleavage reaction. CNP is converted via four different pathways, i.e., aromatic hydroxylation, oxidative dechlorination (path b), reductive dechlorination (path c), and the reduction of the nitro group to amine (path d). Aromatic hydroxylation and oxidative dechlorination are involved during the fungal conversion of NIP (II) to XV and XVI.The fungal conversion of CNP to NIP is via reductive dechlorination (Reddy and Gold, 2000). C. versicolor is capable of dechlorinating 5-chloro-1, 2, 4-trihydroxybenzene to form 2, 4, 6-trichloro-4′-aminodiphenyl ether (III). CNP cytochrome P450-type enzyme is apparently involved in the biotransformation of the herbicide atrazine by *Pleurotus pulmonarius*.

The white-rot basidiomycete *Coriolus versicolor* metabolized chloronitrofen (2, 4, 6-trichloro-4′-nitrodiphenyl ether; CNP) and nitrofen (2, 4-dichloro-4′-nitrodiphenyl ether; NIP). The metabolization resulted in the formation of four metabolites – 2, 4, 6-trichloro-3-hydroxy-4′-nitrodiphenyl ether, 2, 4-dichloro-6-hydroxy-4′-nitrodiphenyl ether, NIP, and 2, 4, 6-trichloro-4′-aminodiphenyl ether – indicating the involvement of hydroxylation, oxidative dechlorination, reductive dechlorination, and nitro-reduction.

2.2.9. Others

The biodegradation of the insecticide dimethoate has been noted in strain *Aspergillus nomius* L3 (Ai et al., 2006; Pinedo-Rivilla et al., 2009). *Stereum hirsutum* showed capacity for degrading methoxychlor to form 2,2-dichloro-1,1-bis(4-methoxyphenyl)ethane, 2,2-dichloro-1,1-bis(methoxyphenyl) ethylene, 2-chloro-1,1-bis(4-methoxyphenyl) ethylene, and 1,1-bis(4-methoxyphenyl)ethylene. *Trametes versicolor, Agaricus augustus, Laetoporeus sulfureus, Gloephyllum trabeum,* and *Ganoderma australe* proved effective in decreasing tribromopenol (TBP) concentrations and biotransforming TBP to tribromoanisole (TBA). The filamentous fungus *Cunninghamella elegans* was able to degrade tributyltin chloride (TBT) giving less toxic compounds, dibutyltin and monobutyltin. *Rhodococcus* sp.was reported to degrade triazines to nitrate. Several bacterial genera, adapted to grow in pesticide contaminated soils, have enzymes involved in the hydrolysis of P-O, P-F, P-S and P-C bonds, found in a wide variety of organophosphorus pesticides (Singh and Walker, 2006).

2.2.9a. Dyes

Bacterial strains degrade azo dyes under aerobic and anaerobic conditions. The azo bond of dye serves as a final electron acceptor and cleaved into aromatic amines. Mixed consortia of bacteria such as *Proteus* sp., *Pseudomonas* sp., *Enterococcus* sp., *Shewanella decolorations* aid in biodegradation and decolorisation of dyes. *Pseudomonas stutzeri* ETL-4 can assist in degradation of the Congo red (Shah, 2014). In fungi, removal of dyes such as Congo red, Bismarck Brown, Brilliant Red K-2BP, Methyl orange, Yellow RR and Remazol Red RRhas been noted in species *Trametes versicolor*, *Phellinus gilvus*, *Pleurotus sajor-caju*, *Pleurotus ostreatus* IBL-02, *Pycnoporus sanguineus*, *Dichomitus squalens*, *Irpex flavus*, *Daedalea flavida*, *Polyporus sanguineus*, *Funalia trogii* ATCC200800, *Ischnoderma resinosum*, *Dichomitus squalens, Phanerochaete chrysosporium* IBL-03, *Coriolus versicolor* IBL-04, *Ganoderma lucidum* IBL-05, *Schizophyllum commune* IBL-06 (Yang et al., 2003; Machado and Matheus, 2006; Freire Andrade et al., 2013) (Table 2.8). The mechanism of degradation involves enzymes of phenol oxidases group such as lignin peroxidase (LiP), manganese peroxidase (MnP) and laccases (Kamida et al., 2005). Biodegradation of sulfonated azo dyes such as Orange I, Acid Orange 7, Acid Orange 8, Great Acid Red, and 4-[(4-hydroxyphenyl) azo]-benzenesulfonic acid by white-rot fungi produced 4-hydroxy-benzenesulfonic acid, 3-methyl-4-hydroxy-benzenesulfonic acid, benzenesulfonic acid, 1,2-naphthoquinone-6-sulfonic acid, and 3-methyl-benzenesulfonic acid as major biotransformation products. The degradation of Mordant Violet 5 by *Pleurotus ostreatus* gave benzenesulfonic acid, 4-hydroxybenzensulfonate, and 1,2-naphthoquinone, while degradation of Disperse Orange 3, Disperse Orange 1, and Disperse Red 1 produced nitrobenzene, 4-nitrophenol and 4-nitroaniline.Rhodamine B accumulation and efflux studies divulged that the bacterial isolate has greatly utilized the efflux pump system to accumulate and siphon out the dye.Some bacterial species also showed capacity to efflux dyes such as Rhodamine B via efflux pump system.

Fungal species such as *Aspergillus flavus*, *A. niger, Helminthosporium* sp, *Mucor* sp, *Penicillium* sp., *Trichoderma viride*, *Myrothecium* sp. IMER1, and *Fusarium* sp. The mushroom fungi *Lentinus conatus*, *Ischnoderma resinosum*, and *Ganoderma lucidum* KMK2 decolorize the dyes such as Remazol Yellow, Remazol Orange and textile dyes viz. reactive black 5, reactive blue 19, reactive Red 22, and reactive yellow 15. *G. lucidum* specifically showed potential for degradation of Remazol Brilliant Blue R (RBBR) and diazo dye Remazol Black-5 (RB-5) (Raghukunar et al., 2008). *Coprinellus xanthothrix* was able to degrade a polyaromatic dye Poly R-478.

Table 2.8. Selected dyes removed by microbes

Dyes	Microbes
Remazol Brilliant Blue R (RBBR)	*Pleurotus ostreatus* BP, *Trogia buccinalis*
Orange G, Amaranth, Orange I, Cuphthalocyanin, Poly R-478, Malachite Green and Crystal Violet	*Dichomitus squalens. Schizophyllum commune* IBL-06
Solar Golden Yellow R, Solar Orange RSN, and Cibacron Red FN-2BL	*Ganoderma lucidum*
Crystal violet, malachite green, and bromophenol blue	*Trametes villosa, Pycnoporus sanguineus*

2.2.9b. Pharmaceuticals

Sphingomonas sp. strain Ibu-2 showed degradation of ibuprofen to produce 1,2-dihydroxy ibuprofen, isobutylcatechol, 5-formyl-2-hydroxy-7-methylocta-2,4-dienoic acid and 2- hydroxy-5-isobutylhexa-2,4-dienedioic acid as intermediate compounds. White-rot fungi *Trametes versicolor, Irpex lacteus, Ganoderma lucidum* and *Phanerochaete chrysosporium* degrade ibuprofen, clofibric acid, and carbamazepine (Marco-Urrea et al., 2009a, b, c; Lagenhoff et al., 2013). Cytochrome P450 enzymes are involved in the oxidation to produce metabolites namely 1-hydroxy ibuprofen and 2-hydroxy ibuprofen. Fungus takes up these as substrates into the cell and that the degradation process occurs intracellularly. Two mono-hydroxylated intermediate products such as 40-hydroxydiclofenac and 5-hydroxydiclofenac (DCFs) were formed via fungal transformation. The degradation of non-steroidal anti-inflammatory drugs diclofenac (DCF) and mefenamic acid (MFA) has been reported by *Phanerochaete sordida* YK-624. Ligninolytic enzymes such as manganese peroxidase (MnP) and laccase effectively remove bisphenol A, nonylphenol, 4-tert-octylphenol, 17b-estradiol, ethinylestradiol, genistein, iso-butylparaben and n-butylparaben (Suzuki et al. 2003; Tamagawa et al. 2005, 2007; Marrco-Urrea et al., 2009b, c; Mizuno et al. 2009).

2.2.9c. Explosives

Both fungi and bacteria possess the ability to reduce explosives (Scheiber et al., 1997). Aerobic bacteria such as *Rhodococcus rhodochrous* strain II Y, *Rhodococcus* sp. strain ON22 *Williumsia* sp. strain KTR4, and *Gordonia* sp. strain KTR9 (Fournier et al., 2002; Thompson et al., 2005), and white-rot fungi reduce and transform nitro aromatics such as TNT into monoamino-dinitrotoluenes, diamino-nitrotoluenes, hydroxylamino-dinitrotoluenes (AmDNT) and degrade it further with the help of laccases, aromatic nitroreductases and phenoloxidases which are further degraded by laccase and MnP. The reduction of a nitro group of the TNT molecule is a successive reaction cascade via ni-troso and hydroxylamino intermediates. Reduction forms hydroxylaminodinitrotoluenes (OHADNT) and aminodinitrotoluenes (ADNT), which are further reduced into diaminonitrotoluenes (DANT). The first step of reduction involves membrane bound nitroreductases. The intermediate compounds are further metabolized by manganese-dependent peroxidase (MnP) or lignin peroxidase LiP (Van Aken et al. 1997). Manganese peroxidase (MnP) cleaves aromatic rings of monoamino-dinitrotoluenes to produce substantial amounts of CO_2. ADNT serve as substrate for MnP (Scheibner et al., 1997). Nitroreductases convert explosives containing*N*-

heterocyclic structures, such as RDX and octahydro-1,3,5,7-tetranitro-1,3,5,7-tetrazocine (HMX) into their respective mono nitroso derivatives. The strains, *Clitocybula dusenii* TMb12 and *Stropharia rugosa-annulata* DSM11372 mineralized TNT to CO_2 (Hughes et al., 1997; van Aken et al., 2004). Micromycetes (deuteromycetes, ascomycetes, zygomycetes) strains convert 2,4,6-trinitrotoluene to AmDNT. MnP acts as an oxidizing agent of the target molecules. A nitroarylreductase enzyme isolated from *Neurospora crassa* reduced range of nitroaromatic compounds including TNT. *Clitocybula dusenii TMb12* and *Stropharia rugosa-annulata* DSM11372 mineralized Micromycetes (deuteromycetes, ascomycetes, zygomy-cetes) produced higher amounts of AmDNT but were unable to mineralize TNT significantly.

Biodegradation of cyclic nitramines include (a) formation of a nitramine free radical and loss of nitro functional groups, (b) reduction of nitro functional groups, (c) direct enzymatic cleavage, (d) a-hydroxylation, or (e) hydride ion transfer. Intermediate compounds formed decompose in water producing nitrite, nitrous oxide, formaldehyde, or formic acid as end products. Enzymatic reactions with the nitramines lead to the reduction of the molecule to nitroso- or hydroxylamino-derivatives, formation of a free radical and loss of a nitro group, direct ring cleavage at N-N or C--N bonds, ex-hydroxylation at C-H bonds, or a hydride ion transfer.

RDX biodegradation occurs under aerobic (Thompson et al., 2005), nitrate-reducing (Bradley and Dinicola, 2005), sulfate reducing, manganese-reducing (Bradley and Dinicola 2005), anaerobic (Hawari et al., 2000), and methanogenic conditions (Adrian et al., 2003). Reduction (two-electron, single electron) involves addition of two or more redox equivalents to RDX leading to reduced nitroso and postulated hydroxylamino- or amino-triazine derivatives (Fournier et al., 2002). Denitration involves the addition of a single electron to RDX generating the anion radical (RDX) followed by ring cleavage. Enzymatic cleavage refers to a direct enzymatic attack at the carbon-nitrogen (C -N), methylene (C-1-1), or nitrogen-nitrogen (N-N) bonds in the molecule. The sequential reduction of RDX occurs through mono-, di-, and trinitroso RDX (MNX, DNX, and TNX). The reduction of these compounds to hydroxylamino derivatives destabilizes the ring leading to hydrolytic cleavage and the fomation of hydrazines, methanol, and formaldehyde. MNX, DNX, and TNX were identified as transient intermediates and methanol and formlaldehyde as end products (Hawari et al., 2000; Halasz et al., 2002). Formaldehyde is subsequently metabolized to formic acid, methanol, carbon dioxide, or methane (Fournier et al., 2002, 2004; Van Aken et al., 2004). The two-electron reduction pathway of RDX degradation has been reported in

Kiehsiella pneumoniae strain SCZ-1, *Clostridium* sp. *Trementus* strain HA W-1, *Shewmwlla halif'axensis* strain HA W-EB4, *Methylohacterium* spp., *Enterobacteria*, *Shetvanclla* sp. 1-IAW-EB2; and *S'hewanella* sp. HAW-EB5 (Zhao et al., 2002, 2004, 2006; van Aken et al., 2004; Fournier et al., 2005; Nyanhongo et al., 2006). The two-electron reduction pathway occurs by oxygen-insensitive type I nitroreductasc enzymes and type I nitroreductase. Nitrate oxidoreductase reduces RDX to ammonium, nitrous oxide, formaldehyde and an intermediate methylenedinitramine (MNX) through ring cleavage product (Hawari et al., 2000; Bhushan et al., 2002; Halasz et al., 2002). The strains utilize RDX as sole carbon, nitrogen, and energy sources in a catabolic pathway. Anaerobic denitration involves a single electron transfer facilitated by oxygen-sensitive type II nitroreductase generating a free-radical anion (RDX) and subsequently releasing a nitro group. The loss of the nitro group destabilizes the molecule leading to ring cleavage and the formation of one molecule of MDNA and unknown intermediate(s). The MDNA further decomposes in water to produce nitrous oxide and formaldehyde (Halasz et al., 2002). Strain *K. pneumoniue* SCZ-1 and *C. hifennentans* HAW-I showed denitration as the main pathway for biodegradation of RDX and MNX. RDX biodegradation catalyzed by a diaphorase enzyme via denitration produces nitrous oxide, nitrite, ammonium, and formaldehyde as noted in *Clostridium kluyvcri*. Studies also showed involvement of a cytochrome P450 enzyme in the transformation of RDX by *Rhodococcus* strains II Y and ON22. The gene responsible for degrading RDX, *xplA,* was isolated from strain I I Y, and it encodes a constitutively expressed, fused flavodoxin-cytochromc P450 enzyme (Seth-Smith ct al., 2002; Rylott et al., 2011). XpA protein assays indicated that NADPH, a natural reductant supported reduction by the enzyme (Rylott et al., 2011). Among fungi, *Phanerochaete chrysosporium* degrade RDX effectively (Reddy et al., 1998; Fournier et al., 2004).

Microbes can also metabolize HMX (Bhushan et al., 2004; Founier et al., 2004). The mechanism of HMX biodegradation is similar to that of RDX as electron donors and acceptors are the same (Adrian et al., 2003; Zhao et al., 2004). Anaerobic fate of HMX produce nitroso derivatives as intermediates such as Octahydro-1-nitroso-3,5. 7-trinitro-1,3,5, 7-tctrazocine (INO-HMX) formed from the two-electron reduction of the nitro groups on the ring. Biotransformation of HMX cleave into intermediates MDNA and bis(hydroxymethyl) nitramine, formaldehyde (trace), formic acid (trace), and nitrous oxide (Hawari et al., 2000; Zhao et al. 2004a, c).

2.3. Other Contaminants

Ligninolytic fungi, *Pleurotus ostreatus and Stereum hirsutum* remove estrogens or anti-androgens such as xenoestrogens bisphenol A (BPA), ethynilestadiol (EE2), bisphenol A (BPA), 17a-ethynilestradiol (EE2), 4-n-nonylphenol (NP), nonylphenol polyethoxilates (surfactants), herbicide linuron and the insecticide dimethoate. The laccase catalyzes the transformation of EDCs to oligomeric coupling products (Cajthaml et al., 2009; Subramanian and Yadav, 2009).*Non-imprinted polymer*particles(NIP) is reported to be degraded by *C. versicolor* via lignin degradative system. The degradation of simple cyanides has also been demonstrated in fungi such as *Trichoderma koningii* TkA, *Trametes versicolor* ATCC 200801, *Phanerochaete chrysosporium* ME 496 and *Pleurotus sajorcaju* (Hossain et al., 2005). Martin et al. (2009) reported the degradation of galaxolide (HHB), and tonalide (AHTN) by *Myrioconium* sp. strain UHH 1-13-18-4 and *Clavariopsis aquatica. Polyporus brumalis* degrade dibutyl phthalate (DBP), the main product being phthalic acid, benzyl alcohol, and hydroxyphenylacetic acid. Degradation of imidazolium compounds (ICs) and quaternary ammonium compounds (QACs) by strains of *Gliocladium roseum*, *Penicillium brevi-compactum*, *P. funiculosum*, *Phialophora fastigiata*, and *Verticillium lecanii* has been reported.

2.3.1. Biosurfactants

Biosurfactants are heterogeneous group of surface-active chemical compounds produced by microbes including bacteria, fungi and yeast (Shivkami et al., 2015). They are amphiphilic molecules with high specificity and stability at extremities of temperature, pH and salt concentration. They enhance solubilization, removal of contaminants and hence lower toxicity of contaminants (Wu et al., 2008; Das and Chandran, 2011). They can act as emulsifying agents by decreasing the surface tension and forming micelles. By reducing surface and interfacial tensions, biosurfactants increase the surface areas of insoluble compounds leading to increased mobility and bioavailability of organic compounds such as hydrocarbons. In Biosurfactants namely rhamnolipids produced by *Pseudomonas sp.* and *Rhodococcus erythropolis* help to amplify the bioavailability of contaminants such as hydrocarbons (Jaysree et al., 2011). Production of glycolipid biosurfactants by *Pseudomonas aeruginosa, P. putida, P. chlororaphis* and *Pseudozyma hubeiensis* has been reported (Abouseoud et al., 2008; Pansiripat et al., 2010; Jaysree et al., 2011; Konishi et al., 2011).

Table 2.9. Bioremediation of organic compounds using endophytic and rhizospheric bacteria*

Compound	Plants used	Microbes used	Reference
PCBs, TCP	*Triticum*spp.	*Herbaspirillum*sp. K1	Mannisto et al. (2001)
Benzoic acids	*Elymus dauricus*	*Pseudomonas aeruginosa*R75 *Pseudomonas savastanoi*CB35	Siciliano et al. (1998)
2,4-D	*Hordeum sativum* *Trifolium pratense* *Lolium perenne* *Pisum sativum* *Hordeum sativum*	*Burkholderia cepacia* *Pseudomonas putida*VM1450 *Burkholderia cepacia* *Burkholderia cepacia*	Jacobsen et al. (1997) Shaw et al. (2004) Germaine et al. (2004) Jacobsen et al. (1997)
Naphthalene	*Pisum sativum*	*Pseudomonas putida*VM1441 (pNAH7)	Germaine et al. (2009)
TCE	*Triticum spp.*	*Pseudomonas fluorescens*	Yee et al. (1998)
PCP	*Lolium perenne*	Indigenous degraders	He et al.2005
Crude oil	*Triticum spp.*	*Azospirillum lipoferum*spp	Muratova et al.(2005) Shaw et al. (2004)
Naphthalene	*Lolium multiflorum*	*Pseudomonas putida* PCL1444	Kuiper et al. (2001, 2004)
PAHs	*Festuca arundinacea*	*Azospirillum brasilense*Cd *Enterobacter cloacae*CAL 2 *Pseudomonas putida*UW3	Huang et al. (2004)
Phenanthracene	*Hordeum sativum*L.	*Pseudomonas*	Ankohina et al. (2013)
Toluene MTBE, BTEX, TCE	*Lupinus luteus* *Populus* *Populus*cv.*Hazendans*and cv.*Hoogvorst*	*Burkholderia cepacia*G4 *Burkholderia cepacia*Bu61 *Pseudomonas*sp.	Barac et al. (2004) Taghavi et al. (2005) Germaine et al. (2004) Porteus-Moore et al. (2006)
TNT, RDX, HMX	*Populus deltoidesnigra*DN34	*Methylobacterium populi*BJ001	Van Aken et al. (2004, 2007)

*Modified from Ryan et al. (2008)

Biosurfactants assist in removal of organic contaminants such as hydrocarbons by increasing substrate bioavailability for microorganisms and interaction with the cell surface, which increases the hydrophobicity of the surface allowing hydrophobic substrates to associate more easily with bacterial cells (Pacwa-Płociniczak et al., 2011).

2.3.2. PGPR and PGPB Degradation/Rhizoremediation

Plant associated endophytic and rhizospheric bacteria contribute to biodegradation of toxic organic compounds in contaminated soil. Naturally occurring plant growth promoting rhizobacteria (PGPR) benefit plants by providing growth promotion (Amora-Lazcano et al., 2010). Plants supply nutrients for growth and multiplication of microorganisms, hence increasing their capacity for degradation. Microbes fix atmospheric nitrogen and supply it to plant, synthesize siderophores, which sequester Fe from the soil and provide it to plant cells. *Pseudomonas* strains possess PGPR activity. PCB degraders associated with rhizosphere growing naturally in a contaminated site have been identified as *Rhodococcus, Luteibacter* and *Williamsia.* The free-living nitrogen fixer *Azospirillum lipoferum* found in the rhizosphere of crop plants degrades pesticides such as malathion. Plant growth promoting bacteria (PGPB) play a vital role in metal mediation (Gao et al., 2010). They also solubilize minerals such as phosphorus making them more readily available for plant growth. PGPB act as biocontrol agents against phytopathogens and stimulate plant growth. Endophytic bacteria inhabit the internal tissues of plants reduces thus reducing the competition between bacterial strains. Porteus Moore *et al.* (2006) isolated 121 endophytic strains from hybrid poplar trees with potential to enhance phytoremediation (Table 2.9). A methylotrophic endophytic bacterium isolated from hybrid poplar trees (*Populus deltoids* X *Populus nigra* DN34) showed capacity for degrading explosives/nitroaromatic compounds such as TNT, RDX and HMX, mineralizing to produce carbon dioxide (van Aken *et al.*, 2004). It also showed capacity of growing on a BTEX-contaminated site and possess potential for degrading VOCs (toluene and naphthalene), chlorinated organic herbicides such as 2,4-D (Porteus Moore *et al., 2006;* Doty, 2008; Germaine*et al.,* 2009*).*

3. Molecular Tools for Enhancing Remediation Capacity

Recombinants developed by alteration in the genetic material of the microbes show enhanced efficiency for degrading wide range of contaminants (Lan et al., 2006; Yuanfan et al., 2010). Genetically engineered endophytic and rhizospheric bacteria show increased efficiency for degradation of toxic compounds. The success of the technology depends upon the microbial strainstability after cloning and high expression of the target gene, tolerance or insensitivity to the contaminant and survival in specific plant rhizosphere is important.

3.1. Inorganic Contaminants

Different approaches have been followed to enhance metal removal or accumulation by microbes. These include i) genes for metal-binding polypeptides by fusion-binding domains to outer-membrane-anchored proteins including metallothioneins, polyhistidines (Valls *et al.,*2000); ii) specific metal transporter genes with MTs in the cytoplasm (Wolfram and Bauerfeind, 2009); iii) synthesis of metabolites like metal bound coenzymes and siderophores involved in the degradation pathway (de Oliveira Martins et al., 2008). Cytoplasmic expression of metal-binding polypeptides assists in cellular detoxification of metals. Protein design and selection of novel metal-binding peptides increase ligand stability and improve the affinity towards heavy metal ions (Yu et al., 2015). Metal-binding peptides increase ligand stability and improve the affinity or selectivity towards heavy metal ions. Metal ligands include peptides with high contents of cysteine or histidine (Martinez et al., 2006).

Bacterial strains like *Escherichia coli*, *Pseudomonas putida*, *Bacillus subtilis* have been genetically modified by overexpressing the genes for phytochelatin synthesis (Table 2.10). The plant enzyme phytochelatin synthase (PCS or glutathione g-glutamylcysteinyltransferase) is involved in the synthesis of phytochelatins (PCs). Co-expression of precursor glutathione (GSH) along with phytochelatins (PC) induces several fold increase in PC finally increasing Cd accumulation. Eukaryotic MTs have been cloned in bacteria. Expressions of MTs on the cell surface i.e., periplasmic space or to outer membrane compartments facilitate the metal accumulation and the efficient desorption of the bound metal to regenerate the system. An *E. coli* strain expressing MT fused to the outer membrane maltose protein (LamB) with a 15-20-fold rise in Cdbinding potential has been developed. Enhanced Cd2þ-binding performance has been achieved by expression of MT as a fusion to the L domain of the IgA protease autotransporter on the surface of *E. coli*, *R. metallidurans* and *Pseudomonas putida*. Genetically engineered bacteria coexpressing the MerT-MerP mercury transporter with MTs or metal-binding peptides in the cytoplasm showed high metal bioaccumulation. The cloning of yeast MTs fused to glutathione S-transferase together with a nickel transporter from *Helicobacter pylori*, produced a 3-fold increase in Ni accumulation in *E. coli*. *Ralstonia eutropha* has been genetically modified to express mouse metallothionein on the cell surface and decrease the toxic effect of the Cd. *Synechococcus* sp. (cynobacterial strains) with the expression of the smtA gene and metal binding protein has been reported. Cytoplasmic production of human MT fused to protein ara B in *E. coli* brought about a 3-5 folds increase in Cd and Cu bioaccumulation. Intracellular or extracellular binding (and thus immobilisation) of the metal with a cognate protein (frequently a metallothionein) leads to transformation of the toxic ion into a less toxic form, and the dissimilatory reduction of the metal. Peptides show high affinity for Cd ions by incorporation of Cys-Gly and Cys-Cys-Gly in *E. coli* as a fusion to the maltose-binding protein. The modified proteins were targeted to the periplasm of *E. coli*, endowing the modified strains with enhanced 109Cd2þ-binding capacity at low metal concentrations. The generation of dimeric and tetrameric human MTs increased metal binding activities when proteins are expressed in *E. coli*. Genetically engineered bacteria coexpressing the MerT-MerP mercury transporter with MTs or metal-binding peptides in the cytoplasm showed an Hg bioaccumulation comparable to that of cells directly expressing the binding peptides on the cell surface.

Metal ligands i.e., peptides with high contents of Cys or His establish coordination bonds with metal ions. Increased Cd2þ and Ni2þ binding was also obtained by expressing poly-His peptides on Gram-positive bacteria such as *Staphylococcusn* species. A peptide shows high affinity for Cd by incorporation of Cys-Gly and Cys-Cys-Gly repeats in *E. coli* as a fusion to the maltose-binding protein. *E. coli* with cysteine- and histidine-rich metal-binding peptides engineered into the LamB protein increased the Cd2þ-binding activity 4-fold. The *E. coli* outer membrane proteins LamB and OmpC have been exploited to surface-exposed poly-His peptides conferring more metal adsorption. Synthetic genes encoding several (Glu-Cys)nGly peptides linked to an lpp-ompA fusion gene has been cloned into *E. coli*.

SRB have been used in the removal of heavy metals. Metals are immobilised from the soil by sulfur-oxidising bacteria such as *Shewanella* and *Geobacter* species. Strains such as *Klebsiella planticola* thrive in high Cd2þ concentrations and precipitate significant amounts of Cd sulfide. The expression of the thiosulfate reductase gene from *Salmonella enterica* into *E. coli* produced modified strain *E. coli* DH5K strains produced significantly more sulfide than the control strains. Overexpression of enzymes serine acetyltransferase and cysteine desulfhydrase in *E. coli* overproduced cysteine and subsequently converted to sulfide. The metal phosphate biomineralisation achieved by the introduction of phoN from *Citrobacter* or related phosphatases is also noted.

Natural resistant pathways for heavy metals have been regulated by metalloregulatory protein in microorganisms. A genetically engineered bacterium *Deinococcus geothemalis* showed Hg reduction due to the expression of mer operon from *E. coli*. Mercury resistant bacteria *Cupriavidus metallidurans* strain MSR33 produced by introducing pTP6 plasmid that provided genes (merB and merG) regulating Hg biodegradation along with the synthesis of organomercurial lyase protein (MerB) and mercuric reductase (MerA). Modification of *Pseudomonas* strain with the pMR68 plasmid with *mer* genes made that strain resistant to Hg. The mechanism of Hg degradation by bacteria (*Klebsiella pneumonia* M426) includes Hg volatilization by reduction of Hg (II) to Hg (0) and precipitation as insoluble Hg is changed to volatile thiol (H_2S). Genetic engineering of *Deinococcus radiodurans* (radiation resistant bacterium) by cloned genes of *tod* and *xyl* operons of *Pseudomonas putida* naturally reduced Cr (IV) to Cr (III) and brought complete toluene (fuel hydrocarbon) degradation (de Oliveira Martins et al., 2008).

Lipopolysaccharides play a major role in metal biosorption in Gram-negative bacteria. *Pseudomonas aeruginosa* PAO1 A-band and B-band lipopolysaccharide mutants differ in their binding specificities towards Fe suggesting high metal specificity. *Arabidopsis thaliana* gene for phytochelatin synthase (PCS; PCSAt) was introduced into *Mesorhizobium huakuii* subsp. *rengei* strain B3 to produce phytochelatins and accumulate Cd. The gene was expressed under the control of bacteroid specific promoter, the nif H gene. Recombinant bacterium, *Alcaligenes eutrophus* AE104 (pEBZ141) removed Cr from industrial wastewater, *Rhodopseudomonas palustris* expressed Hg transport system and metallothionein for Hg removal from heavy metal wastewater. Over-expression of the heterologous PcPCS1 gene in bacterial strains appears to be a promising way to increase the heavy metal accumulation in microbes. The development of soil bacteria expressing the PcPCS1 gene helps to produce PCs that absorb Naþ from saline soils. Expression of the PcPCS1 gene from the pear (*Pyrus calleryana* Dcne.) is induced after Cd and Cu treatments. *E. coli* with over-expressed PcPCS1 had enhanced tolerance to Cd, Cu, and Hg.The transformed *E. coli* cells could tolerate high concentrations of Hg (200 mM), Cd (1.5mM), Cu (2.5mM).

Genes and proteins involved in radionuclide reduction pathways have been elucidated (Methe et al., 2003). Members of family *Geobacteriaceae* have shown potential for radioactive metal reduction (Lloyd *et al.,* 2003). The gene *dcuB* from *G. sulfurreducens* encoding a fumarate transporter was engineered in *G. metallireducens* (Butler *et al.,* 2006). *G. sulfurreducens* showed 100 *c*-type cytochrome gene coding regions in its genome and many of its translated proteins involved with radionuclide reduction pathways. DNA-microarray-mediated analysis revealed 121 genes were upregulated in *Shewanella oneidensis* during U(VI) reduction (Bencheikh-Latmani *et al.,*2005). The expression of the*NiCoT*gene in *Rhodopseudomonas palustris* CGA009 and *Novosphingobium aromaticivorans* F-199 was highly increased when the organisms were grown in the presence of radioactive Co (Raghu *et al.,* 2008). Tian et al. (2000) found a total of 552 differentially regulated proteins on the membrane, including a cytochrome*bd*ubiquinol oxidase involved in the radioresistance of the organism *D. geothermalis*. The cascades of enzymatic proteins in the outer membrane of *S. oneidensis* were found to effectively reduce the levels of radionuclides such as U (Marshall *et al.,* 2006). The radioresistant bacterium, *D. radiodurans*, 2DE and MALDI-TOF MS revealed that 31 radiation-responsive proteins were significantly upregulated, including RecA and PprA (Lu *et al.,* 2009). A genetically engineered *D. radiodurans* strain was created by cloning the *E. coli* gene (*merA*) that provides the ability to utilize carbon and energy from catabolism of toluene and mercury (radioactive contaminants) (Brim *et al.,* 2000). Recombinant strain of *E.coli* was generated with five times the sorption ability for U(VI) radionuclides by altering the transporter genes *nixA* (*Helicobacter pylori*) and *merTP* (*Serratia marcescens*) respectively (Beckwith *et al.,* 2001).

Keasling *et al.* (2000) developed a recombinant by cloning the gene encoding polyphosphate kinase in *P. aeruginosa,* resulting in the precipitation of a complex containing both P and U on the cell surface. Another tool to counteract metal or radionuclide contamination is the generation of bacteria with engineered polyphosphate pathways. The introduction of an inducible polyphosphate kinase into *P. aeruginosa* resulted in the accumulation of large quantities of polyphosphate (Valls and de Lorenzo, 2002). Genetic engineering of peroxidases can improve the catalytic properties and allow enhancement of the H_2O_2 resistance of the enzymes, or the structural modification of commercially available enzymes to enable the oxidation of other compounds with high redox potentials. Efficient production of recombinant CiP, CPO and DyP, and wild-type CPO improve the expression of peroxidases.

Table 2.10. Selected genetically engineered bacteria for the removal of heavy metals

Metals	Genetically Engineered Bacteria	Expressed Gene	Reference
Cd	*Escherichia coli*	SpPCS	Kang et al. (2007)
Cr	*P. putida* strain	Chromate reductase (ChrR)	Ackerley et al. (2004)
Hg	*E. coli* strain	Organomurcurial lyase	Murtaza et al. (2002)
Ni	*P. fluorescens*	Phytochelatin synthase (PCS	Sriprang et al. (2003)
Hg	*E. coli* JM109	Hg^{2+} transporter	Zhao et al. (2005)

3.2. Organic Contaminants

Bacterial strains can be engineered to produce specific enzymes capable of degrading toxic organic substances (Junghanns et al., 2009). Microbes can be engineered via recombinant DNA technology to produce specific enzymes capable of degrading toxic organic pollutants. The rate-limiting steps in metabolic pathways were genetically manipulated to achieve increased degradation rates, or incorporate new metabolic pathways into bacterial strains for the degradation of recalcitrant compounds. Essential genes are carried on a single chromosome but genes specifying enzymes required for the catabolism of substrates may be carried on plasmids in bacteria. Plasmids for degrading various compounds have been identified (Guha et al., 1997). These include i) OCT plasmid, which degrades octane, hexane and decane; ii) XYL plasmid which degrades xylene and toluenes, iii) CAM plasmid that decompose camphor and iv) NAH plasmid, which degrades naphthalene. Genetically modified strains of the rhizospheric bacteria *Pseudomonas fluorescens* F113, i.e., *Pseudomonas fluorescens* F113 *rifbph* (with a single chromosomal insertion of the*bph*operon) and*Pseudomonas fluorescens* F113: 1180 (with a single chromosomal insertion of the *bph* operon under the control of the *Sinorhizobium meliloti* nod regulatory system) have been developed. The modified rhizospheric bacteria colonized roots effectively and degrade organic toxicants such as PCBs more efficiently indicating their potential use in bioremediation technology. Strains of bacteria can degrade fuel hydrocarbons have been genetically transformed to clean oil spillages or treat sewage and effluents.

Plant-associated endophytes are genetically modified to degrade toxic organic compounds. Endophytic bacteria possess genes that degrade contaminants before being reinoculated back into the host plant. Horizontal gene-transfer results in the natural endophyte population having the capacity to degrade environmental pollutants without the need of long-term establishment of the inoculants strain. Genetically engineered strain of *P. putida* has been developedto degrade camphor, naphthalene, xylene, toluene, octanes and hexane. Genes responsible for degradation of compounds such as toluene, chlorobenzene acids, and other halogenated pesticides have also been inserted in microbes. *Pseudomonas* strain capable of oxidizing aliphatic, aromatic, terpenic and polyaromatic hydrocarbons has been successfully produced using a multiplasmid. The genetically engineered microbe such as *Pseudomonas putida* containing XYL, NAH and hybrid plasmid derived by recombinating parts of CAM and OCT with the capacity of degrading camphor, octane, salicylate, naphthalene and other hydrocarbons referred as superbug have been developed. The plasmids TOL are capable ofdegrading toluene and xylene, RA500 for 3, 5-xylene, pAC 25 for 3-cne chlorobenxoate, pKF439 for salicylate toluene. Genetically enhanced endophytic strain of the poplar endophyteis *Pseudomonas putida* VM1441, i.e., *Pseudomonas putida* VM1441 (pNAH7) (Germaine *et al., 2009).* A genetically enhanced endophytic strain *Burkholderia cepacia* G4 showed tolerance to toluene (Barac*et al.,* 2004). The plasmid, pTOM, encodes a pathway for the degradation of toluene was transferred via conjugation to the natural endophyte, providing the genes for toluene degradation. Bacteria (endophytic or rhizospheric) expressing a specific bacterial glutathione-S-transferase (GST) isolated from *Burkholderia xenovorans* LB400, $BphK^{LB400}$ [wildtype and mutant (Ala180Pro)], and dehalogenating toxic chlorinated organic pesticides, chloromequat chloride $BphK^{LB400}$ [wildtype and mutant (Ala180Pro)], have potential for bioremediation of chlorinated organic pollutants.

Germaine *et al.* (2009) reported that a genetically enhanced endophytic strain of the poplar endophyte *Pseudomonas putida* VM1441, i.e.,

*Pseudomonas putida*VM1441 (pNAH7 developed for protection of pea plants from the toxic effects of naphthalene. Inoculation of plants with the strain facilitated higher (40%) naphthalene degradation rate compared with uninoculated plants. Barac *et al.* (2004) reported that a genetically enhanced endophytic strain of the soil bacterium *Burkholderia cepacia* G4 could increase inoculated yellow lupine plant tolerance to toluene, and decrease phytovolatilization of toluene from the plant into the atmosphere by 50-70%. The plasmid, pTOM encodes a pathway for the degradation of toluene, was transferred via conjugation to the natural endophyte, providing the genes for toluene degradation. Taghavi *et al.* (2005) showed that degradative plasmid, pTOM via horizontal gene transfer produced starins with efficient degradation of toluene in poplar plants. Yeast clones expressing individual P450s of *P. chrysosporum* (PcCYPs) transform dioxin. Recombinant white-rot basidiomycetes *Coriolus hirsutus* expressing rat CYP1A1 demonstrated the efficient degradation of 2,7-DCDD through the action of the recombinant rat CYP1A1(Orihara et al., 2005). The introduction of a useful P450 gene enhances the dioxin metabolizing ability of white-rot fungus (Noriyuki et al., 2010). Nagata*et al.* (1993) cloned and sequenced two genes involved in γ-HCH degradation in bacteria. The*lin*A gene encodes Y-HCH dehydrochlorinase which converts γ-HCH to 1,2,4 TCB via γ -PCCH and 1,4-TCDN. The*lin* B gene encodes 1,4-TCDN chlorohydrolase, which converts 1,4-TCDN to 2,4-DDOL via 2,4,5-DNOL. Genetically engineered *P. putida* contained both *lin*A and *lin B* genes. For polychlorinated biphenyls degradation, chromosomally located PCB catabolic genes of *R. eutropha* A5, *Achromobacter* sp. LBS1C1, and *A. denitrificans* JB1 have been transferred into *R. eutropha* CH_3 through natural conjugation.

Most of pesticide-degrading genes responsible for catabolic degradation are located on the chromosomes, plasmids or transposons (Lan et al., 2006). The organophosphate hydrolase gene from *Flavobacterium* sp. and carboxylesterase B1 gene (b1) from *Culex pipiens* were cloned in the coexpression vector and a single microorganism capable of producing enzymes for degradation of organophosphate (parathion), carbamate (pirimicarb) and pyrethroid pesticides (deltamethrim). Methyl parathion hydrolase gene, *mpd*, responsible for hydrolyzing methyl parathion to *p*-nitrophenol and dimethyl phosphorothioate has been cloned from bacterial strains (Yuanfan et al., 2010). A genetically engineered strain, JQL4-5-*mpd*, capable of simultaneously degrading fenpropathrin and methyl parathion was constructed by random insertion of the methyl parathion hydrolase gene (*mpd*) into the chromosome of strain JQL4- 5. This strain proved to beeffective in the bioremediation of multiple pesticides contaminated environments (Yuanfan et al., 2010). Organophosphorus hydrolase (OPH), isolated from both *Flavobacterium* sp. ATCC 27551 and *Pseudomonas diminuta* MG showed potential for hydrolyzing wide range of oxon and thion organophosphate pesticides (Horne et al., 2002). The *mpd* gene encoding an organophosphate degrading protein was isolated from *Plesiomonas* sp. The methyl parathion hydrolase gene (*mpd*) and enhanced green fluorescent protein gene (*egfp*) successfully coexpressed using pETDuet vector in *Escherichia coli* BL21 (DE3). The coexpression produced recombinant protein MPH that showed high enzymatic degradative activity for several OP residues. Human esterases were also expressed in microbes to get enhanced capacity for biodegradation of organophosphates and carbamates. Organophosphate hydrolase (OPH) was coexpressed with CbE, and the mixed enzymes degraded a variety of P-O bond containing OPs (chlorpyrifos, methyl parathion, dichlorvos and phoxim). Bacteria possess a novel triesterase called as parathion hydrolase or organophosphorus acid anhydrase encoded by the highly conserved organophosphate degradation (opd) gene, localized either on dissimilar indigenous plasmids or on the chromosome (Zhang et al., 2005; Singh and Walker, 2006). Parathion hydrolase hydrolyses the characteristic triester bond of OP pesticides and the hydrolytic products are not toxic as they do not inhibit acetyl cholinesterase (AchE) activity in animals. Gram-negative bacteria convert PNP to maleylacetate via hydroquinone, while Gram-positive bacteria such as *Serratia sp.* strain DS001 degrades PNP to 4-nitrocatechol and 1,2,4-benzenetriol (Abdel-Razek et al., 2013). Component "A" of p-nitrophenol hydroxylase in *Serratia sp.* strain DS001 was found to degrade both methyl parathion and its major degradation product p-nitrophenol (PNP). The initial hydrolysis was carried out by parathion hydrolase. Bacterial parathion hydrolases are encoded by the highly conserved organophosphate degradation (*opd*) gene found on large, dissimilar, and indigenous plasmids (Pakala et al., 2006). The oxidative degradation of p-nitrophenol by DS001 was due to initial hydroxylation by 4-nitrophenol 2-hydroxylase. A two-component PNP monooxygenase has been reported in Gram-positive *Bacillus sphaericus* JS 905 and *Rhodococcus opacus* (Kitagawa et al., 2004). P-nitrophenol is converted to nitrocatechol by a Gram-negative organism (Pakala et al., 2006). Bacteria expressing a specific bacterial glutathione-S-transferase (GST) isolated from *Burkholderia xenovorans* LB400, $BphK^{LB400}$ [wildtype and mutant (Ala180Pro)] is capable of dehalogenating toxic chlorinated organic pesticides.

Yang et al. (2006) were successful in cloning the mpd gene from a chlorpyrifos – degrading bacterium. The mpd gene was subsequently cloned in *E. coli* DH5α cells and subsequently expressed in *E. coli* BL21 (DE3). Sequence BLAST result showed that the gene was similar to mpd gene of *Plesimonas sp.* M6, *Pseudomonas putida*, and *Ochrobactrum sp*. MP-4 at the nucleotide level. *Cupriavidus* sp. DT-1 showed capacity for degradation of chlorpyrifos and 3,5,6-trichloro-2-pyridinol (TCP). Chlorpyrifos was first hydrolyzed to TCP, successively dechlorinated to 2-pyridinol, and then subjected to the cleavage of the pyridine ring. The *mpd* gene that encodes the enzyme responsible for chlorpyrifos hydrolysis to TCP was cloned and expressed in *Escherichia coli* BL21 (Lu et al., 2013). Organophosphorus hydrolase (OPH; encoded by the *opd* gene) is the microbial enzyme with the capacity to hydrolyze pesticides. Methyl-parathion hydrolase (MPH; encoded by the *mpd* gene), Are *Pseudaminobacter*, *Achrobacter*, *Brucella* and *Ochrobactrum* genes, they were identified by comparison with the gene *mpd* from *Pleisomonas* sp. M6 strain, the gene for the organophosphorus hydrolase has 996 nucleotides, a typical promoter sequence of the promoter TTGCAA N17 TATACT from *E. coli*. Anatural *p*-nitrophenol (PNP) degrader *Ochrobactrum* sp. strain LL-1 for complete mineralization of dimethyl OPs was developed. The complete mineralization of methyl parathion (MP) involved hydrolysis leading to the generation of PNP and dimethylthiophosphoric acid, and finally PNP degradation to hydroquinone (Zheng et al., 2009). *Streptomyces* sp.HU-S-01, a strain for complete degradation of 3-phenoxybenzoic acid was isolated and characterized (Lin et al., 2011).

Genes encoding the pyrethroids hydrolyzing carboxylesterases isolated from fungi, insect and animal liver were purified, cloned and functionally expressed (Stok et al., 2004; Liang et al., 2005; Wu et al., 2006; Wang et al., 2009). Microbial degradation of pyrethroid compounds involves hydrolysis of carboxyl ester linkage followed by oxidation and dehydrogenation (Liang et al., 2005). Phosphotriesterase (PTE) shows high catalytic activity towards organophosphate pesticides. These enzymes are encoded by gene called opd (organophosphate-degrading). *Flavobacterium* ATCC 27551 contains the opd gene that encodes a PTE (Latifi et al., 2012). The degradation of cypermethrin by HU-S-01 involved two enzymes. The degradation pathway proposed a hydrolytic pathway followed by oxidation and dehydrogenation for allethrin and cypermethrin by *Micrococcus* sp. and *Acidomonas* sp. (Mandar et al., 2005; Tallur et al., 2008). The pod gene associated with degradation of pesticides such as parathion, methylparathion isolated from *Flavobacterium* sp. ATCC27551 and *Pseudomonas diminuta* is well characterized. The gene was expressed in fungus *Phanerochaete chrysosporium*, *Gliocladium virens*. Additionally, insertion of BphKLB400 [wildtype and mutant (Ala180Pro)], into endophytic or rhizospheric bacteria improved potential for bioremediation of chlorinated organic pollutants. RDX biotransfomation by the rabbit liver cytochrome P450 enzyme and the XpA protein was faster under anaerobic than aerobic conditions (Bhushan et al., 2002; Rylottin et al., 2011).

Conclusions

Bacteria and fungi play a vital role in degradation of environmental pollutants. Easy availability, low maintenance, presence of catabolic enzymes and capacity to use pollutants as a growth substrate make them ideal candidates for use in bioremediation technology. Genetically engineering technology improves the potential of microbes to achieve remediation of contaminants through incorporation of genes/proteins capable of degrading contaminants. The genetically engineered microbes show promise for their implementation for *in situ* bioremediation for environmental cleanup via removal from soil, water and sediments.

References

Abdel-Razek, S.A.R.M., Folch-Mallol, J.L., Perezgasga Ciscomani, L., Sánchez-Salinas, E., Castrejón-Godínez, M. L., Ortiz-Hernández, M. L. (2013). Optimization of methyl parathion biodegradation and detoxification by cells in suspension or immobilized on tezontle expressing the *opd* gene. *Journal of Environmental Science and Health B* 48, 449-461.

Abo-Amer, A., Aly, E. (2011). Biodegradation of diazinon by *Serratia marcescens* di101 and its use in bioremediation of contaminated environment. *Journal of Microbiology and Biotechnolog* 21, 71-80.

Abo-Amer, A. E. (2012). Characterization of a strain of *Pseudomonas putida* isolated from agricultural soil that degrades cadusafos (an organophosphorus pesticide). *World Journal of Microbiology and Biotechnology* 28, 805-814.

Abouseoud, M., Maachi, R., Amrane, A., Boudergua, S., Nabi, A. (2008). Evaluation of different carbon and nitrogen sources in production of biosurfactant by *Pseudomonas fluorescens*. *Desalination* 223,143- 151.

Ackerley, D. F., Gonzalez, C. F., Keyhan, M., Blake, R., Matin, A. (2004). Mechanism of chromate reduction by the *Escherichia coli* protein, NfsA, and the role of different chromate reductases in minimizing oxidative stress during chromate reduction. *Envirmental Microbiology* 6, 851–86.

Adenipekun, C.O., Lawal, R. (2012). Uses of mushrooms in bioremediation: A review. *Biotechnology and Molecular Biology Review* 7(3), 62-68.

Adrian, N.R., Arnett, C.M., Hickey, R.F. (2003). Stimulating the anaerobic biodegradation of explosives by the addition of hydrogen or electron donors that produce hydrogen. *Water Research* 37, 3499-3507.

Agarry, S.E., Olu-Arotiowa, O.A., Aremu, M.O., Jimoda, L.A. (2013). Biodegradation of Dichlorovos (organophosphate pesticide) in soil by bacterial isolates. *Journal of Natural Science and Research* 3, 12-16.

Ai, T., Wang, H., Wen, X., Zhang, S., Huang, J. (2006). Microbial remediation of dimethoate-contaminated soil. *Nongye Huanjing Kexue Xuebao* 25(5), 1250-1254.

Aislabie J, Lloyd-Jones G (1995) A review of bacterial-degradation of pesticides. *Australian Journal of Soil Research* 33, 925-942.

Aislabie, J., Richards, N.K., Boul, H.L. (1997). Microbial degradation of DDT and its residues – a review.*New Zealand Journal of Agricultural Research* 40, 269-282.

Aitken, M.D., Stringfellow, W.T., Nagel, R.D., Kazunga, C., Chen, S.H. (1998). Characteristics of phenanthrene-degrading bacteria isolated from soils contaminated with polycyclic aromatic hydrocarbons. *Canadian Journal of Microbiology* 44, 743-752.

Akpor, O.B., Adelani-Akande, T.A., Aderiye, B. (2013). The Effect of temperature on nutrient removal from wastewater by selected fungal species. *International Journal of Current Microbiology and Applied Science* 2(9), 328-340.

Akponah, E. (2013). Bacterial and fungal biosorbents of heavy metals associated with hydrocarbon contaminated rainforest soils of the Niger delta region of Nigeria European *Journal of Experimental Biology* 3(3), 95-100.

Alves, S.A., Albergaria, J.T., Fernandes, D.V., Alvim-Ferraz, M.C., Delerue-Matos, C. (2010). Remediation of soils combining soil vapor extraction and bioremediation: Benzene. *Chemosphere* 80, 823-828.

Amora-Lazcano, E., Guerrero-Zúñiga, L.A., Rodriguez-Tovar, A., Rodriguez-Dorantes, A., Vasquez-Murrieta, M.S. (2010). Rhizospheric plant-microbe interactions that enhance the remediation of contaminated soils. In: *Current Research Technology and Education Topics in Applied Microbiology and Applied Biotechnology* (Mendez-Vilas A Ed.), Formatex 251-256.

Ankohina, T., Kochetkov, V., Zelenkova, N., Balakshina, V., Boronin, A. (2004). Biodegradation of phenanthrene by *Pseudomonas* bacteria bearing rhizospheric plasmids in model plant-microbial associations. *Applied Biochemistry and Microbiology* 40, 568–572.

Arulazhagan, P., Vasudevan, N., Yeom, I.T. (2010). Biodegradation of polycyclic aromatic hydrocarbon by a halotolerant bacterial consortium isolated from marine environment. *International Journal of Environmental Science and Technology* 7 (4), 839-852.

Asgher, M., Bhatti, H., Ashraf, M., Legge, R. (2008). Recent developments in biodegradation of industrial pollutants by white rot fungi and their enzyme system. *Biodegradation* 19, 771–783.

Asha, L.P., Sandeep, R.S. (2013). Review on bioremediation- potential tool for removing environmental pollution. *International Journal of Basic and Applied Chemical Sciences* 3, 2277- 2073.

Awasthi, N., Singh, A.K., Jain, R.K., Khangarot, B.S. (2003). Degradation and detoxification of endosulfan isomers by a defined co-culture of two *Bacillus* strains. *Applied Microbial Biotechnology* 62, 279 - 283.

Bansal, O.P. (2012). Degradation of pesticides. In: Pesticides evaluation of environmental pollution (Rathore H. S., Nollet M.L., Ed.). CRC Press, New York, United States of America, p.47-77.

Barac, T., Taghavi, S., Borremans, B., Provoost, A., Oeyen, L., Colpaert, J.V., Vangronsveld, J., van der Lelie, D. (2004). Engineered endophytic bacteria improve phytoremediation of water-soluble, volatile, organic pollutants. *Nature Biotechnology* 22, 583–588.

Barraga´ n-Huerta, B.E., Costa-pe´rezc, C., Peralta-cruza, J., Barrera-corte´ s, J., Esparza-garci´a, F.,Rodri´guez-va´ zquez, R.(2007). Biodegradation of organochlorine pesticides by bacteria grown in microniches of the porous structure of green bean coffee. *International Journal Biodeterioration Biodegradation* 59, 239-244.

Bass, C., Field, L.M. (2011). Gene amplification and insecticide resistance. *Pest Management Science* 67 (8), 886–890.

Becker, J.G., Seagren, E.A. (2010). Bioremediation of hazardous organics. In: (Mitchell, R; Ji-Dong YG. ed.) *Environmental Microbiology,* Wiley-Blackwell, New Jersey, United States of America p. 177-212.

Beckwith, C.S., McGee, D.J., Mobley, H.L., Riley, L.K. (2001). Cloning, expression, and catalytic activity of *Helicobacter hepaticus* urease. *Infect Immunology* 69, 5914–5920.

Bencheikh-Latmani, R., Williams, S.M., Haucke, L., Criddle, C.S., Wu, L., Zhou, J., Tebo, B.M. (2005). Global transcriptional profiling of *Shewanella oneidensis* MR-1 during Cr(VI) and U(VI) reduction. *Applied Environmental Microbiology* 71, 7453–7460.

Bending, G.D., Friloux, M., Walker, A. (2002). Degradation of contrasting pesticides by white rot fungi and its relationship with ligninolytic potential. *FEMS Microbiology Letters* 212, 59–63.

Bhagobaty, R.K., Joshi, S.R., Malik, A. (2006). Microbial degradation of organophosphorous pesticide: chlorpyrifos.*The Internet Journal of Microbiology*, 4.

Bhalerao, T.S., Puranik, P.R. (2007). Biodegradation of organochlorine pesticide, endosulfan, by a fungal soil isolate, *Aspergillus niger. International Journal of Biodeterioration and Biodegradation* 59, 315–321.

Bhalerao, T.S. (2012). Bioremediation of endosulfan-contaminated soil by using bioaugmentation treatment of fungal inoculant *Aspergillus niger. Turk Journal of Biology 36, 561-567.*

Bhattacharya, S., Das, A., Prashanti, K., Palaniswamy, M., Angayarkanni, J. (2014). Mycoremediation of benzo[a]pyrene by *Pleurotus ostreatus* in the presence of heavy metals and mediators. *Biotechnology* 4(2), 205-211.

Bhosle, N.P., Nasreen, S. (2013). Remediation of cypermethrin-25 EC by microorganisms. *European Journal of Experimental Biology* 3, 144-152.

Bhushan, B., Halasz, A., Spain, J., Thiboutot, S., Ampleman, G., Hawari, J. (2002). Biotransformation of hexahydro-1,3,5-trinitro-1,3,5-triazine catalyzed by a NAD(P)H:nitrate oxidoreductase from *Aspergillus niger. Environmental Science and Technology* 36, 3104–3108.

Bigley, A.N., Raushel, F.M. (2013). Catalytic mechanisms for phosphotriesterases. *Biochimica et Biophysica Acta (BBA)-Proteins and Proteomics* 1, 443-453.

Bradley, P.M., Dinicola, R.S. (2005). RDX (hexahydro-1,3,5-trinitro-1,3,5-triazine) biodegradation in aquifer sediments under manganese-reducing conditions. *Bioremediation Journal* 9, 1–8.

Brim, H., McFarlan, S.C., Fredrickson, J.K., Minton, K.W., Zhai, M., Wackett, L.P., Daly, M.J. (2000).Engineering *Deinococcus radiodurans* for metal remediation in radioactive mixed waste environments. *Nature Biotechnology* 18, 85–90.

Butler, J.E., Glaven, R.H., Esteve-Nunez, A., Nunez, C., Shelobolina, E.S., Bond, D.R., Lovley, D.R. (2006). Genetic characterization of a single bifunctional enzyme for fumarate reduction and succinate oxidation in *Geobacter sulfurreducens* and engineering of fumarate reduction in *Geobacter metallireducens*. *Journal of Bacteriology* 188, 450–455.

Çabuk, A., İlhan, S., Fılık, C., Çalişkan, F. (2005). Pb^{2+} biosorption by pretreated fungal biomass. *Turkish Journal of Biology* 29, 23–28.

Cajthaml, T., Kresinová, Z., Svobodová, K., Sigler, K., Rezanka, T. (2009). Microbial transformation of synthetic estrogen 17α-ethinylestradiol. *Environmental Pollution* 157, 3325–3335.

Camacho-Pérez, B., Ríos-Leal, E., Rinderknecht-Seijas, N., Poggi-Varaldo, H.M. (2012). Enzymes involved in the biodegradation of hexachlorocyclohexane: A mini review. *Journal of Enviroinmental Management* 95, 5306-5318.

Cambria, M.T., Minniti, Z., Librando, V., Cambria, A. (2008). Degradation of polycyclic aromatic hydrocarbons by *Rigidoporus lignosus* and its laccase in the presence of redox mediators. *Applied Biochemistry and Biotechnology* 149, 1-8.

Cameron, M.D., Timofeevski, S., Aust, S.D. (2000). Enzymology of *Phanerochaete chrysosporium* with respect to the degradation of recalcitrant compounds and xenobiotics. *Applied Microbiology and Biotechnology* 54, 751–758.

Carcciolo, A.B., Bottoni, P., Crobe, A., Fava, L., Funari, E., Giuliano, G., Silvestri, C. (2002) Microbial degradation of two carbamate insecticides and their main metabolites in soil. *Chemical Ecology* 18, 245–255.

Collins, G., Foy, C., McHugh, S., O'Flaherty, V. (2005). Anaerobic treatment of 2,4,6-trichlorophenol in an expanded granular sludge bed anaerobic filter (EGSB-AF) bioreactor at 15 0C. *FEMS Microbiology and Ecology* 53(1), 167-178.

Chang, Y.S. (2008).Recent developments in microbial biotransformation and biodegradation of dioxins. *Journal of Molecular Microbiology and Biotechnology* 15, 152–171.

Chanika, E., Georgiadou, D., Soueref, E., Karas, P., Karanasios, E., Nikolaos, G.T., Tzortzakakis, E.A., Karpouzas, D.G. (2011). Isolation of soil bacteria able to hydrolyze both organophosphate and carbamate pesticides. *Bioresource Technology* 102, 3184-3192.

Chaturvedi, A.D., Pal, D., Penta, S., Kumar, A. (2015). Ecotoxic heavy metals transformation by bacteria and fungi in aquatic ecosystem. *World Journal of Microbiology and Biotechnology* 31, 1595–1603.

Chen, S.H., Aitken, M.D. (1999). Salicylate stimulates the degradation of high molecular weight polycyclic aromatic hydrocarbons by *Pseudomonas saccharophila* P15. *Environmental Science and Technology* 33, 435-439.

Chen, S., Liu, C., Peng, C., Peng, C., Liu, H., Hu, M., Zhong, G. (2012). Biodegradation of chlorpyrifos and its hydrolysis product 3,5,6-trichloro-2-pyridinol by a new fungal strain *Cladosporium cladosporioides* Hu-01. *PLoS ONE* 7, 1-12.

Chino-Flores, C., Dantán-González, E., Vázquez-Ramos, A., Tinoco-Valencia, R., Díaz-Méndez, R., Sánchez-Salinas, E.,Castrejón-Godínez, M.L., Ramos-Quintana, F., Ortiz-Hernández, M.L. (2012). Isolation of the opdE gene that encodes for a new hydrolase of *Enterobacter sp.* capable of degrading organophosphorus pesticides. *Biodegradation* 23, 387-397.

Cramer, R.A. Jr, Stajich, J.E., Yamanaka, Y., Dietrich, F.S., Steinbach, W.J. (2006). Phylogenomic analysis of non-ribosomal peptide synthetases in the genus *Aspergillus*. *Perfect Journal of Gene* 383, 24-32.

Crocker, F., Indest, K., Fredrickson, H. (2006). Biodegradation of the cyclic nitramine explosives RDX, HMX, and CL-20. *Applied Microbiology and Biotechnology* 73, 274–290.

Cycon, M., Wojcik, M., Piotrowska-Seget, Z. (2009). Biodegradation of the organophosphorus insecticide diazinon by *Serratia sp.* and *Pseudomonas sp.* and their use in bioremediation of contaminated soil. *Chemosphere* 76, 494-501.

Cycon, M., Zimijowska, A., Piotrowska-Seget, Z. (2014). Enhancement of deltamethrin degradation by soil bioaugmentation with two different strains of *Serratia marcescens*. *International Journal of Environment Science and Technology* 11, 1305–1316.

da Silva, N.A., Birolli, W.G., Seleghim, M.H.R., Porto, A.L.M. (2013). Biodegradation of the organophosphate pesticide profenofos by marine fungi, Chapter 7 In: *Applied Bioremediation - Active and Passive Approaches*, InTech.

Darrah, P.K., Tlalka, M., Ashford, A., Watkinson, S.C., Fricker, M.D. (2006). The vacuole system is a significant intracellular pathway for longitudinal solute transport in basidomycete fungi. *Eukaryotic Cell* 5, 1111–1115.

Das, N., Chandran, P. (2011). Microbial degradation of petroleum hydrocarbon contaminants: An Overview. *Research Biotechnology Research International* 941810, 13.

de Oliveira Martins, P.S., de Almeida, N.F., Ferreira Leite, S.G. (2008). Application of a bacterial extracellular polymeric substance in heavy metal adsorption in a co-contaminated aqueous system. *Brazilian Journal of Microbiology* 39, 780-786.

De Luca, G., de Philip, P., Dermoun, Z., Rousset, M., Vermeglio, A. (2001). Reduction of technetium(VII) by *Desulfovibrio fructosovorans* is mediated by the nickel-iron hydrogenase. *Applied Environmental Microbiology* 67, 4583–4587.

De Schrijver, A., De Mot, R. (1999). Degradation of pesticides by actinomycetes. *Critical Reviews in Microbiology* 25, 85-119.

Dixit, R., Wasiullah, Malaviya, D., Pandiyan, K., Singh, U.B., Sahu, A., Shukla, R., Singh, B.P., Rai, J.P., Sharma, P.K., Lade, H., Paul, D. (2015).Bioremediation of heavy metals from soil and aquatic environment: An overview of principles and criteria of fundamental processes. *Sustainability* 7, 2189-2212.

Doty, S. (2008). Enhancing phytoremediation through the use of transgenics and endophytes. *New Phytologist* 179, 318–333.

Drufovka, K., Danevcˇic, T., Trebše, P., Stopar, D. (2008). Microorganisms trigger chemical degradation of diazinon. *International Journal of Biodeterioration and Biodegradation* 62, 293–296.

Dudhagara, P., Bhalani, S., Bhatt, S., Ghelani, A. (2012). Degradation of organophosphate and organochlorine pesticides in liquid culture by marine isolate *Nocardiopsis sp.* and its bioprospectives. *Journal of Environmental Research and Development* 7, 995-1001.

Dursun, A.Y., Uslu, G., Cuci, Y., Aksu, Z. (2003). Bioaccumulation ofcopper (II), lead (II) and chromium (VI) by growing *Aspergillus niger*. *Process Biochemistry* 38, 1647– 1651.

Durve, A., Naphade, S., Bhot, M., Varghese, J., Chandra, N. (2013). Quantitative evaluation of heavy metal bioaccumulation by microbes. *Journal of Microbiology and Biotechnological Research* 3 (6), 21-32.

Edward, A., Melchias, G., Visvanathan, D., Soni, P.K., Ajithan, C. (2012). Role of efflux pumps in affording organic solvent tolerance to bacterial strains from petroleum contaminated soil.*Middle-East Journal of Scientific Research* 12 (1), 80-87.

Ehlers, G.A., Rose, P.D. (2005). Immobilized white-rot fungal biodegradation of phenol and chlorinated phenol in trickling packed-bed reactors by employing sequencing 20 batch operation. *Bioresource Technology* 96, 1264-1275.

El Bestawy, E., Abu Rass, M., Abdel-Kawi, M.A. (2013). Removal of lead and oil hydrocarbon from oil refining contaminated wastewater using *Pseudomonas spp. Journal of Natural Sciences Research* 3, 112.

El-Sayed, W.S., Ibrahim, M.K., Abu-Shady, M., El-Beih, F., Ohmura, N., Saiki, H., Ando, A. (2003). Isolation and characterization of phenol catabolizing bacteria from a coking plant. *Biosciences, Biotechnolology and Biochemistry* 67(9), 2026-2029.

Ellegaard-Jensena, L., Aamand, J., Kragelund, B.B., Johnsend, A.H., Rosendahl, S. (2013). Degradation of the phenylurea herbicide diuron by soil fungi: different degradation potentials of five *Mortierella* strains endosulfan sulfate. *FEMS Microbiology Letters* 215, 255-259.

Fang, H., Dong, B., Yan, H., Tang, F., Yunlong, Y. (2010). Characterization of a bacterial strain capable of degrading DDT congeners and its use in bioremediation of contaminated soil. *Journal of Hazardous Materials* 184, 281-289.

Finley, S.D., Broadbelt, L.J., Hatzimanikatis, V. (2010). *In silico* feasibility of novel biodegradation pathways for 1,2,4-Trichlorobenzene.*BMC Systems Biology* 4, 4-14.

Fournier, D., Halasz, A., Spain, J., Fiurasek, P., Hawari, J. (2002). Determination of key metabolites during biodegradation of hexahydro-1,3,5-trinitro-1,3,5-triazine with *Rhodococcus sp* strain DN22. *Applied and Environmental Microbiology* 68, 166-172.

Fournier, D., Halasz, A., Thiboutot, S., Ampleman, G., Manno, D., Hawari, J. (2004). Biodegradation of octahydro-1,3,5,7-tetranitro-1,3,5,7-tetrazocine (HMX) by *Phanerochaete chrysosporium*: new insight into the degradation pathway. *Environmental Science and Technology* 38, 4130–4133.

Freire Andrade, M.V., Lima da Silva, K.M., da Silva Siqueira, J.P., Pessoa Wanderley, C.R., dos Araújo, R.S., Marinho, G., Rodrigues, K. (2013) Azo dye degradation by *Phanerochaete chrysosporium* in the medium enriched with nitrogen in the presence of primary co-substrate. *Archives of Biology and Technology* 56, 867-874.

Freitas, A.C., Ferreira, F., Costa, A.M., Pereira, R., Antunes, S.C., Gonçalves, F. (2009). Biological treatment of the effluent from a bleached kraft pulp mill using basidiomycete and zygomycete fungi. *Science of the total Environment* 407, 3282-3289.

Fuchs, G., Boll, M., Heider, J. (2011). Microbial degradation of aromatic compounds — from one strategy to four. *Nature Reviews Microbiology* 9, 803-816.

Fuentes, M.S., Briceño, G.E., Saez, J.M., Benimeli, C.S., Diez, M.C., Amoroso, M.J. (2013). Enhanced removal of a pesticides mixture by single cultures and consortia of free and immobilized *Streptomyces* strains. *Biomediation Research International* 2013, 1-9.

Fulekar, M.H. (2014). Rhizosphere bioremediation of pesticides by microbial consortium and potential microorganism. *International Journal of Current Microbiology and Applied Science* (7), 235-248.

Gao, Y.Z., Cheng, Z.X., Ling, W.T., Huang. (2010). Arbuscular mycorrhizal fungal hyphae contribute to the uptake of polycyclic aromatic hydrocarbons by plant roots. *Bioresource Technology* 101, 6895–6901.

Gao, Y., Chen, S., Hu, M., Hu, Q., Luo, J., Li, Y. (2012). Purification and characterization of a novel Chlorpyrifos hydrolase from *Cladosporium cladosporioides* Hu-01. *PLoS ONE* 7, 1-8.

Gelagutashvili, E. (2013). Comparative study on heavy metals biosorption by different types of bacteria. *Open Journal of Metal* 3, 62-67.

Germaine, K., Keogh, E., Garcia-Cabellos, G., Borremans, B., van der Lelie, D., Barac, T., Oeyen, L., Vangronsveld, J., Moore, F.P., Moore, E.R.B., Campbell, C.B., Ryan, D., Dowling, D.N. (2004). Colonisation of poplar trees by gfp expressing bacterial endophytes. *FEMS Microbiology Ecology* 48, 109–118.

Germaine, K.J., Keogh, E., Ryan, D., Dowling, D.N. (2009). Bacterial endophyte-mediated naphthalene phytoprotection and phytoremediation. *FEMS Microbiology Letters* 296, 226–234.

Ghosh, P.G., Sawant, N.A., Patil, S.N., Aglave, B. (2010). Microbial biodegradation of organophosphate pesticides. *International Journal of Biotechnology and Biochemistry* 6, 871-876.

Goda, S.K., Elsayed, I.E., Khodair, T.A., El-Sayed, W., Mohamed, M.E. (2010). Screening for and isolation and identification of malathion degrading bacteria: cloning and sequencing a gene that potentially encodes the malathion-degrading enzyme, carboxylestrase in soil bacteria. *Biodegradation* 21, 903-913.

Guest, R.K., Smith, D.W. (2002). A potential new role for fungi in a wastewater MBR biological nitrogen reduction system. *Journal of Environmental Engineering and Science* 1(6), 433-437.

Guha, A., Kumari, B., Bora, T.C., Roy, M.K. (1997). Possible involvement of plasmids in degradation of malathion and chlorpyrifos by *Micrococcus* sp. *Folia Microbiology* 42, 574-576.

Guo, L.Q., Shi, D.C., Wang, D.L. (2010). The key physiological response to alkali stress by the alkali-resistant halophyte *Puccinellia tenuiflora* is the accumulation of large quantities of organic acids and into the rhyzosphere. *Journal of Agronomy and Crop Science* 96, 123–135.

Gupta, A., Kaushik, C.P., Kaushik, A. (2000). Degradation of hexachlorocyclohexane (HCH; a, b, g and d) by *Bacillus circulans* and *Bacillus brevis* isolated from soil contaminated with HCH. *Soil Biology and Biochemistry* 32, 1803-1805.

Gur, O., Ozdali, M., Algur, O.F. (2014), Biodegradation of the synthetic pyrethroid insecticide α-cypermethrin by *Stenotrophomonas maltophilia* OG2. *Turk Journal of Biology* 38, 1-6.

Gutnick, D.K., Bach, H. (2000). Engineering bacterial biopolymers for the biosorption of heavy metals; new products and novel formulations. Applied *Microbiology and Biotechnology* 54, 441–460.

Haferburg, G., Merten, D., Buchel, G., Kothe, E. (2007). Biosorption of metal and salt tolerant microbial isolates from a former uranium mining area. Their impact on changes in rare earth element patterns in acid mine drainage. *Journal of Basic Microbiology* 47, 474–484.

Halasz, A., Manno, D., Perreault, N.N., Sabbadin, F., Bruce, N.C., Hawari, J. (2012). Biodegradation of RDX nitroso products MNX and TNX by cytochrome P450 XplA. *Environmental Science and Technology* 46(13), 7245-51.

Hammel, K.E. (1995). Mechanisms for polycyclic aromatic hydrocarbon degradation by ligninolytic fungi. *Environmental Health Perspectives* 103, 41–43.

Harish, R.S.M., Chauhan, J.B. (2013). Biodegradation of organophosphate pesticide by soil fungi. *Advances in Biotechnology* 12, 1-8.

Hata, T., Kawai, S., Okamura, H., Nishida, T. (2010). Removal of diclofenac and mefenamic acid by the white rot fungus *Phanerochaete sordida* YK-624 and identification of their metabolites after fungal transformation. *Biodegradation* 21, 681–689.

Hawari, J., Beaudet, S., Halasz, A., Thiboutot, S., Ampleman, G. (2000). Microbial degradation of explosives: biotransformation versus mineralization. *Applied Microbiology and Biotechnology* 54, 605–618.

Hayatsu, M., Hirano, M., Tokuda, S. (2000). Involvement of two plasmids in Fenitrothion degradation by *Burkholderia sp.* strain NF100. *Applied Environmental Microbiology* 66, 1737–1740.

He, Y., Xu, J.M., Tang, C.X., Wu, Y.P. (2005). Facilitation of pentachlorophenol degradation in the rhizosphere of ryegrass (*Lolium perenne* L.) *Soil Biology and Biochemistry* 37, 2017–2024.

Hegazy, A.K., Emam, M.H. (2011). Accumulation and soil-to-plant transfer of radionuclides in the Nile Delta coastal black sand habitats. *International Journal of Phytoremediation* 13, 140–155.

Hong, L., Zhang, J.J., Wang, S.J., Zhang, X.E., Zhou, N.Y. (2005). Plasmid-borne catabolism of methyl parathion and p-Nitrophenol in *Pseudomonas sp.* strain WBC-3. *Biochemical Biophysical Research Communication* 334, 1107-1114.

Horne, I., Sutherland, T.D., Harcourt, R.L., Russell, R.J., Oakeshott, J.G. (2002). Identification of an opd (organophosphate degradation) gene in an *Agrobacterium* Isolate.*Applied Environmental Microbiology* 68, 3371–3376.

Hossain, S.M., Das, M., Begum, K.M.M.S., Anantharaman, N. (2005). Studies on biodegradation of cyanide (AgCN) using *Phanerochaete chrysosporium*. *Journal of Institution of Engineers* 85, 45-49.

Huang, X.D., El-Alawi, Y., Penrose, D.M., Glick, B.R., Greenberg, B.M. (2004). A multi-process phytoremediation system for removal of polyaromatic hydrocarbons from contaminated soils. *Environmental Pollution* 130, 465–476.

Huang, W., Wang, Z., Choudhary, G., Guo, B., Zhang, J., Ren, D. (2012). Characterization of microbial species in a regenerative bio-filter system for volatile organic compound removal. *HVAC and R Research* 18, 169-178.

Hughes, J.B., Shanks, J., Vanderford, M., Lauritzen, J., Bhadra, R. (1997). Transformation of TNT by aquatic plants and plant tissue cultures. *Environmental Science and Technology* 31, 266–271.

Hussain, S., Siddique, T., Arshad, M., Saleem, M. (2009). Bioremediation and phytoremediation of pesticides: recent Advances.*Critical Reviews in Environmnetal Science and Technology* 39, 843-907.

Hussein, H., Ibrahim, S.F., Kandeel, K., Moawa, H. (2004). Biosorption of heavy metals from waste water using *Pseudomonas sp. Electronic Journal of Biotechnology* 7(1).

Igwo-Ezikpe, M.N., Gbenle, O.G., Ilori, M.O., Okpuzor, J., Osuntoki, A.A. (2010). High molecular weight polycyclic aromatic hydrocarbons biodegradation by bacteria isolated from contaminated soils in Nigeria. *Research Journal of Environmental Sciences* 4, 27-137.

Jacobsen, C.S. (1997). Plant protection and rhizosphere colonization of barley by seed inoculated herbicide degrading *Burkholderia (Pseudomonas) cepacia* DB01 (pR0101) in 2,4-D contaminated soil. *Plant and Soil* 189, 139–144.

Jahangeer, Kumar, V. (2013). An overview on microbial degradation of petroleum hydrocarbon contaminants. *International Journal of Engineering and Technical Research* 1(8), 34-37.

Jain, P.K., Gupta, V.K., Pathak, H., Lowry, M., Jaroli, D.P. (2010). Characterization of 2T engine oil degrading indigenous bacteria, collected from high altitude Mussoorie region of India. *World Journal of Microbiology and Biotechnology* 26, 1419–1426.

Jain, P.K., Lowry, M., Gupta, V.K., Bajpai, V., Sharma, S, Joshi N, Gaur RK (2010) Enterobacter sp. as an engine oil degrader. *Biospectra* 5(1), 115-120.

Jaysree, R.C., Basu, S., Singh, P.P., Ghosal, T., Patra, P.A., Keerthi, Y., Rajendran, N. (2011). Isolation of biosurfactant producing bacteria from environmental samples. *Pharmacology* 3, 1427-1433.

Jena, J., Kumar, R., Dixit, A., Pandey, S., Das, T. (2015). Evaluation of simultaneous nutrient and COD removal with polyhydroxybutyrate (PHB) accumulation using mixed microbial consortia under anoxic condition and their bioinformatics analysis. *PLoS ONE* 10(2), e0116230.

Jha, S., Chauhan, R., Dikshit, S.N. (2014). Fungal biomass as biosorbent for removal of heavy metal from industrial wastewater effluent. *Asian Journal of Plant Sciences* 13, 93-97.

Jilani S, Altaf Khan M (2006) Biodegradation of cypermethrin by Pseudomonas in a batch activated sludge process. *International Journal of Environmnetal Science and Technology* 3, 371-380.

John, S.G., Ruggiero, C.E, Hersman, L.E., Tung, C.S., Neu, M.P. (2001). Siderophore mediated plutonium accumulation by *Microbacterium flavescens* (JG-9).*Environmental Science and Technology* 35, 2942–2948.

Joutey, N.T., Bahafid, W., Sayel, H., El Ghachtouli, N. (2013). Biodegradation: Involved microorganisms and genetically engineered microorganisms. Chapter 11 Agricultural and Biological Sciences. *"Biodegradation - Life of Science,"* Chamy, R., Rosenkranz, F. (Eds.).

Junghanns, C., Pecyna, M.J., Böhm, D., Jehmlich, N., Martin, C., von Bergen, M., Schauer, F., Hofrichter, M., Schlosser, D. (2009). Biochemical and molecular genetic characterization of a novel laccase produced by the aquatic ascomycete *Phoma sp.* UHH 5-1-03. *Applied Microbiology and Biotechnology* 84, 1095–105.

Kafilzadeh, F., Moghtaderi, Y., Jahromi, A.R. (2013). Isolation and identification of cadmium-resistant bacteria in Soltan Abad river sediments and determination of tolerance of bacteria through MIC and MBC.*Pelagia Research Library European Journal of Experimental Biology* 3(5), 268-273.

Kamei, I., Suhara, H., Kondo, R. (2005). Phylogenetical approach to isolation of white-rot fungi capable of degrading polychlorinated dibenzo-p-dioxin. *Applied Microbiology and Biotechnology* 69, 358-366.

Kamida, H.M., Durrant, L.R., Rosim-Monteiro, R.T.,Dutra de Armas, E. (2005). Biodegradao de efluente têxtil por *Pleurotus sajor-caju*. *Química Nova* 28, 629-632.

Kang, S.H., Singh, S., Kim, J.Y., Lee, W., Mulchandani, A., Chen, W. (2007). Bacteria metabolically engineered for enhanced phtochelatin production and cadmium accumulation. *Applied Environmental Microbiology* 73, 6317–6320.

Kapoor, A., Viraraghavan, T., Roy, D. (1999). Removal of heavy metals using the fungus *Aspergillus niger*. *Bioresource Technology* 70, 95–104.

Karayilanoglu, T., Kenar, L., Serdar, M., Kose, S., Aydın, A.T. (2008). Bacterial biodegradation of aldicarb and determination of bacterium which has the most biodegradative effect. *Turk Journal of Biochemistry* 33, 209–214.

Karpouzas, D.G., Singh, B.K. (2006). Microbial degradation of organophosphorus xenobiotics: metabolic pathways and molecular basis. *Advances in Microbiology and Physiology* 15, 119-225.

Karunya, K.S., Reetha, D. (2012). Efficiency of bacterial isolates in the degradation of malathion and parathion. *International Journal of Pharmaceutical Biology and Archives* 3, 659-665.

Kasai, N., Ikushiro, S., Shinkyo, R., Yasuda, K., Hirosue, S., Arisawa, A., Ichinose, H., Wariishi, H., Sakaki, T. (2010). Metabolism of mono- and dichloro-dibenzo-p-dioxins by *Phanerochaete chrysosporium* cytochromes P450. *Applied Microbiology Biotechnology* 86, 773–780.

Keasling, J.D., Van Dien, S.J., Trelstad, P., Renninger, N., McMahon, K. (2000). Application of polyphosphate metabolism to environmental and biotechnological problems. *Biochemistry* 65, 324–331.

Kim, I.S., Ryu, J.Y., Hur, H.G., Gu, M.B., Kim, S.D., Shim, J.H. (2004). *Sphingomonas sp.* strain SB5 degrades carbofuran to a new metabolite by hydrolysis at the furanyl ring. *Journal of Agricultural Food Chemistry* 52, 2309-2314.

Kim, S.J., Koh, D.C., Park, S.J., Cha, I.T., Park, J.W., Na, J.H. (2012). Molecular analysis of spatial variation of iron-reducing bacteria in riverine alluvial aquifers of the Mankyeong. *River Journal of Microbiology* 50, 207–217.

Kitagawa, W., Kimura, N., Kamagata, Y. (2004). A novel p-nitrophenol degradation gene cluster from a gram-positive bacterium, *Rhodococcus opacus* SAO101. *Journal of Bacteriology* 186, 4894–4902.

Konishi, M., Nagahama, T., Fukuoka, T., Morita, T., Imura, T., Kitamoto, D., Hatada, Y. (2011). Yeast extract stimulates production of glycolipid biosurfactants, mannosylerythritol lipids by *Pseudozyma hubeiensis* SY62. *Journal of Bioscience and Bioengineering* 111, 702- 705.

Kuiper, I., Bloemberg, G.V., Lugtenberg, B. (2001). Selection of a plant-bacterium pair as a novel tool for rhizostimulation of polycyclic aromatic hydrocarbon-degrading bacteria. *Molecular Plant Microbe Interactions* 14, 1197–1205.

Kuiper, I., Lagendijk, E.L., Bloemberg, G.V., Lugtenberg, B.J.J. (2004). Rhizoremediation, a beneficial plant-microbe interaction. *Molecular Plant Microbe Interactions* 17, 6–15.

Kumar, R., Singh. S., Singh, O.V. (2007). Bioremediation of radionuclides: emerging technologies. *OMICS* 11, 295–304.

Kumar, K., Devi, S.S., Krishnamurthi, K., Kanade, G.S., Chakrabarti, T. (2007). Environment and isolation of endosulfan degrading and detoxifying bacteria. *Chemosphere* 68, 479-488.

Kwon, G.S., Sohn, H.Y., Shin, K.S., Kim, E., Seo, B.I. (2005). Biodegradation of the organochlorine insecticide, endosulfan and the toxic metabolite, endosulfan sulfate by *Klebsiella oxytoca* KE-8. *Applied Microbiology and Biotechnology* 67, 845-850.

Lakshmi, C., Kumar, M., Khanna, S. (2008). Biotransformation of chlorpyrifos and bioremediation of contaminated soil. *International Journal of Biodeterioration and Biodegradation* 62, 204–209.

Lan, W.S., Gu, J.D., Zhang, J.L., Shen, B.C., Jiang, H., Mulchandani, A., Chen, W., Qiao, C.L. (2006). Coexpression of two detoxifying pesticide-degrading enzymes in a genetically engineered bacterium. *International Journal of Biodeterioration and Biodegradation* 58, 70-76.

Langenhoff, A., Inderfurth, N., Veuskens, T., Schraa, G., Blokland, M., Kujawa-Roeleveld, K., Rijnaarts, H. (2013). Microbial removal of the pharmaceutical compounds ibuprofen and diclofenac from wastewater. *BioMed Research International* http://dx.doi.org/10.1155/2013/325806. Article ID325806, 9 pages.

Latifi, A.M., Khodi, S., Mirzaei, M., Miresmaeili, M., Babavalian, H. (2012). Isolation and characterization of five chlorpyrifos degrading bacteria. *African Journal of Biotechnology* 11(13), 3140-3146.

Lee, S.G., Yoon, B.D., Park, Y. (1998). Isolation of a novel pentachlorophenol- degrading bacterium, *Pseudomonas sp. Journal of Applied Microbiology* 85, 1-8.

Lee, J.W., Lee, S.M., Hong, E.J., Jeung, E.B., Kang, H.Y., Kim, M.K., Choi, I.G. (2006). Estrogenic reduction of styrene monomer degraded by *Phanerochaete chrysosporium* KFRI 20742. *The Journal of Microbiology* 44, 177-184.

Li, W., Dai, Y., Xue, B., Li, Y., Peng, X., Zhang, J., Yan, Y. (2009). Biodegradation and detoxification of endoulfan in aqueous medium and soil by *Achromobacter xylosoxidans* strain CS5. *Journal of Hazardous Materials* 67, 209-216.

Liang, W.Q., Wang, Z.Y., Li, H., Wu, P.C., Hu, J.M., Luo, N., Cao, L.X., Liu, Y.H. (2005). Purification and characterization of a novel pyrethroid hydrolase from *Aspergillus niger* ZD11. *Journal of Agriculture Food Chemistry* 53, 7415-7420.

Lin, J., Reddy, M., Moorthi, V., Qoma, B.E. (2008). Bacterial removal of toxic phenols from an industrial effluent. *African Journal of Biotechnology* 7 (13), 2232-2238.

Lin, Q.S., Chen, S.H., Hu, M.Y., Ul Haq, M.R., Yang, L., Li, H. (2011). Biodegradation of cypermethrin by a newly isolated actinomycetes. HU-S-01 from wastewater sludge. *International Journal of Environment science and Technology* 8, 45-56.

Liu, Y., Lam, M.C., Fang, H.H.P. (2001). Adsorption of heavy metals by EPS of activated sludge *Water Science and Technology* 43, 59–66.

Liu, Z., Hong, Q., Xu, J.H., Wu, J., Zhang, X.Z., Zhang, X.H., Ma, A.Z., Zhu, J., Li. (2003). Cloning, analysis and fusion expression of methyl parathion hydrolase. *Acta Genetica Sinica* 30, 1020-1026.

Liu, F., Hong, M., Liu, D., Li, Y. (2007). Biodegradation of methyl parathion by *Acinetobacter radioresistens* USTB-04. *Journal of Environmental Science* 19(10), 1257-1260.

Lloyd, J.R. (2003). Microbial reduction of metals and radionuclides. *FEMS Microbiology Reviews* 27, 411-425.

Lloyd, J.R., Cole, J.A., Macaskie, L.E. (1997). Reduction and removal of heptavalent technetium from solution by *Escherichia coli. Journal of Bacteriology* 179, 2014–2021.

Lloyd, J.R., Sole, V.A., Van Praagh, C.V., Lovley, D.R. (2000). Direct and Fe(II)-mediated reduction of technetium by Fe(III)-reducing bacteria. *Applied Environmental Microbiology* 66, 3743–3749.

Lloyd, J.R., Mabbett, A., Williams, D.R., Macaskie, L.E. (2001). Metal reduction by sulfate-reducing bacteria: physiological diversity and metal specificity. *Hydrometallurgy* 59, 327–337.

Lloyd, J.R., Leang, C., Hodges Myerson, A.L., Coppi, M.V., Cuifo, S., Methe, B. et al.(2003). Biochemical and genetic characterization of PpcA, a periplasmic c-type cytochrome in *Geobacter sulfurreducens*. *Biochemical Journal* 369, 153–161.

Lovley, D.R. (2003). Dissimilatory Fe(III) and Mn(IV) reduction. *Microbiol Reviews* 55, 259–287.

Lone, M.A., Rafiq, M. (2012). Degradation of dimethoate and pyrethroid by using fungal strains isolated from the rhizosphere of *Juglans regia* in the northern region of Jammu and Kashmir, India. *International Journal of Pharmacy and Biological Sciences* 3, 716 – 723.

Lu, H., Gao, G., Xu, G., Fan, L., Yin, L., Shen, B., Hua, Y. (2009). *Deinococcus radiodurans* PprI switches on DNA damage response and cellular survival networks after radiation damage. *Molecular Cell Proteomics* 8, 481–494.

Lu, P., Li, Q., Liu, H., Feng, Z., Yan, X., Hong, Q., Li, S. (2013). Biodegradation of chlorpyrifos and 3,5,6-trichloro-2-pyridinol by *Cupriavidus sp.* DT-1. *Bioresource Technology* 127, 337-342.

Macaskie, L.E., Bonthrone, K.M., Yong, P., Goddard, D.T. (2000). Enzymically mediated bioprecipitation of uranium by a *Citrobacter sp.:* a concerted role for exocellular lipopolysaccharide and associated phosphatase in biomineral formation. *Microbiology* 146(8), 1855–1867.

Machado, K.M.G., Matheus, D.R. (2006). Biodegradation of remazol Brilliant Blue R by ligninolytic enzymatic complex produced by *Pleurotus ostreatus*. *Brazilian Journal of Microbiology* 37, 468-473.

Mahiudddin, M., Fakhruddin, N.M., Abdullah-al-mahin, M.A.Z., Chowdhury, M.A., Rahman, M.K. (2014). Degradation of the organophosphorus insecticide diazinon by soil bacterial isolate. *The International Journal of Biotechnology* 3(1), 12-23.

Majcherczyk, A., Johannes, C., Huttermann, A.(1998). Oxidation of polycyclic aromatic hydrocarbons (PAH) by laccase of *Trametes versicolor*. *Enzyme Microbial Technology* 22, 335–341.

Malghani, S., Chatterjee, N., Yu, H.X., Luo, Z. (2009). Isolation and identification of profenofos degrading bacteria. *Brazilian Journal of Microbiology* 40, 893-900.

Mandar, P., Manish, J., Dileep, D. (2005). Biodegradation of allethrin, a pyrethroid insecticide, by an *Acidomonas sp. Biotechnology Letters* 27, 1909-1913.

Mannisto, M.K., Tiirola, M.A., Puhakka, J.A. (2001). Degradation of 2,3,4,6-tetraclorophenol at low temperature and low dioxygen concentrations by phylogenetically different groundwater and bioreactor bacteria. *Biodegradation* 12, 291–301.

Marco-Urrea, E., Aranda, E., Caminal, G., Guillén, F. (2009a). Induction of hydroxyl radical production in *Trametes versicolor* to degrade recalcitrant chlorinated hydrocarbons. *Bioresource Technology* 100, 5757–5762.

Marco-Urrea, E., Pérez-Trujillo, M., Vicent, T., Caminal, G. (2009b). Ability of white-rot fungi to remove selected pharmaceuticals and identification of degradation products of ibuprofen by *Trametes versicolor. Chemosphere* 74, 765–772.

Marco-Urrea, E., Pérez-Trujillo, M., Cruz-Morató, C., Caminal, G., Vicent, T. (2010). White-rot fungus-mediated degradation of the analgesic ketoprofen and identification of intermediates by HPLC-DAD-MS and NMR. *Chemosphere* 78, 474–481.

Marshall, M.J., Beliaev, A.S., Dohnalkova, A.C., Kennedy, D.W., Shi, L, Wang, Z, et al. (2006). c-Type cytochrome-dependent formation of U(IV) nanoparticles by *Shewanella oneidensis*. *PLoS Biology* 4, 268.

Martin, C., Corvini, P.F., Vinken, R., Junghanns, C., Krauss, G., Schlosser, D. (2009). Quantification of the influence of extracellular laccase and intracellular reactions on the isomer-specific biotransformation of the xenoestrogen technical nonylphenol by the aquatic hyphomycete *Clavariopsis aquatica. Applied Environmental Microbiology* 75, 4398–4409.

Martinez, R.J., Wang, Y., Raimondo, M.A., Coombs, J.M., Barkay, T., Sobecky, P.A. (2006). Horizontal gene transfer of PIB-type ATPases among bacteria isolated from radionuclide- and metal-contaminated subsurface soils. *Applied Environmental Microbiology* 72, 3111–3118.

Martins, M., Dairou, J., Rodrigues-Lima, F., Dupret, J., Silar, P. (2010). Insights into the phylogeny or arylamine N-acetyltransferases in fungi. *Journal of Molecular Evolution* 71, 141-152.

Megharaj, M., Ramakrishnan, B., Venkateswarlu, K., Sethunathan, N., Naidu, R. (2011).Bioremediation approaches for organic pollutants: A critical perspective. Bioremediation approaches for organic pollutants: A critical perspective. *Environmental International* 37, 1362-1375.

Melchias, E.G., Visvanathan, D., Soni, P.K., Ajithan, C. (2012). Role of efflux pumps in affording organic solvent tolerance to bacterial strains from petroleum contaminated soil middle-east. *Journal of Scientific Research* 12 (1), 80-87.

Mi, S., Mi, O., Berney, E., Elyachiou. (2007). Biodegradation of the carbofuran by a fungus isolated from treated soil. *African Journal of Biotechnology* 6 (4), 419-423.

Mizuno, H., Hirai, H., Kawai, S., Nishida, T. (2009). Removal of estrogenic activity of iso-butylparaben and n-butylparaben by laccase in the presence of 1-hydroxybenzotriazole. *Biodegradation* 20, 533-539.

Mohammadia, A., Nasernejad, B. (2009). Enzymatic degradation of anthracene by the white rot fungus *Phanerochaete chrysosporium* immobilized on sugarcane bagasse. *Journal of Hazardous Material* 161, 534-537.

Murtaza, I., Dutt, A., Ali, A. (2002). Biomolecular engineering of *Escherichia coli* organomercurial lyase gene and its expression. *Indian Journal of Biotechnology* 1, 117–120.

Muratova, A.Y., Turkovskaya, O.V., Antonyuk, L.P., Makarov, O.E., Pozdnyakova, L.I., Ignatov, V.V. (2005). Oil-oxidizing potential of associative rhizobacteria of the genus *Azospirillum*. *Microbiology* 74, 210–215.

Murugesan, A.G., Jeyasanthi, T., Maheswari, S. (2010). Isolation and characterization of cypermethrin utilizing bacteria from brinjal cultivated soil. *African Journal of Microbiol Research* 4 (1), 10-13.

Nagashetti, V., Mahadevaraju, G.K., Muralidhar, T.S., Javed, A., Trivedi, D., Bhusal, K.P. (2013). Biosorption of heavy metals from soil by *Pseudomonas aeruginosa. International Journal of Innovative Technology and Exploring Engineering* 2, 22-24.

Nakamiya, K., Hashimoto. S., Ito, H., Edmonds, J.S., Yasuhara, A., Morita, M. (2005). Degradation of dioxins by cyclic ether degrading fungus, *Cordyceps sinensis.FEMS Microbiology Letters* 248, 17–22.

Ningfeng, W., Minjie. D., Guoyi. L., Xiaoyu. C., Bin. Y., Yunliu. F. (2004). Cloning and expression of ophc2, a new organphosphorus hydrolase gene. *Chinese Science Bulletin* 49, 1245-1249.

Nirmali, S., Subrata, K.D., Bharat, K.C., Patel, R.N., Aqbal, S., Madhuban, G. (2005). Biodegradation of beta-cyûuthrin by *Pseudomonas stutzeri* strain S1. *Biodegradation* 16 (6), 581-589.

N'Guessan, A.L., Vrionis, H.A., Resch, C.T., Long, P.E., Lovley, D.R. (2008). Sustained removal of uranium from contaminated groundwater following stimulation of dissimilatory metal reduction. *Environmental Science and Technology* 42, 2999–3004.

Nyanhongo, G.S., Couto, S.R., Guebitz, G.M. (2006). Coupling of 2,4,6-trinitrotoluene (TNT) metabolites onto humic monomers by a new laccase from *Trametes modesta. Chemosphere* 64, 359–370.

Olawale, A., Akintobi, O.A. (2011). Biodegradation of glyphosate pesticide by bacteria isolated from agricultural soil. *Report Opinion* 3(1), 124-128.

Orihara, K., Yamazaki, T., Shinkyo, R., Sakaki, T., Inouye, K., Tsukamoto, A., Sugiura, J., Shishido, K., (2005). Rat cytochrome P450-mediated transformation of dichlorodibenzo-p-dioxins by recombinant white-rot basidiomycete *Coriolus hirsutus. Applied Microbiology and Biotechnology* 69(1), 22-8.

Ortega, N.O., Nitschke, M., Mouad, A.M., Landgraf, M.D., Rezende, M.O.O., Seleghim, M.H.R., Sette, L.D., Porto, A.L.M. (2011). Isolation of Brazilian marine fungi capable of growing on DDD pesticide. *Biodegradation* 22, 43- 50.

Ortiz-Hernández, M.L., Sánchez-Salinas, E. (2010). Biodegradation of the organophosphate pesticide tetrachlorvinphos by bacteria isolated from agricultural soils in México. *Reviews in International Contamination Ambie* 26, 27-38.

Ortiz-Hernandez, M.,Enrique Sanchez Salinas, E., María Luisa Castrejon Godinez, M.L., Gonzalez, E.D.,Popoca, E.C. (2013). Strategies for pesticide biodegradation: opportunity for waste, soils and water cleaning mechanisms. *Reviews in International Contamination Ambie* 29, 85-104.

Ortiz-Hernández, M.L., Sánchez-Salinas, E., Olvera-Velona, A., Folch-Mallol, J.L. (2011). Pesticides in the Environment: Impacts and its Biodegradation as a Strategy for Residues Treatment. In: *Pesticides-Formulations, Effects, Fate. (Stoytcheva, M. Ed.). InTech. Croatia, pp. 551-574.*

Pacwa-Płociniczak, M., Płaza, G.A., Piotrowska-Seget, Z., Cameotra, S.S. (2011). Environmental applications of biosurfactants: recent advances. *International Journal of Molecular Science* 12(1), 633–654.

Pakala, S.B., Gorla, P., Pinjari, A.B., Krovidi, R.K., Baru, R., Yanamandra, M., Merrick, M., Siddavattam, D. (2007). Biodegradation of methyl parathion and p-nitrophenol: evidence for the presence of a p-nitrophenol 2-hydroxylase in a gram-negative *Serratia sp.* strain ds001. *Applied Microbiology and Biotechnology*73 (6), 1452-62.

Pansiripat, S., Pornsunthorntaweea, O., Rujiravanit, R., Kitiyanana, B., Somboonthanate, P., Chavadej, S. (2010). Biosurfactant production by *Pseudomonas aeruginosa* SP4 using sequencing batch reactors: effect of oil-to-glucose ratio. *Biochemical Engineering Journal* 49, 185-191.

Ramana, K.V., Karanth, N.G. (1989). Production of biosurfactants by the resting cells of *Pseudomonas aeruginosa* CFTR-6. *Biotechnology Letters* 11, 437-442.

Pathak, H., Jain, P.K., Jaroli, D.P., Lowry, M. (2008). Technical note: Degradation of phenanthrene and anthracene by *Pseudomonas strain*, isolated from coastal area. *Bioremediation Journal* 12, 111-116.

Pathak, R.K., Dikshit, A.K. (2012). Effect of various environmental parameters on biosorptive removal of atrazine from water environment. *International Journal of Environmental Science and Development* 3, 289.

Pinedo-Rivella, C., Aleu, J., Collado, I.G. (2009). Pollutants biodegradation by fungi. *Current Organic Chemistry* 13, 1194–1214.

Porteus-Moore, F., Barac, T., Borremans, B., Oeyen, L., Vangronsveld, J., van der Lelie, D., Campbell, D., Moore, E.R.B. (2006). Endophytic bacterial diversity in poplar trees growing on a BTEX-contaminated site: the characterization of isolates with potential to enhance phytoremediation. Systems *Applied Microbiology* 29, 539–556.

Prakash, D., Gabani, P.,Chandel, A.K., Ronen, Z.,Singh, O.V. (2013). Bioremediation: a genuine technology to remediate radionuclides from the environment. *Microbial Biotechnology* 6(4), 349–360.

Premjet, S., Bunthong, O., Tachibana, S., Premjet, D. (2009). Screening of fungi from natural sources in thailand for degradation of polychlorinated hydrocarbons. *American-Eurasian Journal of Agricultural and Environmental Sciences* 5, 466-472.

Rachitra, H.R., Sowmya, H.S., Jyothi, V.A. (2013). Biosorption of heavy metals by fungi and bacteria. *Journal of Biotechnology and Biosafety* 1, 37-43.

Raghu, G., Balaji, V., Venkateswaran, G., Rodrigue, A., Maruthi Mohan, P. (2008) Bioremediation of trace cobalt from simulated spent decontamination solutions of nuclear power reactors using *E. coli* expressing NiCoT genes. *Applied Microbiology and Biotechnology* 81, 571–578.

Raghukumar, C., D'Souza-Ticlo, D., Verma, A. (2008). Treatment of colored effluents with lignin-degrading enzymes: an emerging role of marine-derived fungi. *Critical Reviews in Microbiology* 34, 189–206.

Ramadevi, C., Nath, M.M., Prasad, M.G. (2012). Mycodegradation of malathion by a soil fungal isolate, *Aspergillus niger*. *International Journal of Basic and Applied Chemical Sciences* 2, 2277-2073.

Reddy, G.V.B., Sollewijn Gelpke, M.D., Gadd, M.H. (1998). Degradation of 2,4,6-Trichlorophenol by *Phanerochaete chrysosporium*: involvement of reductive dechlorination. *Journal of Bacteriology* 19, 5159-5164.

Reddy, G.V.B., Gold, M.H. (2000). Degradation of pentachlorophenol by *Phanerochaete chrysosporium*: intermediates and reactions involved. *Microbiology* 146, 405–413.

Reddy, T.B., Reddy, A.V.R., Reddy, Y.V.R. (2012). Biodegradation of organochlorine pesticides endosulfan in soil and water. Environments. *Der Pharma Chemica* 4 (3), 1016-1018.

Rhodes, C.J. (2013). Applications of bioremediation and phytoremediation. *Science Progress* 96(4), 417-427.

Rivera, A., Quintero, D., Hernández, F., Romero, O., Villareal Espino Barros, O., Gil, C. B. (2015). Bacterial isolation ability to metabolize organic matter. *Research Journal of Recent Sciences* 4(6), 99-101.

Rocha, L.C., Ferreira, H.V., Pimenta, E.F., Berlinck, R.G.S., Seleghim, M.H.R., Javaroti, D.C.D., Sette, L.D., Bonugli, R.C., Porto, A.L.M. (2009). Bioreduction of α-chloroacetophenone by whole cells of marine fungi. *Biotechnology Letters* 3, 1559-1563.

Romero, M.C., Reinoso, E.H., Kiernan, A.M., Urrutia, M.I.(2006). Chlorinated biphenyl degradation by wild yeasts pre-cultured in biphasic systems. *Electronic Journal of Biotechnology* 9.

Ryan, R.P., Germaine, K., Franks, A., Ryan, D.J., Dowling, D.N. (2008). Bacterial endophytes; recent developments and applications. *FEMS Microbiology Ecology* 278, 1–9.

Rylott, E.L., Lorenz, A., Bruce, N.C. (2011). Biodegradation and biotransformation of explosives. *Current Opinion in Biotechnology* 22, 434-444.

Sasaki, K., Morikawa, H., Kisibe, T., Takeno, K., Mikami, A., Harada, T., Ohta, M. (2013). Simultaneous removal of cesium and strontium using a photosynthetic bacterium, *Rhodobacter sphaeroides* SSI immobilized on porous ceramic made from waste glass. *Advances in Bioscience and Biotechnology* 4, 6-13.

Scheibner, K., Hofrichter, M., Herre, A., Michels, J., Fritsche, W. (1997). Screening for fungi intensively mineralizing 2,4,6-trinitrotoluene. *Applied Microbiology and Biotechnology* 47, 452-457.

Scott, C., Pandey, G., Hartley, C.J., Jackson, C.J., Cheesman, M.J., Taylor, M.C., Pandey, R., Khurana, J.L., Teese, M., Coppin, C.W., Weir, K.M., Jain, R.K., Lal, R., Russell, R.J., Oakeshott, J.G. (2008). The enzymatic basis for pesticide bioremediation. *Indian Journal of Microbiology* 48, 65-79.

Seo, J., Jeon, J., Kim, S., Kang, S., Han, J., Hur, H. (2007). Fungal biodegradation of carbofuran and carbofuran phenol by the fungus *Mucor ramannianus*: identification of metabolites.*Water Science Technology* 55, 163–167.

Seth-Smith, H.M.B., Rosser, S.J., Basran, A., Travis, E.R., Dabbs, E.R., Nicklin, S., Bruce, N.C. (2002). Cloning, sequencing, and characterization of the hexahydro- 1,3,5-trinitro-1,3,5-triazine degradation gene cluster from *Rhodococcus rhodochrous*. *Applied and Environmental Microbiology* 68, 4764-4771.

Shah, M.P. (2014). Exploited application of *Pseudomonas stutzeri* ETL-4 in microbial degradation of Congo red. *Opinion in Biotechnology* 10, 3(1):2.

Sharma, A., Mishra, M., Sheet, S., Thite, M. (2013). Role of microbes as cleaning degrading industrial wastes for environmental sustainability- A Review. *Recent Research in Science and Technology* 5(5), 21-25.

Shaw, L.J., Burns, R.G. (2004). Enhanced mineralization of [U14-C] 2,4-dichlorophenoxyacetic acid in the soil from the rhizosphere of *Trifolium pratense*. *Applied Environmental Microbiology* 70, 4766–4774.

Shen, Y.J., Lu, P., Mei, H., Yu, H.J., Hong, Q., Li, S.P. (2010). Isolation of a methyl-parathion degrading strain *Stenotrophomonas sp.* SMSP-1 and cloning of the ophc2 gene. *Biodegradation* 21, 785-792.

Siciliano, S.D., Goldie, H., Germida, J.J. (1998) Enzymatic activity in root exudates of Dahurian wild rye (*Elymus dauricus*) that degrades 2-chlorobenzoic acid. *Journal of Agriculture and Food Chemistry* 46, 5–7.

Singh, B.K., Allan, S., Walker, A., Alun, J., Morgan, W., Denis, J. (2004). Biodegradation of chlorpyrifos by *Enterobacter* strain B-14 and its use in bioremediation of contaminated soils. *Applied Environmental Microbiology* 70, 4855-4863.

Singh, B.K., Kuhad, R.C.H., Singh, A., Lal, R. (1999). Biochemical and molecular basis of pesticide degradation by microorganisms. *Critical Reviews in Biotechnology* 19, 197–225.

Singh, B.K., Walker, A. (2006). Microbial degradation of organophosphorus compounds. *FEMS Microbiology Reviews* 30, 428-471.

Sinha, P.K., Sinha, A.,Das, M. (2011). Microbial removal of phenol and p-chlorophenol from industrial waste water using *Rhodococcus sp.* RSP8 and its growth kinetic modeling.*Journal of Water Resource and Protection* 3, 634-642.

Sivakami, R., Saravanan, S., Premkishore, G. (2015). Bacterial biosurfactants and their role in heavy metal degradation. *International Journal of Current Research and Academic Review* 3, 272-281.

Soriano, J.M., Jiménez, B., Font, G., Moltó, J.C. (2001). Analysis of carbamate pesticides and their metabolites in water by solid phase extraction and liquid chromatography: A Review.*Critical Reviews in Analytical Chemistry* 31, 19-52.

Soares, A., Guieysse, B., Mattiasson, B. (2003). Biodegradation of nonylphenol in a continuous packed-bed bioreactor. *Biotechnology Letters* 25, 927–933.

Sriprang, R., Hayashi, M., Ono, H., Takagi, M., Hirata, K., Murooka, Y. (2003). Enhanced accumulation of Cd^{2+} by a *Mesorhizobium sp.* transformed with a gene from *Arabidopsis thaliana* coding for phytochelatin synthase. *Applied Environmental Microbiology* 69, 79–796.

Stok, J.E., Huang, H.Z., Jones, P.D., Wheelock, C.E., Morisseau, C., Hammock, B.D. (2004). Identification, expression, and purification of a pyrethroid-hydrolyzing carboxylesterase from mouse liver microsomes. *Journal of Biological Chemistry* 279, 29863–29869.

Subramanian, V., Yadav, J.S. (2009). Role of P450 monooxygenases in the degradation of the endocrine-disrupting chemical nonylphenol by the white rot fungus *Phanerochaete chrysosporium*. *Applied Environmental Microbiology* 75, 5570–5580.

Suzuki, K., Hirai, H., Murata, H., Nishida, T. (2003). Removal of estrogenic activities of 17β-estradiol and ethinylestradiol by ligninolytic enzymes from white rot fungi. *Water Research* 37, 1972-1975.

Tallur, P.N., Megadi, V.B., Ninnekar, H.Z. (2008). Biodegradation of cypermethrin by *Micrococcus sp.* strain CPN 1. *Biodegradation* 19 (1), 77–82.

Tamagawa, Y., Hirai, H., Kawai, S., Nishida, T. (2005). Removal of estrogenic activity of endocrine-disrupting genistein by ligninolytic enzymes from white rot fungi. *FEMS Microbiology Letters* 244, 93-98.

Tamagawa, Y., Hirai, H., Kawai, S., Nishida, T. (2007). Removal of estrogenic activity of 4-tert-octylphenol by ligninolytic enzymes from white rot fungi. *Environmental Toxicology* 17 (22), 281-286.

Taghavi, S., Barac, T., Greenberg, B., Borremans, B., Vangronsveld, J., van der Lelie, D. (2005). Horizontal gene transfer to endogenous endophytic bacteria from poplar improves phytoremediation of toluene. *Applied Environmental Microbiology* 71, 8500–8505.

Tawfik, Z., Abu-shady, M., Haytham, M. (2005). Uranium uptake by some locally isolated and some reference bacterial species. *Acta Pharmaceutica* 5, 93–105.

Thabit, T., Medhat, M.A., El-naggar, A.H. (2013). Malathion degradation by soil isolated bacteria and detection of degradation products by GC-MS. *International Journal of Environmental Sciences* 3, 1467-1476.

Theriot, C.M., Grunden, A.M. (2010). Hydrolysis of organophosphorus compounds by microbial enzymes. *Applied Microbiology and Biology* 89, 35-43.

Thompson, K.T., Crocker, F.H., Fredrickson, H.L. (2005). Mineralization of the cyclic nitramine explosive hexahydro-1,3,5-trinitro-1,3,5-triazine by *Gordonia* and *Williamsia* spp. *Applied and Environmental Microbiology* 71, 8265-8272.

Throne-Holst, M., Wentzel, A., Ellingsen, T.E., Kotlas, H.K., Zotchev, S.B. (2007). Identification of novel genes involved in long-chain n-alkane degradation by *Acinetobacter sp.* strain DSM 17874. *Applied and Environmental Microbiology* 73, 3327–3332.

Ulfig, K., Przysta, W., Pfaza, G., Miksch, K. (2006). Biodegradation of petroleum hydrocarbons by keratinolytic fungi I. Twardowska et al. (eds.), *Soil and Water Pollution Monitoring, Protection and Remediation*, 3–23. Springer.

Valls, M., de Lorenzo, V. (2002). Exploiting the genetic and biochemical capacities of bacteria for the remediation of heavy metal pollution. *FEMS Microbiology Reviews* 26 327-338.

Valls M, Atrian S, de Lorenzo V, Fernandez LA (2000a). Engineering a mouse metallothionein on the cell surface of *Ralstonia eutropha* CH34 for immobilization of heavy metals in soil. *Nature Biotechnology* 18, 661–665.

Valls, M., de Lorenzo, V., Gonzalez-Duarte, R., Atrian, S. (2000b). Engineering outer-membrane proteins in *Pseudomonas putida* for enhanced heavy-metal bioadsorption. *Journal of Inorganic Biochemistry* 79, 219–223.

Van Aken, B., Hofrichter, M., Scheibner, K., Hatakka, A.I., Naveau, H., Agathos, S.N. (1999). Transformation and mineralization of 2,4,6-trinitrotoluene (TNT) by manganese peroxidase from the white-rot basidiomycete *Phlebia radiate*. *Biodegradation* 10, 83–91.

Van Aken, B., Yoon, J.M., Schnoor, J.L. (2004). Biodegradation of nitro-substituted explosives 2,4,6-trinitrotoluene, hexahydro-1,3,5-trinitro-1,3,5-triazine, an octahydro-1,3,5,7-tetranitro-1,3,5-tetrazocine by a phytosymbiotic *Methylobacterium sp* associated with poplar tissues (*Populus deltoides x nigra* DN34). *Applied and Environmental Microbiology* 70, 508-51.

van Beek, T.A., Kuster, B., Claassen, F.W., Tienvieri, T., Bertaud, F., Lenon, G., Petit-Conil, M., Sierra-Alvarez, R. (2007). Fungal bio-treatment of spruce wood with *Trametes versicolor* for pitch control: Influence on extractive contents, pulping process parameters, paper quality and effluent toxicity. *Bioresource Technology* 98, 302-311.

Wang, B.Z., Guo, P., Hang, B.J., Li, L., He, J., Li, S.P. (2009). Cloning of a novel pyrethroid-hydrolyzing carboxylesterase gene from *Sphingobium sp.* strain JZ-1 and characterization of the gene product. *Applied Environmental Microbiology* 75 (17), 5496-5500.

Wang, J., Chen, C. (2009). Biosorbents for heavy metals removal and their future. *Biotechnology Advances* 27, 195–226.

Weelink, S.A.B., van Eekert, M.H.A., Stams, A.J.M. (2010). Degradation of BTEX by anaerobic bacteria: physiology and application. *Reviews in Environmental Science and BioTechnology* 9, 359-385.

Wick, L.Y., Remer, R., Wurz, B., Reichenbach, J., Braun, S., Schafer, F., Harms, H. (2007). Effect of fungal hyphae on the access of bacteria to phenanthrene in soil. *Environmental Science and Technology* 41, 500–505.

Wildung, R.E., Gorby, Y.A., Krupka, K.M., Hess, N.J., Li, S.W., Plymale, A.E., et al. (2000). Effect of electron donor and solution chemistry on products of dissimilatory reduction of technetium by *Shewanella putrefaciens*. *Applied Environmental Microbiology* 66, 2451–2460.

Wolfram, L., Bauerfeind, P. (2009). Activities of urease and nickel uptake of *Helicobacter pyloriproteins* are media- and host-dependent. *Helicobactertia* 14, 264–270.

Wolski, E.A., Murialdo, S.E., Gonzalez, J.F. (2006). Effect of pH and inoculum size on pentachlorophenol degradation by *Pseudomonas sp. Water SA* 32. 93-98.

Wu, P.C., Liu, Y.H., Wang, Z.Y., Zhang, X.Y., Li, H., Liang, W.Q., Luo, N., Hu, J.M., Lu, J.Q., Luan, T.G., Cao, L.X. (2006). Molecular cloning, purification, and biochemical characterization of a novel pyrethroid-hydrolyzing esterase from *Klebsiella sp.* strain ZD112. *Journal of Agriculture andFood Chemistry* 54,836–842.

Wu, P., Ji, X., Song, X., Shen, Y. (2014). Nutrient removal from municipal wastewater and microbial community analysis of a combined ABR-MBR (CAMBR) Process. *CLEAN – Soil, Air, Water* 42, 53–759.

Wu, N.Y., Zhang, S.Z., Huang, H.L., Christie, P. (2008). Enhanced dissipation of phenanthrene in spiked soil by arbuscular mycorrhizal alfalfa combined with a non-ionic surfactant amendment. *Science of the Total Environment* 394, 230–236.

Xiao, P., Mori, T., Kamei, I., Kondo, R.(2011). Metabolism of organochlorine pesticide heptachlor and its metabolite heptachlor epoxide by white-rot fungi belonging to genus *Phlebia.Microbiology Letters* 314, 140-146.

Xie, S., Yang, J., Chen, C., Zhang, X., Wang, Q., Zhang, C. (2008). Study on biosorption kinetics and thermodynamics of uranium by *Citrobacter freudii*. *Journal of Environmental Radioactivity* 99, 126–133.

Xu, G.M., Li, Y.Y., Zheng, W., Peng, X., Li, W., Yan, Y.C. (2007). Mineralization of chlorpyrifos by co-culture of *Serratia* and *Trichosporon spp. Biotechnology Letters* 29, 1469–1473.

Yadav, J.S., Loper, J.C. (2000). Cytochrome P450 oxidoreductase gene and its differentially terminated cDNAs from the white rot fungus *Phanerochaete chrysosporium*. *Current Genetics* 37, 65–73.

Yang, Q., Yang, M., Pritsch, K., Yediler, A., Kettrup, A. (2003). Decolorization of synthetic dyes and production of manganese dependent peroxidase by new fungal isolates. *Biotechnology Letters* 25, 709 -715.

Yang, L., Zhao, Y.H., Zhang, B.X., Yang, C.H., Zhang, X. (2005). Isolation and characterization of a chlorpyrifos and 3, 5, 6- trichloro-2-pyridinol degrading bacterium. *FEMS Microbiology Letters* 251, 67-73.

Yang, C., Liu. N., Guo, X., Qiao, C. (2006). Cloning of mpd gene from a chlorpyrifos-degrading bacterium and use of this strain in bioremediation of contaminated soil. *FEMS Microbiology Letters* 265, 118-125.

Ye, D., Siddiqi, M.A., Maccubbin, A.E., Kumar, S., Sikka, H.C. (1996). Degradation of polynuclear hydrocarbons by *Sphingomonas paucimobilis*. *Environmental Science and Technology* 30, 136-142.

Yee, D.C., Maynard, J.A., Wood, T.K. (1998). Rhizoremediation of trichloroethylene by a recombinant, root-colonising *Pseudomonas fluorescens* stain expressing toluene ortho-monooxygenase constitutively. *Applied Environmental Microbiology* 64, 112–118.

Yu, H.L., Jing, L., Chang, Y. (2015). Enhanced tolerance and accumulation of heavy metal ions by engineered *Escherichia coli* expressing *Pyrus calleryana* phytochelatin synthase. *Journal of Basic Microbiology* 55, 398–405.

Yuanfan, H., Jin, Z., Qing, H., Qian, W., Jiandong, J., Shunpeng, L. (2010). Characterization of a fenpropathrin-degrading strain and construction of a genetically engineered microorganism for simultaneous degradation of methyl parathion and fenpropathrin. *Journal of Environmental Management* 91, 2295-2300.

Zhang, R., Cui, Z., Jaindong, J., Jain, H., Xiangyang, G., Sphunpeng, L. (2005). Diversity of organophosphorus pesticide degrading bacteria in a polluted soil and conservation of their organophosphorus hydrolase genes. *Canadian Journal of Microbiology* 51, 337–343.

Zhao, J.S., Halasz, A., Paquet, L., Beaulieu, C., Hawari, J. (2002). Biodegradation of hexahydro-1,3,5-trinitro-1,3,5-triazine and its mononitroso derivative hexahydro-1-nitroso-3,5-dinitro-1,3,5-triazine by *Klebsiella pneumoniae* strain SCZ-1 isolated from an anaerobic sludge. *Applied and Environmental Microbiology* 68, 5336-5341.

Zhao, J.S., Greer, C.W., Thiboutot, S., Ampleman, G., Hawari, J. (2004a). Biodegradation of the nitramine explosives hexahydro-1,3,5-trinitro-1,3,5-triazine and octahydro- 1,3,5,7-tetranitro-1,3,5,7-tetrazocine in cold marine sediment under anaerobic and oligotrophic conditions. *Canadian Journal of Microbiology* 50, 91- 96.

Zhao, J.S., Paquet, L., Halasz, A., Manno, D., Hawari, M. (2004b). Metabolism of octahydro-1,3,5,7-tetranitro-1,3,5,7-tetrazocine by *Clostridium bifermentans* strain HAW-1 and several other H2 producing fermentative anaerobic bacteria. *FEMS Microbiology Letters* 237, 65-72.

Zhao, J.S., Spain, J., Thiboutot, S., Ampleman, G., Greer, C., Hawari, J. (2004c). Phylogeny of cyclic nitramine-degrading psychrophilic bacteria in marine sediment and their potential role in the natural attenuation of explosives. *FEMS Microbiology Ecology* 49, 349-357.

Zhao, X. W., Zhou, M. H., Li, Q.B., Lu, Y.H., He, N., Sun, D.H., Deng, X. (2005). Simultaneous mercury bioaccumulation and cell propagation by genetically engineered *Escherichia coli. Process Biochemistry* 40, 1611–1616.

Zhao, J. S., Manno, D., Leggiadro, C., O'Neill, D., Hawari, J. (2006) *Shewanella halifaxensis* sp nov., a novel obligately respiratory and denitrifying psychrophile. *International Journal of Systematic and Evolutionary Microbiology* 56, 205-2.

Zheng, Y. Z., Lan, W.S., Qiao, C.L., Mulchandani, A., Chen, W. (2007). Decontamination of vegetables sprayed with organophosphate pesticides by organophosphorus hydrolase and carboxylesterase (B1). *Applied Biochemistry and Biotechnology* 136, 233-241.

Zheng, Y., Liu, D., Xu, S., Yuan, Y., Xiong, L. (2009). Kinetics and mechanisms of p-nitrophenol biodegradation by *Pseudomonas aeruginosa* HS-D38. *Journal of Environmental Science* 21, 1194-1199.

In: Chemical Pollution Control … ISBN: 978-1-53611-034-0
Editor: Naser A. Anjum

Chapter 3

ARBUSCULAR MYCORRHIZAL FUNGI AND THE REMEDIATION OF SOILS CONTAMINATED WITH HYDROCARBONS

Marcia Toro[1,*], Rita Gamarra[2], Liliana López[3] and Carmen Infante[3]

[1]Laboratorio de Ecología de Agroecosistemas,
Instituto de Zoología y Ecología Tropical, Facultad de Ciencias,
Universidad Central de Venezuela, Caracas, Venezuela
[2]Universidad Centro-occidental Lisandro Alvarado,
Estado Lara, Venezuela
[3]Instituto de Ciencias de La Tierra, Facultad de Ciencias,
Universidad Central de Venezuela, Caracas, Venezuela

ABSTRACT

Phytoremediation is the use of plants to recover contaminated soils; is a nondestructive and cost-effective *in situ* technology to clean up contaminated soils. In the case of contamination with petroleum hydrocarbons, plants enhance microbial degradation of the contaminant in the rhizosphere. Venezuela is one of the largest oil producers in the

---

* Corresponding author: Marcia Toro, Laboratorio de Ecología de Agroecosistemas, Instituto de Zoología and Ecología Tropical, Facultad de Ciencias, Universidad Central de Venezuela, Caracas 1041-A, Venezuela; E-mail: marcia.toro@ciens.ucv.

world. For the rehabilitation of oil-contaminated sites, phytoremediation, using arbuscular mycorrhizal fungi to remediate soils contaminated with hydrocarbons have high potential and represents a promising technology whereby plants are used to enhance biodegradation processes in soil. The feasibility of using plant-Glomeromycota fungi association in the biorestoration of contaminated soil was studied in an oil pit, located in Anzoategui state, Venezuela. The results and discussion presented in this chapter aim to highlight the role of mycorrhizal symbiosis as a promising tool for phytoremediation of soils contaminated with hydrocarbons; rhizodegradation seemed to be the most feasible mechanism.

Keywords: hydrocarbons, phytoremediation, arbuscular mycorrhizae, oily sludge

1. INTRODUCTION

Contamination of the environment with oil or its derivatives is one of major problems affecting the world for its apparent deleterious effects on humanity, fauna and flora. Particularly in Venezuela, this problem exists since mid-twentieth century, when started the first exploratory oil drilling activity (Delon, 2001). Today, with intensive oil extraction and transport, chances of accidents and mishaps, hauling spills and environmental release of large amounts of toxic have increased. According to Venezuelan Ministry of Environment, in 2001 a total of 17,000 oil pits were estimated in Venezuela, of which 12,000 have not received adequate treatments that could prevent further damage to environment and health of surrounding population; for this reason, it becomes environmental liability for oil industry and nation (Rojas, 2006).

A pit is an excavation of land, with an approximate capacity between 100 to 100,000 m^3, built to receive the waste products of drilling fluids process with bentonite mud, hydrocarbons, water, gravel and chemicals (Madrid and Cataldi, 2002). However, the growing awareness worldwide about the impact on the environment and public health has encouraged development and use of methods and technologies to reduce lethal effects. Is essential to suggest mechanisms of cleaning as bioremediation, a viable and attractive biological technique for restoring hydrocarbon-contaminated soils.

Phytoremediation is a kind of bioremediation based on the use of a plant or combination of plants with microorganisms in the rhizosphere, to remove, degrade or immobilize contaminants located in soils, sediments and surface water (Perez et al., 2002). A very interesting case of association or combination of plants with soil microorganisms are arbuscular mycorrhiza (AM), which are mutualistic symbiotic associations established between the roots of most of the plants and fungi from Glomeromycota Phylum (Alarcon and Ferrera, 2000). In Venezuela, Hernandez and Mager (2003) and Merkl et al. (2004), have selected several plant species with phytoremediating potential in hydrocarbon-contaminated soils. Until present the use of indigenous microorganisms as mycorrhizal fungi for phytoremediation, is scarce in the country.

The use of AM with phytoremediation purposes is a task that must be rigorously investigated; in this work we show results indicating the feasibility of using indigenous biological resources, such as Glomeromycota fungi associated with plants to clean up contaminated environments.

2. PHYTOREMEDIATION

Phytoremediation is an emerging technique that involves the use of plants and their associated microbiota, soil amendments, organic matter and/or fertilizers, which together remove, take or immobilize contaminants from the ecosystem (Hou et al., 2001). As an agronomic practice it may be economic, especially in areas where contamination is not severe (Garcia et al., 2000).

A number of researchers including Merkl et al. (2005) and Infante (2007) have provided several benefits involved in the use of phytoremediation. Major benefits are listed below:

- Inexpensive compared to other techniques.
- Contaminated soil need not be excavated, reducing implementation costs.
- Soil can be reused after treatment.
- Environmental acceptance.
- Can be used in the recovery of soil and groundwater in the same ecosystem.
- Applicable to gasoline, explosives, solvents and/or heavy metals contaminated soils.

According to Cunningham et al. (1997), this technique may present obstacles when applied, such as location of pollutant beyond rooting depth and/or phytotoxic effects on the plant. Another limiting aspect is vagueness regarding length of the process, therefore to overcome these constraints, interactions with microorganisms are necessary to study and understand processes involved in phytoremediation.

2.1. Processes Involved in Phytoremediation

Phytoextraction or Phytoaccumulation: The process of phytoextraction or phytoaccumulation is the absorption of pollutants by the roots, and involves the ability of some plants to accumulate contaminants in their roots, stems or leaves. This mechanism has been extensively studied in plants that accumulate metals (Banks et al., 2000) and radioactive materials (Dushenkov, 2003). When plant extracts metals, it is harvested for incineration or burial (Infante, 2007).

Phytostabilization: Herein, plants use to develop a dense root system, which allows to reduce bioavailability and mobility of contaminants, preventing its transport to underground layers or the atmosphere (Joner and Leyval, 2003).

Phytodegradation: This is the transformation of organic pollutants into simpler molecules. In cases, degradation products may contribute to plant growth (Pilon, 2005).

Phytovolatilization: This occurs as growing plants absorb water along with soluble organic contaminants. Some of the contaminants may reach the leaves and evaporate or volatilize into the atmosphere (Lopez et al., 2005).

Phytostimulation or Rhizodegradation: Herein, the root system releases enzymes that initiate oxidation of organic contaminants and contributes to stimulate microbial activity to oxidate, degrade and mineralize soil contaminants (Joner and Leyval, 2003). The root effect on degradation of organic compounds via stimulation of microbial activity in the rhizosphere, and is called rhizodegradation (Siciliano et al., 2001).

3. Interactions of Microorganisms with Phytoremediating Plants

In natural ecosystems, microorganisms live near plant roots in stable equilibrium, in an area called rhizosphere (Figure 3.1), where there is a gradient of microorganisms from root surface, with maximum value at its surface (rhizoplane) to 1-2 mm where microorganisms are significantly lower or bulk soil (Strong and Phillips, 2001).

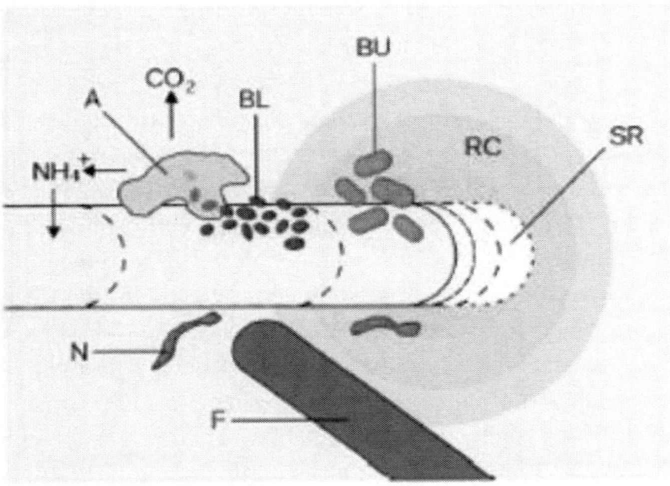

Figure 3.1. Schematic representation of the diversity of organisms interacting among themselves and with the root, ultimately forming the community of the rhizosphere. (A = amoeba eating bacteria; BL = Bacteria limited by energy; BU = Bacteria not limited by energy; RC = Carbohydrates derived from root; SR = sloughed root hair cells; F = fungal hyphae N = Nematode) (Strong and Phillips, 2001).

Efficiency of soil phytoremediation is often related to microorganism´s activity in the rhizosphere (Muratova et al., 2003). However, pollutants break the balance in the rhizosphere and can trigger microorganism-plant interactions which may favor or disfavor phytoremediation process. Toxic contaminant levels may cause adverse effects to plants; however, at low concentrations, hydrocarbons may stimulate growth of certain plant species, because rhizosphere provides favorable nutritional conditions and aeration to increase populations and diversity of microbial flora (Perez et al., 2002).

3.1. Plants as Phytoremediation Process Components

Numerous plant species have been studied as phytoremediators of heavy metals and contaminated soils. Many of them metabolize or immobilize petroleum hydrocarbons through different processes, while stimulates populations of microorganisms in the rhizosphere (Perez et al., 2002). Several studies tested the ability of plants to remediate slightly contaminated soils. List and Alexander (2000), studied phytoremediation of nine species growing in 78.4 ppm pyrene contaminated soil, obtaining a removal of 74% in eight weeks. Huang et al. (2005) achieved 50% removal in soil contaminated with 50,000 ppm of total petroleum hydrocarbons (TPH) with *Festuca arundinacea*.

Amaranthus tricolar Linn, *Brassica parachinensis* Bailey, *Raphnus sativus* L., *Ipomoea aquatica* Forsk, *Glycine max* Merr, *Phaseolus vulgaris* L., *Brassica chinensis* L., *Brassica oleracea* L., *Spinacia oleracea* L., *Solanum melongena* L. and *Lolium multiflorum* Lam have shown the ability to grow in xenobiotic contaminated soils. These plants were able to remedy 0.01% to 0.24% phenanthrene and pyrene contaminated soils (Gao and Zhu, 2004). Mattina and Dykas (2000) observed that *Cucurbita pepo* L. removed and accumulated a wide range of persistent organic pollutants in their roots and aerial parts, including chlorinated as dichlorodiphenyl trichloroethane (DDT). White et al. (2003) noted the accumulation of certain polychlorinated biphenyls (PCBs) and some polycyclic aromatic hydrocarbons (PAHs) in the same species. However, pollutant content in plant is variable and depends on phytoremediating ability of cultivar (White et al., 2000). *Cyperus laxus* is proposed by Escalante et al. (2005) as a model plant for phytoremediating tropical soils in flooded conditions; it achieved 90% removal after six months, compared to control soil artificially contaminated with 5000 ppm of TPH (Total Petroleum Hydrocarbons).

Siciliano et al. (2003) pointed *Festuca arundinacea* as good degrader of oil hydrocarbons, both in flooded and dry soils as it showed a rate of degradation up to 38 mg hydrocarbon/kg.month; this figure was twice of found in contaminated soil without plant. In Venezuela, Merkl et al. (2004) listed fifty-seven native species growing in petroleum contaminated soils, including 18 Fabaceae, 19 grasses and 17 herbaceous, an important data basis to study phytoremediation in the tropics. More recently, Brandt et al. (2006) evaluated the potential of Vetiver (*Vetiveria zizanioides* L.) to tolerate presence of heavy oils. Comparison of plant growth and soil parameters in different treatments showed that Vetiver is able to reproduce in the presence of the contaminant, and reduce its levels in soil. The authors suggest Vetiver cultivation in areas contaminated with hydrocarbons.

3.2. Glomeromycota Fungi-Interaction with Soil Biotic Components

Glomeromycota fungi are part of the rhizosphere and form the mycorrhizosphere, term referring to the soil area affected by mycorrhizal association, where microbial communities grow and change dynamically in time (Sylvia, 1999). The mycorrhizosphere is the soil area around mycorrhizal roots; the soil closer to the hyphae of the mycorrhizal fungal mycelium external to the root is called the hyphosphere or mycosphere.

Microorganism populations grow in the mycorrhizosphere and interact with mycorrhizal fungi (e.g., phosphate solubilizing bacteria), due to both root and fungi exudates (Bending et al., 2006). Glomeromycota fungi interact stimulating or inhibiting certain groups of microorganisms, which in turn may promote or counteract development of mycorrhizae (Azcón et al., 2002). Johansson et al. (2004), express these relationships in the following figure, which highlights the interactions established during a mycorrhizal symbiosis, both in natural and artificial ecosystems. These effects confirm that soil biotic community exerts an impact on the establishment of mycorrhizae because they can control dynamics of its population, root colonization and both internal development and external hyphae (Bending et al., 2006). These interactions may be affected by the effect of pollutants such as hydrocarbons.

3.2.1. Effect of Petroleum Hydrocarbons on Glomeromycota Fungi

AM fungi and also ectomycorrhizal associations (formed by Basidiomycetes or Ascomycetes fungi and roots of around 2% of woody/plant species) are involved in remediation of oil-contaminated soils (Tedersoo et al., 2010). Genney et al. (2004) demonstrated degradative capacity of *Thelephora terrestris* and *Laccaria laccata* pure cultures on polycyclic aromatic hydrocarbons. Ectomycorrhizal fungi *Suillus bovinus* can form hydrophobic substances in its mycelium, which contacts Polycyclic Aromatic Hydrocarbons (PAHs) and promotes its degradation (Joner et al., 2006). Hernandez et al. (2000) suggested that although the AM symbiosis can be adversely affected in the initial stages of colonization by the presence of hydrocarbons reducing the number of species of fungi in soil, plant morphometric parameters tend to improve at the end of its life cycle, which could be adaptation of fungi to such conditions. One possible advantage of AM in hydrocarbons contaminated soils, is mycorrhized plants can best survive under water stress (Joner and Leyval, 2001), thanks to root external hyphae that captures more water under drought conditions.

Joner and Leyval (2001) suggest the establishment of AM associations together with N-fixing symbiotic associations in hydrocarbon contaminated soils can help reduce polluting effect of hydrocarbon, as these associations contribute absorbing mineral nutrients from soil and plant growth. Acquisition of phosphate is vital for the plant, for its key role in biological systems; it is known that mycorrhizal plants take phosphate more efficiently than single roots. Improving phosphorus nutrition helps plant in nitrogen acquisition.

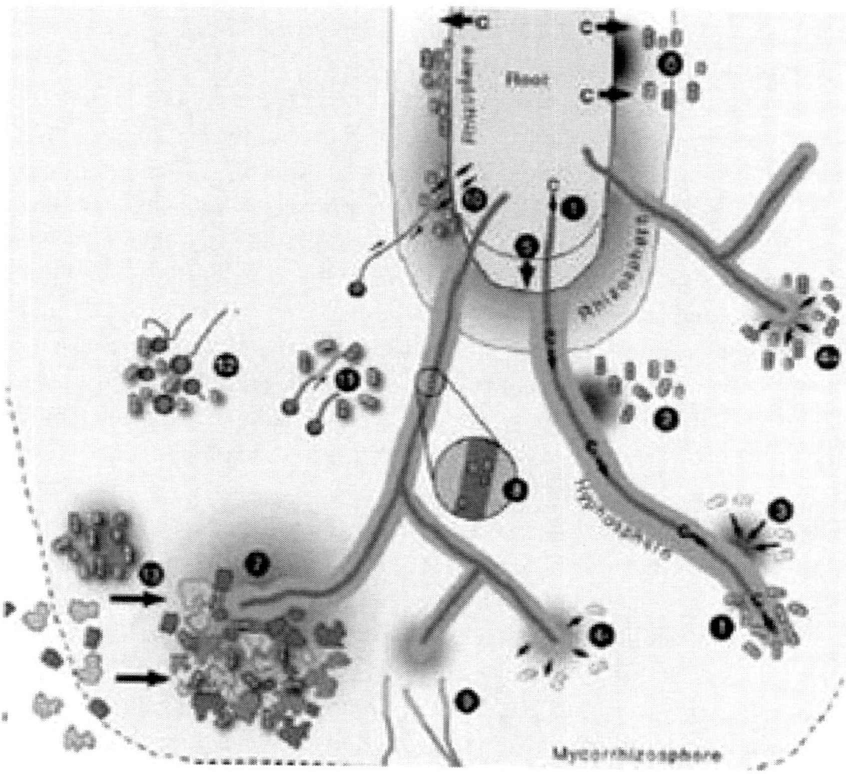

Figure 3.2. Schematic representation of the mycorhizospheric relationships. Processes involved include: 1) Supply of energy C-components via hyphae; 2) pH changes; 3) Competition for nutrients; 4) Inhibitory or stimulating compounds; 5) Stimulated root growth; 6) Changes in root exudation; 7) Changes in root structure; 8) Endophytic bacteria in AM fungal structures; 9) Effect of pathogenic fungi; 10) Effect on receptivity of roots to AM/fungus recognition; 11) Effects on fungal growth; and 12) Modification of soil chemistry (Johansson et al., 2004).

One of the first promising evidences attributable to AM fungi in hydrocarbons contaminated soils was obtained by Volante et al. (2005), assessing the effect of AM on the persistence of benzene, toluene, ethylbenzene and xylene (BTEX). They not only showed sharp reductions in the concentration of BTEX in soil, but also showed these reductions can vary with AM fungal species and nature of BTEX.

To determine the potential of these fungi in remediation of hydrocarbons contaminated soils Verdin et al. (2006) conducted *in vitro* studies to evaluate the effect of anthracene on development of symbiosis, and its role on transformation of the contaminant. They found decrease in intra radical fungal mycelium, sporulation and percentage of mycorrhizal colonization, but the presence of the symbiosis was clearly positive; besides a decrease in concentration of the pollutant, they detected traces of both radicals in plant cells and intra radical fungal hyphae.

Strategies involved in AM biodegradation of hydrocarbons have not been elucidated at present. Verdin et al. (2006) suggest that accumulation of anthracene was result of bidirectional exchange along coenocytic mycelium mediated by enzymes and transport of lipids from extra radical hyphae towards intra radical hyphal and vice versa; is interpreted as anthracene translocation through lipid bodies in fungal hyphae.

4. Effect of Arbuscular Mycorrhizae in Phytoremediation of *Brachiaria brizantha*

The feasibility of using plant-Glomeromycota fungi association in the biorestoration of contaminated soil was studied in an oil pit, located in Anzoategui state, Venezuela (Gamarra, 2014). The vegetation and soil of the area affected by the oil spill was studied. Chemical and biological analyses of the rhizospheric soil of predominant plants were made, and samples of the oily sludge were taken. Among the predominant plant species, *Fimbristylis cymosa*, was the most abundant, with a frequency of 65.33%.

Rhizospheric soil study associated with 8 dominating plant species in the oil pit showed Glomeromycota fungal spores; among these, 86.21% were similar to *Glomus* genera; 10.34% were similar to *Acaulospora* genera, and 3.45% were identified as possible *Entrophospora* genera. These samples were used to prepare an inoculum of native Glomeromycota species to be used experimentally.

In greenhouse experiments carried out in Venezuela, using *Brachiaria brizanta* inoculated with native Glomeromycota fungi or *Acaulospora mellea* species and 4% of oily sludge. Gamarra (2014) measured oil degradation capacity of the plant in presence or absence of hydrocarbon. The following table shows that hydrocarbon adversely affected plant growth, however, inoculating Glomeromycota fungi significantly favored biomass production both in presence or absence of hydrocarbon.

Shoot/root ratio average values in plants grown in uncontaminated and contaminated soil were 1.04 and 0.75 respectively; this parameter may have reduced due to application of oily sludge (Table 3.1). This suggests aerial part of plant is most sensitive to hydrocarbon action so beneficial effects of inoculation were observed in biomass root production; inoculation helped roots grew more. Hernandez and Mager (2003) also observed reduced shoot/root ratio as result of hydrocarbon application. Advantages of AM plants are uptake and transfer of water and nutrients (Camargo, 2002), increasing plant biomass (Van der Heijden, 2002), immobilization of high concentrations of non-essential metals in root preventing to pass plant aerial part (Leyval et al., 2002) and blocking entry of pathogens by stimulating growth of rhizobacteria in the mycorrhizosphere (Lioussanne, 2010). One or several factors simultaneously could influence inoculated plants in this work.

Table 3.1. Dry matter of the stem, root, total biomass (g) and stem/root ratio of *Brachiaria brizantha* growing with and without 4% of hydrocarbon and inoculated or not with a mixture of Glomeromycota native species or *Acaulospora mellea* after 3 months of treatment

Soil treatment	Inocula	Shoot biomass (g)	Root biomass (g)	Shoot/Root Ratio
Without Hidrocarbon (oily sludge)	No inoculum	45.15 a	46.88 c	1.04 a
	Native inoculum	33.02 b	63.89 ab	0.53 bcd
	A. mellea	43.80 a	76,44 a	0.60 bc
With Hidrocarbon (oily sludge)	No inoculum	18.11 c	28.09 d	0.75 b
	Native inoculum	18.39 c	54.48 bc	0.35 d
	A. mellea	24.01 c	56.91 bc	0.43 cd

Significance simple effect (soil) *** *** **
Significance simple effect (inoculum) ** *** ***
Interaction significance (soil * inoculum)* ns ns
C.V. 20,82 23,55 25,40
Mean separation in columns by Duncan test ($P \leq 0,05$) ns
*Non-significant, * $\leq 0,05$; ** $\leq 0,01$; *** $\leq 0,001$*
Values followed by the same letter do not differ significantly

The following table shows initial and final values of oils and greases or TPH (Total petroleum hydrocarbon) in experiments with *B. brizantha* (Gamarra, 2014) (Table 3.2). First column indicates initial percentage of oils and greases (4%, equivalent to 40000 mg/kg soil). Second column indicates oils and greases content at the end of the experiment for each treatment; third column indicates difference between initial and final content of oils and greases.

Table 3.2. Initial and final concentration of oils and greases in different treatments of inoculation with AM fungi of *Brachiaria brizantha* after a growing period of three months

Treatments	Initial Oils and Greases (%)	Total oils and Greases (%)	Difference Oils and Greases (%)
Brachiaria brizantha	4	2.83	1.17
B. brizantha + native inoculum	4	0.81	3.19
B. brizantha + *A. mellea*	4	2.42	1.83
Soil without plant	4	3.81	0.19

Results indicate that all treatments contained hydrocarbons remaining after three months of experimentation, and reduction of oils and greases, where the greatest reduction (3.19%) was noted in *B. brizantha* inoculated with native inoculum. Only in this case less than 1% limit prescribed by Venezuelan environmental regulations for appropriate cleaning was achieved or the maximum allowed TPH levels in soil (Republic of Venezuela, 1998).

4.1. Estimated Biodegradation Processes

Phytoremediation was verified based on the percentage of TPH degradation achieved in AM inoculation treatments for *B. brizantha*, as seen in Table 3.3. Inoculation with native fungi resulted in greater degradation rate (79.75%), followed by *A. mellea* inoculation (45.63%) and treatment without inoculum (29.25%). Rates of hydrocarbon degradation in AM plants may be related to mycorhizospheric effect and microbial activity in the rhizosphere. Different microbial groups, bacteria and fungi were detected in these rhizospheres, data not shown (Gamarra, 2014). Also a fraction of this degradation could be due to natural volatilization of hydrocarbon product of high temperature where the experiment was made (±37 monthly average).

Merkl et al. (2006) indicate degradation of petrolized crude oil in experiments with *B. brizantha* may be related to fungal consortia different to Glomeromycota fungi, reinforcing the point made by Malachowska and Kalka (2010) who suggest efficiency of AM is related with rhizospheric microorganisms. Gamal (2005) suggests inoculation with Glomeromycota fungi can improve rhizosphere sanitation and healthy plant growth, improving microbial activity, growth and tolerance of plants to oil, so plants would be sink for pollutants.

Table 3.3. TPH biodegradation rate in AM fungal inoculation treatments with of *Brachiaria brizantha* after growing three-months period

Tretatments	% TPH Degradation
Plant not inoculated	29.25 c
Inoculated with native AM	79.75 a
Inoculated with *A. mellea*	45.63 b
C.V. 17,88	

Mean separation in columns by Duncan test ($P \leq 0.05$). Equal letters indicate no significant differences between treatments

CONCLUSION

Results of this study suggest that soil remediation mechanism that would prevail in this study may be Rhizodegradation, according to which microbial populations and their activity increase by root exudates. If so, an increase of organic pollutants biodegradation from soil leads to co-metabolism of pollutants in the rhizosphere. It also revealed the outstanding role of Glomeromycota fungi and associated microflora in biodegradation of oily sludge in artificially contaminated soil. The highlighted values cumulatively indicate the Rhizodegradation of petrolized sludge in the rhizosphere of *Brachiaria brizanta*. These results also evidence the important role of arbuscular mycorrhizae in phytoremediation of oil spills or other hydrocarbon debris in tropical polluted soils, which represents a huge potential in the search of efficient and innovative options for remediation of hydrocarbons contaminated soils.

ACKNOWLEDGMENTS

The authors wish to express their gratitude to translator Mrs. Auxi Toro for reviewing the manuscript in English.

REFERENCES

Alarcón, A., Ferrera, R. (2000). Ecología, Fisiología y Biotecnología de la micorriza arbuscular. Mundi-prensa. Universidad Autónoma de Chapingo [Ecology, Physiology and Biotechnology of arbuscular mycorrhiza. Mundi-press. Autonomous University of Chapingo]. México, 227 pp.

Azcón-Aguilar, C., Palenzuela, J., Roldan, A. (2002). Analysis of the mycorrhizal potential in the rizosphere of representative plant species from desertification-threatened Mediterranean shrublands. *Applied Soil Ecology* 21, 1-9.

Banks, M., Govindaraju, R. S., Schwab, A. P., Kulakow, P. (2000). Effect of soil depth an root surface area on phytoremediation efficiency. In: *Phytoremediation of Hydrocarbon-Contaminated Soil* (Fiorenza, S., Oubre, C.L., and Herb Ward, C. Eds.). Boca Raton, FL, CRC. pp. 29-44.

Bending, G., Aspray, T., Whipps, J. (2006). Significance of microbial interactions in the mycorrhizosphere. *Advances in Applied Microbiology* 60, 97-132.

Brandt, R., Merkl, N., Schultze-Kraft, R., Infante C. (2006). Potential of vetiver (*Vetiveria zizanoides* L. Nash) for phytoremediation of hydrocarbon contaminated soils in Venezuela. *International Journal of Phytoremediation* 8, 273-284.

Camargo, S. (2002). Dispersal, distribution and establishment of arbuscular mycorrhizal fungi: a review. *Boletin de la Sociedad Botánica de México* 71, 33-44.

Córdova, A., Infante, C., Lugo, P., López, L. (2011). Comparación de la eficiencia de extracción con solventes de diferente polaridad, para un crudo mediano a dos concentraciones en un suelo sometido a biorremediación [Comparison of extraction efficiency with solvents of different polarity, for a medium crude at two concentrations in a soil subjected to bioremediation]. *Paper presented at the biannual meeting of Venezuelan Society of Chemistry*, Vargas, Venezuela.

Cunningham, S., Shann, J., Crowley, D., Anderson, T. (1997). Phytoremediation of contaminated soil and water. In: *Phytoremediation of Soil and Water contaminants* (Kruger, E.L., Anderson, T. A. and Coats, J. R. Eds.) Am. Chem. Soc. Washington DC. pp. 2-17.

Delon, F. (2001). Hidrocarburos origen y acumulación. Petróleos de Venezuela, SA. Programa de educación petrolera [Hydrocarbons origin and accumulation. Petróleos de Venezuela, SA. Oil education program]. *Litografía Beta*. pp 14.

Dushenkov, S. (2003). Trends in phytoremediation of radionuclides. *Plant and Soil* 249, 167-175.

Escalante, E., Gallegos, M., Favela, E., Gutiérrez, M. (2005). Improvement of the hydrocarbon phytoremediation rate by *Cyperus laxus* Lam. inoculated with a microbial consortium in a model system. *Chemosphere* 59, 405-413.

Gamal, R. (2005). Role of arbuscular mycorrhizal fungi in phytoremediation of soil rhizosphere spiked with poly aromatic hydrocarbons. *Mycobiology* 33(1), 41-50.

Gamarra, R. (2014). Factibiidad de uso de la asociación planta-hongos Glomeromycota en la biorestauración de un suelo contaminado con hidrocarburo. Trabajo de Grado Doctorado en Ciencias del Suelo. Postgrado de Ciencias del Suelo. Facultad de Agonomía [Feasibility of using the plant-fungal association Glomeromycota in the bioresorption of a soil contaminated with hydrocarbon. PhD Degree in Soil Science. Postgraduate in Soil Science. Faculty of Agronomy]. Universidad Central de Venezuela [Central University of Venezuela], Maracay, pp. 235.

Gao Y., L. Zhu. (2004). Plant uptake, accumulation and translocation of phenanthrene and pyrene in soils. *Chemosphere* (55), 1169-1178.

García, G., Ferrera, R., Almaráz, J., Rodríguez, R. (2000). Colonización micorrízica en gramíneas creciendo en un suelo contaminado con hidrocarburos. Ecología, fisiología y biotecnología de la micorriza arbuscular. Alarcón Ediciones [Mycorrhizal colonization in grasses growing on soil contaminated with hydrocarbons. Ecology, physiology and biotechnology of arbuscular mycorrhiza. Alarcón Editions]. *Colegio de Postgraduados (México)*: 221-227.

Genney, D., Alexander, I., Killham, K., Meharg, A. (2004). Degradation of the polycyclic aromatic hydrocarbon (PAH) fluorene is retarded in a Scots pine ectomycorrhizosphere. *New Phytologist* 163, 641-649.

Hernández, E., Ferrera, R., Fernandez, L., Rodriguez, R. (2000). Ocurrencia de la micorriza arbuscular y bacterias fijadoras de nitrógeno atmosférico en suelos contaminados por hidrocarburos. Ecología, fisiología y biotecnología de la micorriza arbuscular. Alarcón ediciones [Occurrence of arbuscular mycorrhiza and bacteria fixing atmospheric nitrogen in soils contaminated by hydrocarbons. Ecology, physiology and biotechnology of arbuscular mycorrhiza. Alarcón editions]. *Colegio de Postgraduados (México):* 213-220.

Hernández, I., Mager, D. (2003). Uso de *Panicum maximum* y *Brachiaria brizantha* para fitorremediar suelos contaminados con un crudo de petróleo liviano [Use of Panicum maximum and Brachiaria brizantha for phytoremediation of soils contaminated with light petroleum]. *Bioagro* 15(3), 149-144.

Hou F., Milke, M., Leung, D., Macpherson, D. (2001). Variations in phytoremediation performance with diesel-contaminated soil. *Environmental Technology* 22(2), 215-222.

Huang, X. D., El-Alawi, Y., Gurska, J., Glick, B. R., Greenberg, B. M. (2005). A multiprocess phytoremediation system for decontamination of persistent total petroleum hydrocarbons (TPHs) from soils. *Microchemical Journal* 81(1), 139-147.

Infante C. (2007). Biorremediación en suelos contaminados con hidrocarburos. Guía de apoyo de Curso UCV. Postgrado de Ciencias del Suelo. Universidad Central de Venezuela [Bioremediation in soils contaminated with hydrocarbons. UCV Course Support Guide. Postgraduate in Soil Science. Central University of Venezuela].

Infante, C., Hernández, I., López, L., Toro, M. (2012). Petroleum hydrocarbon-contaminated soils in Venezuela. In: *Phytotechnologies: Remediation of Environmental Contaminants*. Anjun, N. A., Pereira, M. E., Ahamad, I., Duarte, A., Umar, S., Khan, N.A. (Eds.), Boca Raton, FL (US). CRC Press/Taylor and Francis Group, Boca Raton, FL (US).

Johansson, J., Leslie, P., Roger, F. (2004). Microbial interactions in the mycorrhizosphere and their significance for sustainable agriculture. *FEMS Microbiology Ecology* 48, 1-13.

Joner, E., Leyval, C. (2001). Influence of arbuscular mycorrhiza on clover and ryegrass grown together in a soil spiked with polycyclic aromatic hydrocarbons. *Mycorrhiza* 10:155-159.

Joner, E., Leyval C. (2003). Rhizosphere gradients of polycyclic aromatic hydrocarbon (PAH) dissipation in two industrial soils and the impact of arbuscular mycorrhiza. *Environmental Science Technology* 37, 2371-2375.

Joner, E., Leyval, C., Colpaert, J. (2006). Ectomycorrhizas impede phytoremediation of polycyclic aromatic hydrocarbons (PAHs) both within and beyond the rhizosphere. *Environmental Pollution* 142, 34-38.

Leyval, C., Joner, E., del Val, C., Haselwandter, K. (2002). Potential of arbuscular mycorrhizal for bioremediation. In: *Mycorrhizal Technology in Agriculture* (Gianinazzi, S., Schüepp, H., Barea, J.M. and Haselwandter, K. Eds.). Birkhaüser Verlag, Basel, Switzerland, pp. 175-186.

Lioussanne, L. (2010). Review. The role of the arbuscular mycorrhiza-associated rhizobacteria in the biocontrol of soilborne phytopathogens. *Spanish Journal of Agricultural Research* 8(S1), S51-S61.

Liste, H. H., Alexander, M. (2000). Plant-promoted pyrene degradation in soil. *Chemosphere* 40, 7-10.

López, S., Gallegos, M., Pérez, L. J., Gutiérrez, M. (2005). Mecanismos de fitorremediación de suelos contaminados conmoléculas orgánicas xenobióticas [Mechanisms of phytoremediation of soils contaminated with xenobiotic organic molecules]. *Revista Internacional de Contaminación Ambiental* 21(2), 91-100.

Malachowska, A., Kalka, J. (2010). Influence of mycorrhizal fungi on remediation of soil contaminated by petroleum hydrocarbons. *Fresenius Environmental Bulletin* 19(12b), 3217-3223.

Madrid, M., Cataldi, A. (2002). Caracterización de fosas petroleras y sitios contaminados por crudo a través de métodos geofísicos y sensores geoquímicas *in situ*. XI Congreso venezolano de geofísica. Sociedad venezolana de ingenieros geofísicos. Noviembre 2002 [Characterization of oil wells and sites contaminated by crude oil through geophysical methods and in situ geochemical sensors. XI Venezuelan Congress of Geophysics. Venezuelan society of geophysical engineers. November 2002].

Mattina, M., Dykas, L. (2000). Chlordane uptake and its translocation in food crops. *Journal Agriculture Food and Chemical* 48, 1909-1915.

Merkl, N., Schultze, R., Infante, C. (2004). Phytoremediation of petroleum-contamined soils in the tropics-pre-selection of plant species from eastern Venezuela. *Journal of Applied Botany and Food Quality* 78, 185-192.

Merkl, N., Schultze, R., Infante, C. (2005). Phytoremediation in the tropics-influence of heavy crude oil on root morphological characteristics of graminoids. *Enviromental Pollution* 138(1), 86-91.

Merkl, N., Schultze, R., Arias, M. (2006). Effect of the tropical grass *Brachiaria brizantha* (Hochst. ex A. Rich.) Stapf on microbial population and activity in petroleum-contaminated soil. *Microbiological Research* 161, 80-91.

Muratova A., Thorsten, H., Neeru, N., Helmut, W., Turkovskaya, O., Kuschk, P., Jahn R., Moder, M. (2003). Plant-rhizosphere-microflora association during phytoremediation of PAH-contaminated soil. *Int. J. Phytoremediation* 5, 137-151.

Pérez, J., García, E., Esparza, F. (2002). Papel ecológico de la flora rizosférica en fitorremediación [Ecological role of rhizospheric flora in phytoremediation]. *Avance y Perspectiva* 21, 297-300.

Peters, K., Walters, C., Moldowan, E. (2005). The biomarker guide. Biomarkers and isotopes in the environment and human history. Cambridge, University Press Cambridge. 354 p.

Pilon, E. (2005). Phytoremediation. *Annual Review of Plant Biology* 56, 15-39.

República de Venezuela (1998). Normas para el control de la recuperación de materiales peligrosos y el manejo de desechos peligrosos. *Decreto 2635.* Gaceta Nº 5245 [Standards for the control of the recovery of hazardous materials and the management of hazardous wastes. Decree 2635. Gazette No. 5245]. Caracas.

Rojas, S. (2006). Tratamiento de crudo proveniente de fosas de desecho hacienda uso de tecnologías aplicadas para el mejoramiento de crudos pesados. Trabajo Especial de Grado. Universidad Central de Venezuela [Treatment of crude oil from waste pits using technologies applied for the improvement of heavy crude oil. Degree thesis. Central University of Venezuela], Caracas. 215 p.

Siciliano, S., Germida, J., Banks, K., Creer, C. (2003). Changes in microbial community composition and function during a polyaromatic hydrocarbon. *Applied and Environmental Microbiology* 69, 483-489.

Smith, S., Read, D. (2008). *Mycorrhizal Symbiosis*, Elseiver Edition. Third Edition, London, Academic Press. 785 p.

Soto, J. (2001). Quimica organica II: hidrocarburos y sus derivados halógenos. Editorial Síntesis [Organic Chemistry II: hydrocarbons and their halogenated derivatives. Editorial Synthesis]. Madrid. 544 p.

Strong, D., Phillips, D. (2001). Notes from the underground. Communication and control in the rhizospher. *Plant Physiology* 127, 727-30.

Sylvia, D. (1999). Mycorrhizal symbioses. In: *Principles and applications of soil microbiology*. Sylvia, D., Fuhrmann, J., Hartel, P., Zuberer, D. (Eds.), Upper Saddle River, NJ, Prentice Hall. pp. 408-426.

Tedersoo, L., May, T., Smith, M. (2010). Ectomycorrhizal lifestyle in fungi: global diversity, distribution, and evolution of phylogenetic lineages. *Mycorrhiza* 20, 217-263.

Verdin, A., Sahraoui, L. J., Fontaine, A., Grandmougin, F., Durand, R. (2006). Effects of anthracene on development of an arbuscular mycorrhizal fungus and contribution of the symbiotic association to pollutant dissipation. *Mycorrhiza* 16, 397-405.

Volante, A., Lingua, G., Cesaro, P., Cresta, A., Puppo, M., Ariati, L., Berta, G. (2005). Influence of three species of arbuscular mycorrhizal fungi on the persistence of aromatic hydrocarbons in contaminated substrates. *Mycorrhiza* 16, 43-50.

White, J., Wang, X., Gent, P., Iannucci, W., Eitzer, B., Schultes, N., Arienzo, M., Mattina, M. (2003). Subspecies-level variation in the phytoextraction of weathered *p,p_*-DDE by *Cucurbita pepo. Environmental Science Technology* 37, 4368-4373.

White, J., Parrish, Z., Iseleyen, M., Gent, M., Iannucci, W., Eitzer, B., Kelsey, J., Mattina M. (2000). Influence of citric acid amendments on the availability of weathered PCBs to plant and earthworm species. *International Journal of Phytoremediation* 8, 63-79.

Van der Heijden, M. (2002). Arbuscular mycorrhizal fungi as a determinant of plant diversity: in search of underlying mechanisms and general principles. *Ecological Studies* 157, 243-265.

In: Chemical Pollution Control …
Editor: Naser A. Anjum

ISBN: 978-1-53611-034-0

Chapter 4

MICROBIAL REMEDIATION OF TRACE METALS FROM AQUATIC SYSTEM: AN OVERVIEW

Baljinder Singh* and Kashmir Singh
Department of Biotechnology, Panjab University,
Chandigarh 160014, Punjab, India

ABSTRACT

Covering 75% of the earth surface, water is the most important and vital molecule of our planet. However, the pollution of water (aquatic) system with varied pollutants including trace metals is inevitable due to high industrial growth. Trace metals entering through industrial sources to the aquatic system can be fatal for both aquatic organisms and major organisms (animals and humans) closely associated with on this system. Microorganisms have immense inherent capability to control the trace metal levels in aquatic system. This chapter aims to overview trace metal pollution in aquatic system, and to discuss the role and underlying mechanisms of varied microorganisms-mediated control (sorption, degradation, transformation and metabolism) of major trace metals in aquatic system.

Keywords: aquatic pollution, trace metals, microbial remediation

* Corresponding author: Baljinder Singh, Department of Biotechnology, Panjab University, Chandigarh 160014, Punjab, India; E-mail: gilljwms2@gmail.com, sbaljinder@pu.ac.in.

1. INTRODUCTION

Safe drinking water is a necessity for the humankind and more than one billion people all over the world do not have safe water supply (Sekhon and Singh, 2013; Singh et al., 2016). Water covers 75% of earth surface and stands second to none in terms of its importance in the lives of organisms including animals and humans. However, rapid human population and consequent high growth in industries have significantly impacted water (aquatic) system through various ways. Different industries-sourced trace metals are being continuously added to and impacting aquatic system (Zhou et al., 2008; Sharma and Singh, 2013; Sheykhi and Moore, 2016). Originating mainly from mining, millrun, metallurgy, plating, chemical plant, curry and paper making industries, discharge of trace metals-laden wastewater into aquatic system and agricultural activities contribute the predominant causes of aquatic system pollution (Coelho et al., 2015). Trace metals toxicity in microorganisms, aquatic environments, plant and animals have been recognized for a long time (Sekhon and Singh, 2013). These metals are not only toxic and carcinogenic but also can cause serious health hazards after their bioaccumulation in the various food chains. Among the major efforts for decontaminating contaminated sites including aquatic system, bioremediation processes for the removal of trace metals removal have yielded promising results compared to physicochemical methods (Anjum et al., 2012). Owing to their adaptability to extreme conditions, microorganisms have been acting as a major tool to remediate varied trace metals through concentrating, removing, and recovering trace metals from contaminated aquatic systems (reviewed by Coelho et al., 2015).

Herein, trace metal pollution in aquatic system has been overviewed, and the major role and underlying mechanisms of varied microorganisms-mediated control (sorption, degradation, transformation and metabolism) of trace metals in aquatic system have been briefly discussed. The outcomes of the chapter can be useful in developing safer and economic methods for cleanup of trace metal contaminated aquatic system.

2. MICROBIAL REMEDIATION OF TRACE METALS

As the most basic level of trace metals is just atoms, degradation and metabolism are not possible. Instead, microorganisms have developed the capabilities to protect themselves from heavy metal toxicity by various mechanisms, such as adsorption, uptake, methylation, oxidation and reduction. Microbes and their diverse metabolic enzymes are typically employed for safe removal of environment contaminants, either through direct destruction or indirectly through transformation of the contaminant to a safer intermediate. Algae, bacteria, fungi, and yeast are the major microorganisms with biological activity, and can be used in their naturally occurring forms. Literature survey on the subject revealed major (70%) application of microorganisms for Cr, Cu, Cd, and Pb. However, the microorganism-mediated remediation of As and Hg was also reported (reviewed by Coelho et al., 2015).

2.1. Major Mechanisms Underlying Microbial Trace Metal-Remediation

Analysis of bacterial genetic and physiological systems has revealed that they may possess one or more of the four main types of mechanisms to protect against contaminants. 1) Efflux systems: These types of mechanisms export toxic metal ions to the outside of the cell. These may be non-ATPase or ATPase-linked and are generally highly specific for a particular heavy metal ion. 2) Accumulation and complexation: to prevent the exposure of essential cellular components to the contaminant and may be a result of the presence of metallothioneins or cysteine-rich proteins. 3) Reduction: The contaminating ions are enzymatically reduced upon entry to the cell and the less toxic, reduced ions are exported from the cell into the environment. 4) Alteration of cellular components: Some microorganisms may adapt to the presence of heavy metals by altering the sensitivity of cellular components, which may be achieved through mutations that decrease the sensitivity but do not alter basic function or by increasing the production of the sensitive cellular component.

Several microorganisms have been identified for their biosorption properties and different types of biomass have shown high levels of metal uptake (Tables 4.1 and 4.2). The ability of microorganisms to break down heavy metals has been assumed to be a recent trait, selected by the presence of these. Bacteria, which have the ability to export metals out of their cell, reduce damage to the organism by lowering the cellular concentration. It is therefore likely that microbes are presented with metals in water and food and may play a role in protecting the host from their adsorption. Mechanisms by which microorganisms act on heavy metals include biosorption (metal sorption to cell surface by physicochemical mechanisms), bioleaching (heavy metal mobilization through the excretion of organic acids or methylation reactions), biomineralization (heavy metal immobilization through the formation of insoluble sulfides or polymeric complexes), intracellular accumulation, and enzyme-catalyzed transformation (redox reactions). Mechanisms involved in the biosorption process include transport across cell membrane, complexation, ion exchange, precipitation, and physical adsorption. Metabolism-independent metal binding to the cell walls and external surfaces is the only mechanism present in the case of non-living biomass.

Compared to eukaryotes more sensitivity to metal toxicity was reported in prokaryotes (Perpetuo et al., 2011). However, a partial dependence of their eukaryotic or prokaryotic nature was evidenced for the interaction of most microorganisms with trace metals (reviewed by Coelho et al., 2015). Notably, the requirement of enzymatic attack on the pollutants/trace metals has been considered necessary for the conversion of these pollutants/trace metals to harmless products (Sharma, 2012). The possible modes of interaction are (a) Active metal extrusion, (b) various metal-binding peptides-mediated intracellular chelation (in eukaryotes), and (c) toxicity-reduction via transformation of metals into other chemical species are among the potential modes of interaction (reviewed by Coelho et al., 2015). Bacterial metallothioneins (MTs), a family of metal-chelating proteins that are functionally homologous to the short (approx. 60 amino acids) cysteine-rich eukaryotic metal-binding proteins is typical mechanism for regulating intracellular metal ion concentrations in some species (Blindauer et al. 2001). However, production of MTs as the main mechanism of tolerance to metals is exceptional in the bacterial world. In fact, resistance mechanisms - including active efflux and sequestration or transformation to other chemical species - become functional at concentrations above the homeostatic or nontoxic levels. Resistance mechanisms are often plasmid encoded, but in some instances, the genes are found on the chromosome (Table 4.3), suggesting an important evolutionary pressure to keep these genes; examples include mercury resistance in *Bacillus*, cadmium efflux in *Bacillus*, and arsenic efflux in *E. coli*. The first complete bacterial genome sequence, 1.8 megabase pairs (Mbp) from *Haemophilis influenzae* (Fleischmann et al. 1995), has listed homologs for arsC, merP, and merT that involved in arsenic and mercury resistance.

Table 4.1. Bioremediation of heavy metals using microorganisms

Microorganisms	Heavy metals	% uptake	Mechanism study	References
Bacillus sp. L14	Cu, Cd, Pb	Specifically uptake 75.78%, 80.48%, 21.25% of Cd (II), Pb (II) and Cu (II) under the initial concentration of 10 mg/L.	Remediation efficiencies may be greatly promotedThrough inhibiting the activities of ATPase	Guo et al. (2010)
Firmicutes,Gammaproteobacteria Actinobacteria,Bacteroidetes and Betaproteobacteria	U	Representative strains removed more than90% and 53% of U from 100 μM and 2 mM uranyl nitrate solutions, respectively, at pH 3.5 within 10 min of exposureand the activity was retained until 24 h	76% of characterized isolates possessed phosphatase enzyme and53% had PIB-type ATPase genes	Kumar et al. (2013)
Ochrobactrum anthropic	Cr		Reduction of Cr(VI)by cell debris of O. anthropi is enzyme-mediated	Cheng et al. (2010)
Caulobactercrescentus	Cd	Removed 94.3~99.9% of the Cd(II)	The recombinant strain JS4022/p723–6H, expressing RsaA-6His fusion protein was examined for its ability to sequester Cd(II) from the bacterial growth medium.	Patel et al. (2010)
Deinococcusradiodurans	Cd		DR1 strain carrying therecombinant EC20 (synthetic gene) demonstrated 2.5-fold higher toleranceto Cd^{2+}	Chaturvedi and Archana (2014)
Pseudomonas Plecoglossicida	Cd		Biosorption: Ionic exchange, electrostaticInteraction	Guo et al. (2012)
B. cereus and *B. pumilus*	Pb	22.1mg/g and 28.06 mg/g	Biosorption	Colak et al. (2011)
1. Streptomyces lunalinharesii *2. Arthrobacter sp.* *3. Anoxybacillusamylolyticus* *4. Bacillus cereus RC-1* *5. Corynebacterium glutamicum*	1. Cu and Zn 2. As 3. Cd, Cu, Co 4. Cd 5. Cd			

Table 4.1. (Continued)

Microorganisms	Heavy metals	% uptake	Mechanism study	References
6. *Pseudomonas aeruginosa CA207Ni, Burkholderiacepacia AL96Co, Corynebacterium kutscheri FL108Hg, Rhodococcussp AL03N*				
Klebsiellaneumoniae, Bacillus cereus, Bacillus pumilus, Pseudomonaveronii, Pseudomonafluorescens, Klebsiellaoxytoca, Klebsiellaornithinolytica	Pb, Zn, Ag		Biosorption	Munoz et al. (2012)
Bacillus sp. FM1	Cr	Completely reduced 100 mg/L Cr(VI) within 48 h	Reduction	Masood and Malik (2011)
Pseudomonas aeruginosa, Bacillus cereus, Bacillus subtilis, Bacillus megaterium, Kocuriarosea, Micrococcus luteus, Staphylococcus aureus	Cd, Zn, Pb	Maximum MICs of Cd and Zn for bacteria were 0.821 mM and 1.471 mM, respectively,	Bioremediation	Mohamed and Abo-Amer (2012)
Stenotrophomonas,Microbacterium, Acinetobacter,Pseudomonas and *Enterobacter*	U, V, Ni, Mn, Cu			Islam and Sar (2011)
Pseudomonas sp. B50A	Hg	Removed 86% of the mercury	Reductase activity	Giovanella et al. (2015)
Escherichia coli P4	Cd	Remove 18.8, 37, and 56% Cd+2 from aqueous medium after 48, 96, and 144 h, respectively	Biosorption: glutathione (GSH) and non-protein thiol (cysteine) levels which were increased up to 130 and 48%	Khan et al. (2015)
Staphylococcus, Bacillus, Micrococcus	Cr			Roychowdhury et al. (2015)
Cupriavidussp ATHA3, Klebsiellaoxytoca ATHA6 and Methylobacteriumsp ATHA7.	Ni	*K oxytoca* ATHA6 could decrease 83 mg/mL of nickel from the medium after 3 days.		Alboghobeish et al. (2014)

Table 4.2. Summary of major microorganisms, their potential to uptake trace metals and mechanism involved

Microorganisms	Heavy metals	% Uptake	Mechanism study	References
Bacillus sp. L14	Cu, Cd, Pb	Specific uptake (75.78%, 80.48%, and 21.25%) of Cd (II), Pb (II) and Cu (II) under the initial concentration of 10 mg/L.	Remediation efficiencies may be greatly promotedthrough inhibiting the activities of ATPase	Guo et al. (2010)
Firmicutes,Gammaproteobacteria Actinobacteria,Bacteroidetes and Betaproteobacteria	U	Representative strains removed more than90% and 53% of U from 100 μM and 2 mM uranyl nitrate solutions, respectively, at pH 3.5 within 10 min of exposureand the activity was retained until 24 h	76% of characterized isolates possessed phosphatase enzyme and53% had PIB-type ATPase genes	Kumar et al. (2013)
Ochrobactrum anthropic	Cr		Reduction of Cr(VI)by cell debris of O. anthropi is enzyme-mediated	Cheng et al. (2010)
B. cereus and *B. pumilus*	Pb	22.1mg/g and 28.06 mg/g	Biosorption	Colak et al. (2011)
1. Streptomyces lunalinharesii *2. Arthrobacter sp.* *3. Anoxybacillusamylolyticus* *4. Bacillus cereus RC-1* *5. Corynebacterium glutamicum* *6. Pseudomonas aeruginosa CA207Ni, Burkholderiacepacia AL96Co, Corynebacterium kutscheri FL108Hg,Rhodococcussp AL03N*	1. Cu and Zn 2. As 3. Cd, Cu, Co 4. Cd 5. Cd			

Table 4.2. (Continued)

Microorganisms	Heavy metals	% Uptake	Mechanism study	References
Caulobactercrescentus	Cd	Removed 94.3~99.9% of the Cd(II)	The recombinant strain JS4022/p723–6H, expressing RsaA-6His fusion protein was examined for its ability to sequester Cd(II) from the bacterial growth medium.	Patel et al. (2010)
Deinococcusradiodurans	Cd		DR1 strain carrying therecombinant EC20 (synthetic gene) demonstrated 2.5-fold higher toleranceto Cd^{2+}	Chaturvedi and Archana (2014)
Pseudomonas Plecoglossicida	Cd		Biosorption: Ionic exchange, electrostaticinteraction	Guo et al. (2012)

Table 4.3. Genes involved in the tolerance of heavy metals in microorganisms

Microorganisms	Heavy metal	Location of genes	Operon organization	Operon features	References
Pseudomonas strain K-62 ,*Xanthomonas* sp.	Hg	Plasmid	pPB Tn*5044*	2 operons- *merRTPAD* *merRBTPCAD,merRTPsigYAD*	Reniero et al. (1998)
Corynebacteriumglutamicum	As	Plasmid	*ars1* and *ars2*	*arsABC*	Ordonez et al. (2005)
Staphylococcus aureus	Cd	Plasmid	CadC	-	Corbisier et al. (1993)
Escherichia coli.	Cu	Chromosome	cutC and cutF (nlpF)	-	Gupta et al. (1995)
Geobacter species and *Rhodoferaxferrireducens.*	As	Chromosome	arrA	the presence of As(V) increased the transcription of arrA in cultures of*Geobacterlovleyi*	Giloteaux et al. (2013)

Several bacteria can mediate the methylation of Pb, Hg, Se, As, Tn and Sn by transfer a methyl group to the metals, resulting in methylated metal compounds, which may differ in volatility, solubility and toxicity (Bolan et al. 2014). Some bacteria can methylate methylmercury, forming volatile dimethylmercury which further can be enzymatically reduced to volatile metallic mercury, Hg^0 (Gadd, 2010). For example, mercury, Hg(II) can be biomethylated by a number of different bacterial species *Alcaligenes*faecalis, *Bacillus*pumilus, *Bacillus* sp., *P. aeruginosa* and *Brevibacteriumiodinium* to gaseous methyl mercury.

Sulfate-reducing bacteria (SRB) produce sulfide and immobilize toxic ions as metal sulfides (White and Gadd, 2000). Reduction of a metal to a lower redox state occurred by indirect reductive metal precipitation mechanisms, *e.g.*, in SRB systems, where reduction of Cr(VI) can be a result of indirect reduction by Fe^{2+} and the produced sulfide. Sulfur and sulfate-reducing bacteria are geochemically important in reductive precipitation of toxic metals such as U (VI) and Cr(VI). Multi-heme cytochrome c proteins mediate this process. *Desulfotoma culumreducens (*sulfate-reducing bacteria), share physiological properties of both sulfate- and metal-reducing groups of bacteria, and can grow with Cr(VI), Mn(IV), Fe(III) and U(IV) as sole electron acceptors.

CONCLUSION

Many studies on isolation and characterization of microbes capable of resisting heavy metals tolerance have been reported throughout the world. However, limited information is available on identification of genes responsible for resistance of heavy metals in microbes. The complete detoxification of heavy metals has not been studied; their assimilation by microbes remains an open field of study that has largely been unexplored. A great deal of research is needed to understand molecular basis of microbial heavy metals resistance sequences. For designing an effective bioremediation strategy, knowledge of heavy metals resistance and their generic deviations is essential. Molecular biology techniques have provided a useful handle to modify the genomic structure of the microorganisms to improve their qualitative and quantitative bioremeidation capabilities. Genetic engineering can be used to integrate genes into bacteria to enhance heavy metals resistance and accumulation. However, very little literature exists on the expression of heavy metal resistance genes. By understanding the physiological characteristics of these isolates, along with the structure and regulation of genes for both heavy metal resistances, we may gain information that will be useful in the design of biosensor and biosorption systems. Carrying out this work in the laboratory will aid to obtain the information that may allow for these isolates to be further considered for the design of on-site experiments. There is a need to develop transgenic bacterial system that effectively express heavy metals resistance gene for bioremediation. When considering the use of bacteria to detoxify heavy metals it is not only necessary to determine if they possess the genes responsible for this, but also whether these genes are functional and how they are expressed when exposed to heavy metals. The outcome of study will be helpful for tolerance of heavy metals in human and plant as similar study can be further conducted in gut associated bacteria and plant symbiotic bacteria that will provide protection to intestinal health and phytoremediation of trace metals.

REFERENCES

Alboghobeish, H., Tahmourespour, A., Doudi, M. (2014). The study of Nickel Resistant Bacteria (NiRB) isolated from wastewaters polluted with different industrial sources. *Journal of Environmental Health Science and Engineering* 12, 44.

Altinozlu, H., Karagoz, A., Polat, T., Ünver, I. (2012). Nickel hyperaccumulation by natural plants in Turkish serpentine soils. *Turkish Journal of Botany* 36, 269-280.

Anjum, N. A., Pereira, M. E., Ahmad, I., Duarte, A. C., Umar, S., Khan, N. A. (2012). *Phytotechnologies: Remediation of Environmental Contaminants.* Boca Raton, FL (USA). CRC Press/Taylor and Francis Group, Boca Raton, FL (USA).

Bani, A., Pavlova, D., Echevarria, G., Mullaj, A., Reeves, R.D., Morel, J. L., Sulçe, S. (2010). Nickel hyperaccumulation by the species of Alyssum and Thlaspi (Brassicaceae) from the ultramafic soils of the Balkans. *Botanica Serbia* 34, 3-14.

Blindauer, C. A., Harrison, M. D., Parkinson, J. A., Robinson, A. K., Cavet, J. S., Robinson, N. J., Sadler, P. J. (2001). A metallothionein containing a zinc ¢nger within a four-metal cluster protects a bacterium from zinc toxicity. *Proceedings of the National Academy of Sciences USA* 98, 9593-9598.

Chaturvedi, R., Archana, G. (2014). Cytosolic expression of synthetic phytochelatin and bacterial metallothionein genes in *Deinococcus radiodurans* R1 for enhanced tolerance and bioaccumulation of cadmium. *Biometals* 27, 471-482.

Cheng, Y., Yan, F., Huang, F., Chu, W., Pan, D., Chen, Z., Zheng, J., Yu, M., Lin, Z., Wu, Z. (2010). Bioremediation of Cr(VI) and immobilization as Cr(III) by *Ochrobactrum anthropi. Environmental Science and Technology* 44, 6357-6363.

Cho-Ruk, K., Kurukote, J., Supprung, P.,Vetayasuporn, S. (2006). Perennial plants in the phytoremediation of lead contaminated soils. *Biotechnology* 5, 1-4.

Coelho, L. M., Rezende, H. C., Coelho, L. M., de Sousa, P. A., Melo, D. F., Coelho, N. M. (2015). Bioremediation of polluted waters using microorganisms. In: *Advances in Bioremediation of Wastewater and Polluted Soil*. Shiomi, N. (Ed.), InTech, doi:10.5772/60770.

Colak, F., Atar, N., Yazıcıoglu, D., Olgun, A. (2011). Biosorption of lead from aqueous solutions by *Bacillus* strains possessing heavy-metal resistance. *Chemical Engineering Journal* 173, 422-428.

Corbisier, P., Ji, G., Nuyts, G., Mergeay, M., Silver, S. (1993). LuxAB gene fusions with arsenic and cadmium resistance operons of *Staphylococcus aureus* plasmid p1258. *FEMS Microbiology Letters* 110, 231-238.

Ebrahimpour, M., Mushrifah, I. (2008). Heavy metal concentrations (Cd, Cu and Pb) in five aquatic plant species in TasikChini, Malaysia. *Environmental Geology* 54, 689-698.

Fleischmann, R. D., Adams, M. D., White, O., Clayton, R. A., Kirkness, E. F., Kerlavage, A. R., Bult, C. J., Tomb, J. F., et al. (1995). Whole-genome random sequencing and assembly of *Haemophilus influenzae* Rd. *Science* 269, 496-512.

Garcia-Salgado, S., Garcia-Casillas, D., Quijano-Nieto, M. A., Bonilla-Simon, M. M. (2012). Arsenic and heavy metal uptake and accumulation in native plant species from soils polluted by mining activities. *Water, Air and Soil Pollution* 223, 559-572.

Giloteaux, L., Holmes, D. E., et al. (2013). Characterization and transcription of arsenic respiration and resistance genes during *in situ* uranium bioremediation. *ISME Journal* 7(2), 370-383.

Giovanella, P., Cabral, L., Bento, F. M., Gianello, C., Camargo, F. A. (2015). Mercury (II) removal by resistant bacterial isolates and mercuric (II) reductase activity in a new strain of *Pseudomonas* sp. B50A. *Nature Biotechnology* 33(1), 216-23.

Guo, H., Luo, S., Chen, L., Xiao, X., Xi, Q., Wei, W., et al. (2010). Bioremediation of heavy metals by growing hyperaccumulaor endophytic bacterium *Bacillus* sp. L14. *Bioresource Technology* 101, 8599-8605.

Guo, J., Zheng, X. D., Chen, Q. B., Zhang, L., Xu, X. P. (2012). Biosorption of Cd(II) from aqueous solution by *Pseudomonas plecoglossicida. Current Microbiology* 65, 350–355.

Gupta, M., Sharma, P., Sarin, N. B., Sinha, A. K. (2009). Differential response of arsenic stress in two varieties of *Brassica juncea* L.*Chemosphere* 74, 1201–1208.

Gupta, S. D., Lee, B. T. O., Camakaris, J., Wu, H. C. (1995). Identification of *cutC* and *cutF* (*nlpF*) genes involved in copper tolerance in *Escherichia coli.Journal of Bacteriology* 177, 4207-4215.

Huang, F., Dang, Z., Guo, C., Lu., Guc, G. N., Liu, R. R.& Zhang, H. J. (2013). Biosorption of Cd(II) by live and dead cells of *Bacillus cereus* RC-1 isolated from cadmium-contaminated soil. *Colloids and Surfaces. B,Biointerfaces* 107, 11–18.

Islam, E., Sar, P. (2011). Culture-dependent and -independent molecular analysis of the bacterial community within uranium ore. *Journal of Basic Microbiology* 4, 372–384.

Khan, Z., Nisar, M. A., Hussain, S. Z., Arshad, M. N., Rehman, A. (2015). Cadmium resistance mechanism in *Escherichia coli* P4 and its potential use to bioremediate environmental cadmium. *Applied Microbiology and Biotechnology* 99(24), 10745-57.

Kumar, R., Nongkhlaw, M., Acharya, C., Joshi, S. R. (2013). Uranium (U)-tolerant bacterial diversity from U ore deposit of Domiasiat in North-East India and its prospective utilisation in bioremediation. *Microbes and Environment* 28(1), 33–41.

Li, Y. J., Dhankher, O. P., Carreira, L., Lee, D., et al. (2004). Overexpression of phytochelatin synthase in *Arabidopsis* leads to enhanced arsenic tolerance and cadmium hypersensitivity. *Plant and Cell Physiology* 45, 1787–1797.

Mao, J., Won, S. W., Yun, Y. S. (2013). Development of poly(acrylic acid)-modified bacterial biomass as a high performance biosorbent for removal of Cd(II) from aqueous solution. *Industrial and Engineering Chemistry Research* 52, 6446–6452.

Masood, F., Malik, A. (2011). Hexavalent chromium reduction by *Bacillus* sp. strain FM1 isolated from heavy-metal contaminated soil. *Bulletin of Environmental Contamination and Toxicology* 86, 114–119.

Mohamed, R. M., Abo-Amer, A. E. (2012). Isolation and characterization of heavy metal resistant microbes from roadside soil and phylloplane. *Journal of Basic Microbiology* 52, 53–65.

Moreno, F. N., Anderson, C. W. N., Stewart, R. B., Robinson, B. H. (2008). Phytofiltration of mercury-contaminated water: volatilisation and plant-accumulation aspects. *Environmental and Experimental Botany* 62, 78–85.

Munoz, A. J., Ruiz, E., Abriouel, H., et al. (2012). Heavy metal tolerance of microorganisms isolated from wastewaters: identification and evaluation of its potential for biosorption. *Chemical Engineering Journal* 210, 325–332.

Ordonez, E., Letek, M., Valbuena, N., Gil, J. A., Mateos, L. M. (2005). Analysis of genes involved in arsenic resistance in *Corynebacterium glutamicum* ATCC 13032. *Applied Environmental Microbiology* 71, 6206–6215.

Oyetibo, G. O., Ilori, M. O., Obayori, O. S., Amund, O. O. (2014). Equilibrium studies of cadmium biosorption –nonviable bacterial strains isolated from polluted sites. *International Biodeterioration& Biodegradation* 91, 37–44.

Ozdemir, S., Knç, E., Poli, A., Nicolaus, B. (2013). Biosorption of heavy metals (Cd2+, Cu^{2+}, Co^{2+}, and Mn^{2+}) by thermophilic bacteria, *Geobacillusthermantarcticus* and *Anoxybacillusamylolyticus:* equilibrium and kinetic studies. *Bioremediation Journal* 17, 86–96.

Patel, J., Zhang, Q., McKay, R. M., Vincent, R., Xu, Z. (2010). Genetic engineering of *Caulobactercrescentus*for removal of cadmium from water. *Applied Biochemistry and Biotechnology* 160, 232–243.

Perpetuo, E. A., Souza, C. B., Nascimento, C. A. O. (2011). Engineering bacteria for bioremediation. In: Carpi A. (ed.) Progress in Molecular and Environmental Bioengineering—From Analysis and Modeling to Technology Applications. Rijeka: InTech., 605–632.

Prasad, K. S., Ramanathan, A. L., Paul, J., Subramanian, V., Prasad, R. (2013). Biosorption of arsenite (As^{+3}) and arsenate (As^{+5}) from aqueous solution by *Arthrobacter*sp. biomass. *Environmental Technology* 34, 2701–2708.

Rakhshaee, R., Giahi, M., Pourahmad, A. (2009). Studying effect of cell wall's carboxyl–carboxylate ratio change of *Lemna minor* to remove heavy metals from aqueous solution. *Journal of Hazardous Materials* 163, 165–173.

Reniero, D., Mozzon, E., Galli, E., Barbieri, P. (1998). Two aberrant mercury resistance transposons in the *Pseudomonas stutzeri* plasmid pPB. *Gene* 208, 37–42.

Roychowdhury, R., Mukherjee, P., Roy, M. (2015). Identification of chromium resistant bacteria from dry fly ash sample of Mejia MTPS thermal power plant, West Bengal, India. *Bulletin of Environmental Contamination and Toxicology* 99, 10745–10757.

Sekhon, G.S., Singh, B. (2013). Estimation of heavy metals in the groundwater of Patiala district of Punjab, India. *Earth Resource* 1, 1–4.

Sharma, R., Singh, B. (2013). Removal of Ni (II) ions from aqueous solutions using mod-ified rice straw in a fixed bed column. *Bioresource Technology* 146, 519–524.

Sakakibara, M., Ohmori, Y., Ha, N.T. H., Sano, S., Sera, K. (2011). Phytoremediation of heavy metal contaminated water and sediment by *Eleocharisacicularis*. *Clean: Soil, Air, Water* 39, 735–741.

Sharma, S. (2012). Bioremediation: features, strategies and applications. *Asian Journal of Pharmacy and Life Science* 2 (2), 202–213.

Sheykhi, V., Moore, F. (2016). Environmental risk assessment of heavy metals pollution in aquatic ecosystem—A case study: Sediment of Kor River, Iran. *Human and Ecological Risk Assessment: An International Journal* 22, 899-910.

Singh, B., Thakur, V., Bhatia, G., Verma, D., Singh K (2016). Eco-friendly and cost-effective use of rice straw in the form of fixed bed column to remove water pollutants. *Journal of Bioremediation and Biodegradation* 7, 6. doi:10.4172/2155-6199.1000374.

Skinner, K., Wright, N., Porter-Goff, E. (2007). Mercury uptake and accumulation by four species of aquatic plants. *Environmental Pollution* 145, 234–237.

Srivastava, M., Ma, L. Q., Rathinasabapathi, B., Srivastava, P. (2009). Effects of selenium on arsenic uptake in arsenic hyperaccumulator *Pteris vittata* L. *Bioresource Technology* 100, 1115–1121.

Veneu, D. M., Torem, M. L., Pino, G. A. H. (2013). Fundamental aspects of copper and zinc removal from aqueous solutions using a *Streptomyces lunalinharesii* strain. *Minerals Engineering* 48, 44–50.

White, C., Gadd, G. M. (2000). Copper accumulation by sulfatereducing bacterial bio¢lms. *FEMS Microbiology Letters* 183, 313-318.

Zhang, X., Xia, H., Li, Z., Zhuang, P. and Gao, B. (2011). Identification of a new potential Cd hyperaccumulator *Solanum photeinocarpum* by soil seed bank-metal concentration gradient method. *Journal of Hazardous Materials* 189, 414–419.

Zhou, Q., Zhang, J., Fu, J., Shi, J., Jiang, G. (2008). Biomonitoring: An appealing tool for assessment of metal pollution in the aquatic ecosystem. *Analytica Chimica Acta* 606, 135–150.

In: Chemical Pollution Control …
Editor: Naser A. Anjum

ISBN: 978-1-53611-034-0

Chapter 5

BIOCHEMICAL TALK IN THE RHIZOSPHERIC MICROBIAL COMMUNITY FOR PHYTOREMEDIATION

Mazhar Rafique, Khansa Haque, Tehmina Hussain, Amna and Hassan Javed Chaudhary*

Department of Plant Sciences, Quaid-i-Azam University, Islamabad, Pakistan

ABSTRACT

The contamination of different environmental compartments due to anthropogenic and natural sources is increasing day by day because of increase in population, industrialization and urbanization. It is imperative and crying need of the time to explore sustainable measures to control the levels and impacts of environmental contaminants. This chapter aims to overview the belowground processes associated with the plant rhizosphere covering the interactions between plants and microbes, their chemical communications, the biological processes they sustain, and the costs and benefits to plants associated with these interactions particularly in phytoremediation.

Keywords: chemical pollution, phytoremediation, microbial community

* Corresponding Author: Hassan Javed Chaudhary, Department of Plant Sciences, Quaid-i-Azam University, Islamabad, Pakistan; E-mail: hassan_javed@yahoo.com

1. Rhizosphere - A Stage for the Microbial Activity

The soil is a living environment that supports diverse communities of micro- and macro-organisms and is often considered a 'black box'. The zone of soil, which lies immediately adjacent to the roots, and that supports high levels of bacterial activity is known as the rhizosphere (Hiltner, 1904). However, most recently the term has been broadened to include both the volume of soil influenced by the root and the root tissues colonized by micro-organisms (Pinton et al., 2001). The microorganisms in the rhizosphere influence plants in a variety of ways, such as by affecting plant growth, nutrition, development, their susceptibility to disease, resistance to heavy metals, and by affecting their capacity to degrade xenobiotics (Barea et al., 2005).

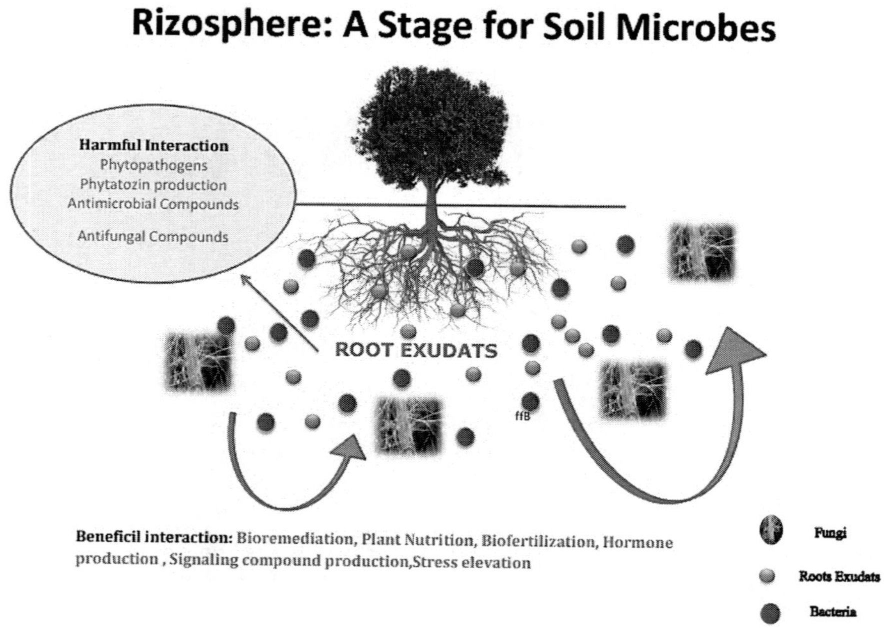

Figure 5.1. Schematic representation of major the processes occurring in the rhizosphere and their important roles.

Rhizosphere is not a single layer but it can be alienated into several distinct zones including endorhizosphere, the rhizoplane and the ectorhizosphere (Lynch, 1987). In addition to bacteria, beneficial fungi (known as mycorrhizal fungi) also occur in the rhizosphere (Lindermann, 1988). Loss of carbon compounds from roots drives the enhanced development of microbial populations (Grayston et al., 1996). Changes in the relative diversity and functional characteristics of microbial communities have unpredictable consequences for nutrient cycling processes, structure and productivity of natural plant communities, which may also impact on the yield of crops (Figure 5.1).

1.1. Plant-Microbe Interactions in the Rhizosphere

The presence of the bacteria in the rhizosphere can affect plants by many ways including growth promotion and pathogenesis. Therefore, the plant-bacteria interaction can be considered beneficial, harmful or neutral (Bais et al., 2006).

1.1.1. Beneficial Interaction

Beneficial interactions between plants and microorganisms are frequent process in the rhizosphere, where the nutrient abundance not only favor the microorganisms' development, improving plant growth and helping the plant to overcome biotic or abiotic stress (Bloemberg et al., 2001; Zamioudis and Pieterse, 2011). The best-known example of beneficial microorganisms is the mycorrhizal fungi that form symbiosis with approximately 80% of all terrestrial plant species by delivering nutrients for the plants in exchange of photosynthates. Beneficial interactions also occur between symbiotic bacteria belonging to *Rhizobium* genus and leguminous plants in which the *Rhizobium* bacteria fix atmospheric nitrogen for the plant. Besides symbiotic beneficial association, free-living rhizosphere microorganisms that include plant growth promoting rhizobacteria (PGPR) can positively affect the plant growth (Zamioudis and Pieterse, 2011). The benefits of plant-PGPR interaction enhanced seed germination rate, root growth, leaf area, chlorophyll content, protein content, hydraulic activity, tolerance to abiotic stress, shoot and root weights, biocontrol, delayed senescence and yield (Adesemoye and Kloepper, 2009, Compant et al., 2010). Beneficial bacteria are used as inoculants with the goal of application as phytostimulators (such as auxin-producing, root-elongating *Azospirillum*), rhizoremediators (pollutant degraders which use root exudate as their carbon source) and biopesticides (Lugtenberg et al., 2002). Direct mechanisms used by rhizobacteria for plant growth promotion are discussed below:

1.1.1a. Nitrogen Fixation

Nitrogen is an essential but limiting factor for plants growth and structure (Vance, 2001; Bhattacharyya and Jha, 2012; Ahemad and Kibret, 2014). The nitrogen fixing bacteria such as *Azoarcus, Azospirillum, Herbaspirillum, Azotobacter, Burkholderia, Gluconobacter, Pseudomonas, Enterobacter* have ability for change of nitrogen to ammonia by using a complex enzyme system called nitrogenase (Bloemberg and Lugtenberg, 2001; Richardson et al., 2009; Bhattacharyya and Jha, 2012; Ahemad and Kibret, 2014).

1.1.1b. Phytohormones Production

For growth and development of plants, phytohormones such as auxins, gibberellins, cytokinin, ethylene and abscisic acid play a very important role, contributing to the coordination of some important physiological processes increasing the plant resistance to environment factors, inducing or suppressing the expression of some genes, synthesis of enzymes, pigments and metabolites (Tsavkelova et al., 2006). Symbiotic and free living bacteria are known for their ability to produce phytohormones, influencing plants hormonal balance (Tsavkelova et al., 2006).

1.1.1c. Auxin Production

The optimal auxin concentration is needed for the plant growth. If the concentration is below the optimal levels, it has no effects on plants, whereas higher concentrations inhibit the growth. The endogenous pool of auxin is influenced by exogenous factors such as bacteria capable of auxin production. The inhibitory effect of the auxin can be the result of the interaction of IAA with 1-aminocyclopropan-1-carboxylate synthase (Felici et al., 2008). The production of high amounts of IAA by bacteria along with the endogenously produced plant IAA activates ACC synthase, leading to production of ACC, a precursor of ethylene. Ethylene is an inhibitory hormone of root growth, especially of the primary root length. However, there are bacteria such as *Pseudomonas putida* GR12-2 that can produce ACC deaminase, which converts ACC to ammonium and 2-ketobutirate lowering the ethylene concentration and also the inhibitory effect on the root growth (Spaepen et al., 2007). More than 80% of the soil bacteria are capable of auxin production bacteria such as: *Azospirillum* spp., *Agrobacterium* spp., *Azotobacter* spp., *Alcaligenes* spp., *Enterobacter* spp., *Erwinia* spp., *Acetobacter* spp., *Rhizobium* spp., *Bradyrhizobium* spp., and *Herbaspirillum* spp. (Tsavkelova et al., 2006; Gamalero and Glick, 2011). The synthesis of this hormone by rhizobacteria is frequently dependent on the presence of IAA precursor in the root exudates (Kravchenko et al., 2004). As in plants, the auxin biosynthesis in bacteria relies on an amino acid called tryptophan (Kravchenko et al., 2004; Tsavkelova et al., 2006; Spaepen et al., 2007; Gamalero and Glick, 2011). The tryptophan concentration in exudates differs strongly among plants, the amount of IAA produced by bacteria depending on this characteristic. For example, the inoculation of seeds with the auxin-generating *P. fluorescens* WCS365 resulted in a significant increase in the root weight of radish, but not in the root and shoot weight of cucumber, sweet pepper or tomato. The explanation is that radish roots produces at least nine times more tryptophan in its exudates per seedling than cucumber, sweet pepper, or tomato (Lugtenberg and Kamilova, 2009).

1.1.1d. Gibberellins Production

It has been identified 136 gibberellins produced by plants, 28 by fungi that belong to 7 species and only 4 by bacteria (GA1, GA3, GA4 and GA20) (Joo et al., 2005). Regarding their role in fungi and bacteria it seems that gibberellins are secondary metabolites that play an important role as signaling factors towards the host plant (Bottini et al., 2004). The best-studied gibberellins that are widely distributed in nature and exhibited maximum biological activity are GA1, GA3, GA4 and GA7 (Tsavkelova et al., 2006). They are detected in many bacterial genera such as: *Azotobacter, Arthrobacter, Azospirillum, Pseudomonas, Bacillus, Acinetobacter, Flavobacterium, Micrococcus, Agrobacterium, Clostridium, Rhizobium, Burkholderia, and Xanthomonas*.

1.1.1e. Cytokinin Production

Microorganisms are capable of synthesizing various cytokinin derivate such as zeatin, kinetin, isopentenyladenine (Tsavkelova et al., 2006). The gene responsible of cytokinin production was first described in *A. tumefaciens* (Gamalero and Glick 2011), but it can also be produced by *Rhizobium, Azotobacter, Azospirillum, Arthrobacter, Bacillus, Proteus, Escherichia, Klebsiella* (Tsavkelova et al., 2006; Hussain and Hasnain, 2009). The factors that can lead to a cytokinin accumulation are related to the presence of bacteria or to some environmental stresses as drought (Gamalero and Glick, 2011).

1.1.1f. Ethylene

High amounts of ethylene are synthesized in the condition of environmental stresses such as extreme temperatures, flooding, drought, the presence of toxic metals, organic pollutants, radiation, wounding, insect predation, high salt, and various pathogens (Gamalero and Glick, 2011). The negative effects of the ethylene can be reduced due to the presence of rhizobacteria that possess the enzyme ACC deaminase which have ability to degrade its immediate precursor, L-aminocyclopropane-L-carboxylic acid (Felici et al., 2008), found in the root exudates and used as a carbon source. Under such stressed environment re-uptake by roots is inhibited and the level of the ethylene inside the roots is highly reduced (Van, 2007). Genes responsible for ACC deaminase synthesis have been identified in many bacterial genera such as *Azospirillum, Rhizobium, Agrobacterium, Achromobacter, Burkholderia, Ralstonia, Pseudomonas* and *Enterobacter* (Gamalero and Glick, 2011).

1.1.1g. Phosphate Solubilization

Most stable organic phosphate synthesized by both plants and microorganisms is inositol phosphate. Other organic phosphates present in soil are in the form of phosphomonoesters, phosphodiesters including phospholipids and nucleic acids and phosphotriesters (Rodriguez and Fraga, 1999). Most of the phosphates solubilizing bacteria are found in the rhizosphere, where the metabolic activity is higher than in the bulk soil (Khan et al., 2009; Rodriguez and Fraga, 1999). Bacterial genera capable of phosphate solubilization or mineralization include *Pseudomonas, Bacillus, Rhizobium, Burkholderia, Enterobacter, Klebsiella, Serratia, Xanthomonas, Azotobacter, Azospirillum, Citrobacter,* and *Proteus* etc. (Chaiharn and Lumyong, 2011; Gamalero and Glick, 2011).

1.1.1h. Siderophore Production

The favorable form of iron for absorption by plant roots is more reduced ferrous (Fe^{2+}) ion (Vessey, 2003). The importance of soil bacteria in improving iron nutrition of plants is demonstrated in the case of exposure to heavy metals. Metals mobility in the soil can be influenced by microbial metabolites, especially by siderophores that can bind to magnesium, manganese, and chromium (III), gallium (III), cadmium, copper, nickel, arsenic, lead and zinc and radionuclides, such as plutonium (IV) as well as to iron (Gamalero and Glick, 2011).

1.1.1i. Competition

Beneficial rhizobacteria can compete with pathogens for nutrients and a suitable niche to be colonized. Bacteria that colonize the plant roots consume the nutrients derived the exudates limiting the amount of available nutrients for pathogens. This mechanism is often used by *P. fluorescens* due to its nutritional versatility and high rate of multiplication in the rhizosphere (Podile and Kishore, 2006).

1.1.1j. Enzymatic Action

Production of enzymes by PGPRs such as chitinase, cellulose, β-1,3 glucanase, protease or lipase, that induce lysis of fungal cell walls represents another mechanism of suppressing the growth of soil borne pathogens and simultaneously enhancing physiological processes.

1.1.1k. Induced Systematic Resistance

Rhizobacteria can stimulate indirectly the plant growth by increasing the resistance against pathogens by triggering the induced systemic resistance. Beside the local defense response (production of reactive oxygen species), plants can also initiate a systemic response in case of pathogenic attack.

1.1.2. Harmful Interactions

Root exudates can attract both beneficial microorganism and pathogenic population that can have negative effects on plant physiology (Morgan et al., 2005; Nihorimbere et al., 2011). Only few groups of bacteria are pathogenic for plants such as *Ralstonia solanacearum* L. which can cause bacterial wilt of tomato, *A. tumefaciens* known as crown gall agent, *Pantoaea stewartii* - cause of Stewart's wilt of corn, *Xanthomonas campestris* - a vascular pathogen that causes black rot of cabbage and other cruciferous plants etc. (Von et al., 2003).

2. Microbial Assistance in the Phytoremediation

The activities of rhizospheric microbes and plant's ability to cope with soil contaminations are closely related. The microbial activity in the immediate vicinity of the root seems to offer a favorable environment for co-metabolism of soil-bound and recalcitrant chemicals (Walton and Anderson, 1990; Shann, 1995). The microbial transformations of organic compounds are usually not driven by energy needs, but a necessity to reduce toxicity for which microbes may have to suffer an energy deficit. Thus, the processes may be helped and driven by the abundant energy that is provided by root exudates. Such stimulation of soil microbial communities by root exudates also benefits plants through increased availability of soil-bound nutrients and degradation of phytotoxic soil contaminants (El and Makhadmeh, 2001).

The microbial populations in a plant rhizosphere are usually composed of diverse communities rather than a single strain (Anderson and Coats, 1995). Such diversity can further promote remediation of contaminated soils as degradation of organic compounds may require several organisms with distinctive enzyme systems (Meharg and Cairney, 2000). The composition of microbial community in a rhizosphere is also known to differ both qualitatively and quantitatively from non-rhizospheric soil. It is, however, worth mentioning that the true extent of microbial diversity in soil or rhizosphere remains to be established.

It appears from a number of studies that rhizospheric bacteria are predominantly gram-negative (Anderson and Coats, 1995). The prevalence of gram-negative species has been speculated to be due to their abilities to efficiently utilize growth substrates available in the rhizosphere, and to better cope with toxic environments due to the presence of detoxifying enzymes (Anderson and Coats, 1995). The genera commonly found in rhizosphere include *Rhizobium*, *Azotobacter* and *Pseudomonas* (Hoagland and Williams, 1985), as well as a number of root-associated fungi, including those involved in the formation of symbiotic mycorrhizae. The microbial population and diversity in a rhizosphere may, however, vary with plant species and environmental factors such as temperature, moisture and nutrient status of the soil. For example, the genera that have been identified in barley rhizospheres include *Pseudomonas*, *Arthrobacter*, *Nocardia*, *Streptomyces*, *Penicillium* and *Trichoderma* (Krol, 1999). Activities of a number of exoenzymes have been reported in plant rhizosphere (Naseby and Lynch, 2002). These include cellulase, pectinase, chitinase (Krol, 1999; Hanzlikova et al., 1988), arylsulfatase (Knauff et al., 2003), glutathione-S-transferase (Zablotowicz et al., 1995) and peroxidase (Criquet et al., 2000). Such enzymes may play an important role in the detoxification of soil contaminants. Glutathione-Stransferase activity has been found in gram-negative isolates of rhizobacteria such as *Pseudomonadaceae* and *Enterobacteriaceae* that could dechlorinate chloronitrobenzene and certain herbicides (Zablotowicz et al., 1994). Soils contaminated with organic chemicals often also contain toxic heavy metals. In addition to providing a carbon source for rhizospheric microbes, root exudates in this case may also take part in direct detoxification by forming chelates with the metal ions. Under such conditions, mycorrhizae, the symbiotic root-fungi, may play a major role in phytoremediation as it improve plant establishment and survival on soils contaminated with both heavy metals (Shetty et al., 1994) and organic compounds (Leyval and Binet, 1998). Recent studies on PAH-polluted soil, with high levels of experimentally added Zn, have shown that ectomycorrhizae are able to tolerate mixed (PAH and heavy metal) pollutions. The presence of metals or mycorrhizae alone, however, showed a reduction in the rates of PAH degradation, but such reduction may have been due to low availability of mineral nutrients (Bending, 2003), which may not be relevant to other mixed pollutions. The use of mycorrhizae may give plants several advantages including protection against harmful conditions in the soil due to low nutrient levels, drought, and presence of toxic pollutants. The mycorrhizal hyphae network is more extensive than a root system, which can increase uptake of pollutants due to ability to penetrate relatively small soil pores. Many persistent organic chemicals such as 2, 4-D, atrazine, and PCBs can be degraded by mycorrhizal fungi (Donnelly and Fletcher, 1994; Anderson and Coats, 1995). Some ectomycorrhizal species are particularly efficient in degrading recalcitrant organic compounds, such as BTEX, TNT and PAHs (Meharg et al., 1997; Sarand et al., 1999; Dittmann et al., 2002), and their capacity is retained even under symbiotic conditions in soil (Sarand et al., 1999; Heinonsalo et al., 2000).

3. Factors Affecting Soil-Microbial Community (Biotic and Abiotic) in Remediation

The exudation of organic compounds by roots is influenced by either biotic (for example by soil microbial uptake) (Kuzyakov et al., 2003) or by abiotic processes (Hees et al., 2003). Some of the factors influencing exudation are listed as thus explained.

3.1. Plant Species

The amount, range and balance of compounds in root exudates vary for different plant species. Some worker found differences between wheat and barley (*Hordeum vulgare* L.) root exudates with respect to certain sugars (galactose, glucose and rhamnose), whereas other sugars occurred in similar amounts in exudates of both plants (Vancura, 1964). The specificity of root exudates from different plants in stimulating only certain groups of organisms; the cysts of potato eelworm (*Heterodera rostochiensis* L.) hatched when supplied the root washings of potato (*Solanum tuberosum* L.), tomato and some other solanaceous plants, but not the washings of beet (*Beta vulgaris*), rape (*Brassica napus* L.), lupin (*Lupinus lilosus* L.), mustard (*Brassica* sp.) or oats (Xiaoe et al., 2005).

3.2. Root Age

Research on peas and oats indicated that more amino acids and sugars exuded during the first 10 days of growth than during the second 10 days (Rovira, 1956). Vancura and Hovadik, (1965) study found 3- pyrazolylalanine in root exudate of cucumber (*Cucumis sativus* L.) only during the early seeding stage. In case of tomato and red pepper (*Capsicum anznumm* L.), they found that tyrosine occurred in the exudate only at fruiting and not at any other stage of growth.

3.3. Microorganisms

Microorganisms may affect the permeability of root cells, metabolism of roots, absorption and excretion of certain compounds in root exudates. Filtrates of cultures of some bacteria and fungi and also some antibiotics, increased the exudation of scopoletin (6 methoxy -7 hydroxycoumarin) by oat roots (Blaylock et al., 1997). Norman (1961) found that certain polypeptide antibiotics, for example, polymyxin that is formed by *Bacillus polymyxa* from soil, altered cell permeability and increased leakage. There are two main difficulties in interpreting the significance of their results, which show that culture filtrates or products increase the leakiness of plant roots. Consideration of the significance of the rhizosphere population in modifying exudation must involve a concept of micro-ecology, with a wide variety of organisms occupying different "niches on the roots and only those plant cells in the immediate vicinity of "exudation- promoting" organisms are likely to be affected. Microorganisms also influenced the exudation of organic materials into soil. A supplementary study showed that the exudation from wheat roots into synthetic soil was increased at least four-fold by microorganisms (Norman, 1961). The magnitude of the effects of microorganisms upon exudation no doubt will depend on the species colonizing the roots (Albert, 1969). Some other plant biotic factors like developmental status, shoot herbivory, photosynthesis, supply of carbon from shoot to root, evaporation, transpiration, nutrient deficiency, root architecture, cytosolic concentration, membrane permeability, membrane electrochemical potential, release of microbial signal, allelochemical release, mycorrhizae, nodulation and some soil biotic factor are also influenced by the root exudation.

3.4. Temperature

The release of amino acids especially, asparagine from roots of tomato and subterranean clover (*Trifolium subterraneum* L.) increased with rising temperature (Rovira, 1959). Some worker found more amino acids in exudates from strawberry plants (*Fragatia vesca* L.) grown at 5 to 10°C than at 20 to 30°C; influenced the pathogenicity which attacks strawberries at low soil temperatures (Husain and Mckeen, 1963; Hale et al., 1978).

3.5. Light

The light intensity at which plants grow is known to affect the amount and balance of compounds exuded into nutrient solution by the roots of plants such as tomato and subterranean clover (Rovira, 1959). Clover grown at full daylight intensity exuded more serine, glutamic acid, and c-alanine than plants grown in 60% shade. In tomato, shading reduced the levels of aspartic acid, glutamic acids, phenylalanine and leucine in exudate. Adjacent to these abiotic factors, few others such as moisture, humidity, wind speed and light intensity, elevated CO_2 pesticides, available space, atmospheric nitrogen deposition, ozone, physical disturbance, fire, irrigation, erosion, altitude and latitude are also influencing the exudation (Torsivik et al., 1996). Some soil abiotic factors resembling compaction, soil type, salinity, soil pH, metal toxicity, water availability, organic matter, cation and anion exchange, drainage, aeration, rooting depth, soil texture, soil structure and redox-potential stimulate the release of organic chemical from plant root (Ross et al., 2000; Rangarajan et al., 2001).

4. Root Exudates and Other Chemicals: Major Players in the Microbes-Assisted Remediation

Plants depend on the ability of roots to communicate with microbes. The converse is also true; many bacteria and fungi are dependent on associations with plants that are often regulated by root exudates. A chemotactic response towards root-secreted organic and amino acids is the first step in root colonization (Zheng and Sinclair, 1996). Several studies show the chemo attraction of various PGPR such as *P. fluorescens*, *Azospirillum spp.*, *Azotobacter chroococcum* and *Bacillus* species toward whole roots exudates (Gafny et al., 1986; Heinrich and Hess, 1985; Zheng and Sinclair, 1996; De Weert et al., 2002; Sood, 2003; Yaryura et al., 2008; Tan et al., 2013). In fact, plant roots represent important ecological niches that support microbial communities, which can influence the plant growth by their beneficial or deleterious effects (Stefan, 2008; Raaijmakers et al., 2009). Research suggested that root exudates might act as messengers that communicate and initiate biological and physical interactions between roots and soil organisms (Walker et al., 2003). The specific structure and diversity of the rhizosphere bacterial community varies between plant species (Smalla et al., 2001). Moreover, and different root zones on the same plant can support distinct bacterial communities, reflecting qualitative and quantitative differences in root exudation (Yang and Crowley, 2000). In addition, soil type has a key role in determining the specific dominant bacteria colonizing the rhizosphere (Marschner et al., 2001). The structure of rhizosphere bacterial communities can also be influenced by root infection by pathogenic bacteria, which promote greater bacterial community variability compared with healthy roots (Yang et al., 2001). The driving force for plant – microorganism interactions is represented by the organic compounds released by roots in a few mm of thickness of the surrounding soil, called rhizosphere (Nihorimbere et al., 2011, Walker et al., 2003). Broadly, in the rhizosphere there are three different, but interacting components: the rhizosphere (soil) or ectorhizosphere, the rhizoplane and the root itself or endorhizosphere. Thus, the rhizosphere or ectorhizosphere is the zone of soil whose chemical composition is influenced by the roots exudates, which can stimulate or inhibit the microbial community and its activity. More than 40% of plants photosynthates are secreted into the rhizosphere (Berendsen et al., 2012). Root exudation includes water, sugars and polysaccharides, amino acids, organic acids, fatty acids, sterols, growth factors, enzymes, flavonoids and proteins (Nicholas, 2007; Stefan, 2008) which serves as signals and growth substrates for beneficial or pathogenic microbial partners (Bakker et al., 2012). Depending on their molecular weight, root exudates are often divided into two classes of compounds: low-molecular weight compounds as amino acids, organic acids, sugars, phenolics and high-molecular weight exudates such as polysaccharides and proteins (Bais et al., 2006), many of which can induce chemotactic response in PGPR (Bais et al., 2008).

Not remarkably, specific compounds identified in root exudates have been shown to play roles in root–microbe interactions. For example, isoflavonoids and flavonoids present in the root exudates of a variety of leguminous plants activate the Rhizobium genes responsible for the nodulation process and might be responsible for vesicular-arbuscular mycorrhiza (VAM) colonization (Peters et al., 1986; Becard et al., 1995; Trieu et al., 1997). Characteristics of flavonoid profiles in root exudates differ considerably among legumes (Philips, 2000), and this specificity enables mutualists and beneficial bacteria such as rhizobia to distinguish their hosts from other legumes (Mithofer, 2002). The release of carbon compounds from plants into the soil results in greater microbial populations in the rhizosphere relative to the bulk soil, and increased microbial biomass and activity (Lynch, 1987; Bending, 2003). The root exudates composition and concentration can vary with the plant species, cultivars, plant's age, soil properties, level of biotic and abiotic stress etc. (Nicholas, 2007; Nihorimbere et al., 2011; Chapparo et al., 2012). Moreover, in maize root exudates, the N deficiency reduced the release of amino acids, P deficiency stimulated the release of gamma-aminobutyric acid and carbohydrates, K-deficiency led to a less sugars release and Fe deficiency increased the release of glutamate, glucose, ribitol and citrate (Carvalhais et al., 2011). As a consequence, plant roots exudates have a strong influence on microbial population, which can differ between the same plant species growing in the same soil. It has been reported that root of *A. thaliana* infected with the pathogen *P. syringae* start secreting malic acid, which in turn act as a strong attractant for the beneficial *B. subtilis* (Rudrappa et al., 2008). Similarly, cucumber roots infected by *Fu. oxysporum* secrete more citric and fumaric acid, stimulating the chemotactic response of the protective *B. amyloliquefaciens* strain SQR9 toward the plant (Liu et al., 2014). These two examples elegantly illustrate how a plant can modulate its root-associated microbial community via attracting molecules to help it cope with environmental stresses.

Hereunder, efforts have been made to briefly discuss major composition, chemistry and the potential mechanisms underlying the root exudation.

4.1. Composition of Root Exudates

Roots also exude a variety of low-molecular weight organic compounds. These include sugars and simple polysaccharides (such as arabinose, fructose, glucose, maltose, mannose, oligosaccharides), amino acids (such as arginine, asparagine, aspartic, cysteine, cystine, glutamine), organic acids (such as acetic, ascorbic, benzoic, ferulic, malic acids), and phenolic compounds. Some of these compounds, especially the phenolic, influence the growth and development of surrounding plants and soil microorganisms. In addition, higher-molecular weight compounds such as flavonoids, enzymes, fatty acids, growth regulators, nucleotides, tannins, carbohydrates, steroids, terpenoids, alkaloids, polyacetylenes, and vitamins are released in large quantities (Rovira, 1969; Hale et al., 1978; Curl and Truelove, 1986; Fan et al., 1997a; Uren, 2000). Varieties of these compounds are involved in either primary or secondary plant metabolic processes or also plant defense. Root exudates contain water soluble, insoluble, and volatile compounds including sugars, amino acids, organic acids, nucleotides, flavanones, phenolic compounds and certain enzymes (Park et al., 2002). The root exudates mediate acquisition of minerals by plants; stimulate microbial growth in the rhizosphere, and change pH, water flux, and availability of oxygen in soil around the roots (Morgan et al., 2005).

Living root hairs as well as actively growing primary and secondary roots release large quantities of root exudates. Depending upon the plant species and environmental conditions encountered, the living root system varies widely in terms of root architecture, including the number and size of root hairs. Plant root exudates are primarily composed of carbon-containing compounds derived from products of photosynthesis. The non-carbon- containing compounds present include the ubiquitous H^+, inorganic ions, water, and electrons. Although their quantities are lower in root exudates than those of carbon-containing compounds and their presence is significant (Uren, 2000). Estimated amounts and proportions of photosynthate devoted to root and root products vary considerably (Uren, 2000). In the seedling stage, root exudates may represent between 30 and 40% of the plant's photosynthetic productivity (Whipps, 1990). In young plants, the release of these molecules is quite significant and may account for up to 30% of total dry matter production

(Sauerbeck et al., 1981; Whipps, 1990). It has been well documented in a number of species such as sorghum and wheat that root exudation decreases with age of plants and increases with soil stress such as compaction, drought, and low nutrient supply (Brady and Weil, 1999; Brimecombe et al., 2001). Though roots produce nearly every major type of compound found in plants, they generally do not produce chlorophyll and specific compounds linked to photosynthesis (Uren, 2000).

4.2. Chemistry of Root Exudates

The amount of root exudates produced varies with the plant species, cultivar, age, and stress factors (Uren, 2000). It has been described previously, significant quantities of organic compounds are released at the surface of young roots (Rovira, 1969). High molecular weight polysaccharides secreted by root cap cells and epidermal cells at the level of the apical zone form a substance referred to as mucigel when mixed with microbial cells and clay particles. The role of mucigel appears to be associated with lubrication of root movement through the soil, and enhancement of interactions between roots and soil, especially in dry soils where soil contact is diminished. Mucigel can also serve as a means of protection against toxins that provides a suitable environment for the growth and development of microorganisms.

4.3. Mechanisms of Root exudation

Root exudates are released in three ways from living root systems. These processes of release are briefly described below.

4.3.1. Diffusion

Low molecular weight organic compounds such as sugars, amino acids, carboxylic acids, and phenolic are released using a passive process involving steep concentration gradients between the cytoplasm of intact root cells and the soil. Membrane permeability determines if direct diffusion through the lipid bilayer of the plasmalemma is possible. This condition depends on the physiological state of the root cell and on the polarity of the compounds to be exuded. Permeation of lipophilic exudates is generally facilitated by this method (Guern et al., 1987). Root exudation of amino acids and sugars occurs mainly passively via diffusion and may be enhanced under stress.

4.3.2. Ion Channel

Some chemicals such as specific carboxylates (*e.g.*, citrate, malate and oxalate), which are normally exuded in high concentrations, cannot diffuse through root membranes when the root is under specific stress such as nutritional deficiency or toxicity. In these cases, anion channels mediate the controlled release of these products by roots.

4.3.3. Vesicle Transport

Transport of high-molecular- weight compounds generally involves vesicular transport (Battey and Blackbourn, 1993). Transport of mucilage polysaccharides across the root cap is mediated by Golgi vesicles, while proteins such as ectoenzymes (acid phosphatase, peroxidase) are synthesized at the level of membrane-bound polysomes and enter the lumen of the endoplasmic reticulum using vectorial segregation (Neumann and Romheld, 2000). Through the Golgi apparatus, proteins are separated from those destined for vacuolar compartmentation, and those transported to the plasmalemma by transfer vesicles (Chrispeels, 1991; Chrispeels and Raikhel, 1992). Extracellular and intracellular calcium levels influence processes that involve exocytosis (Marschner, 1995). Phenolic (Rougier, 1981; Gagnon et al., 1992), phytosiderophores (Nishizawa and Mori, 1987) as well as high-molecular-weight compounds, are stored and released using vesicles, but the exact mechanisms utilized remain unknown. Chemicals that occurred in root exudates, when released in large quantities, generally enter the rhizosphere and are directed to physical (sorption), chemical (metal oxidation) and biological (microbial degradation) processes in the soil (Cheng, 1995; Dalton, 1999; Huang et al., 1999; Inderjit and Dakshini, 1999). The biological activity of chemicals in the rhizosphere may be altered rapidly in terms of their efficacy because of chemical oxidation, microbial breakdown or immobilization by irreversible binding to soil particles. This alteration in activity can occur before the compound(s) in question reach a biological target (Lehmann et al., 1987; Cheng, 1995).

4.4. Root Exudation-Impact on Soil Microorganisms and Root-Microbe Communication

Root exudates also act as antimicrobials against rhizospheric microflora and provides the defense to plants. Collectively, plants produce a compositionally diverse array of 100 000 different low-molecular-mass natural products, known as secondary metabolites (Dixon, 2001). The rich diversity of secondary metabolites arises partly because of selection for improved defense mechanisms against a broad array of microbes, insects and plants. *Bacillus cereus* strain known to promote tobacco growth showed alterations in its secreted proteome in the presence of tobacco root-exudates and groundnut root-exudates. In response to the plant's exudates, *B. cereus* can modify its chemical composition of surface, suggesting an adaptation that can favor strong root colonization (Dutta et al., 2013). These studies determined that the plant presence intensely influence the bacterial function with which it associates, promoting cellular processes as diverse as secondary metabolites biosynthesis, metabolism, transport and motility.

Bacteria-secreted molecules mediate the positive effects of PGPRs on the plant. Among these molecules, we find auxin IAA, a phytohormone involved in plant growth and development produced both by phytopathogenic bacteria and PGPRs. Interestingly, the plant can affect PGPR's IAA production by secreting tryptophan in its root exudates. Tryptophan is a precursor of IAA biosynthesis, which will activate gene transcription, upregulating IAA production in the beneficial *P. putida* and *A. brasilense* (Patten and Glick, 2002; Ona et al., 2005; Rothballer et al., 2005; Ryu and Patten, 2008). *Pseudomonas putida* strain PCL1444 degrades the pollutant naphthalene, which can be toxic to the plant. Grass cv. Barmultra can positively influence this activity since the degradation genes are most expressed in presence of salicylic acid, which could come from the plant, and of the major carbon sources present in the grass' exudates (Kuiper et al., 2002). These examples illustrate that the plant not only stimulates its own colonization by PGPR, but also favors the activation of the plant-beneficial bacterial activities.

5. Signaling Molecules for Biochemical Talk in the Rhizosphere

The structure of rhizosphere microbial communities differs from that of the bulk soil (Marschner et al., 2004). Plants are able to shape their microbiome (Zamioudis and Pieterse, 2011). Though, the dominant population of the rhizosphere is made up by species belonging to Proteobacteria and Actinobacteria (Nihorimbere et al., 2011). Because of the significant amounts of rhizodeposits (10 to 250 mg/g root) microbial biomass and activity are generally much higher in the rhizosphere, than in the bulk soil [FRANS et al.,

2007). The concentration of bacteria per gram of soil can reach here between 10^{10} and 10^{12} cells. Therefore, plant roots, through their rhizodeposits, can be seen as initiators of cross talked with soil microbes, which in turn produce signals that initiate colonization (Nihorimbere et al., 2011). Rhizosphere colonization is an important aspect of plant-bacteria interactions involving different steps: movement of microbes to root surface, adsorption, anchoring and gene expression (Frans et al., 2007).

A widespread mechanism used by bacteria to sense and respond to their environment stimuli such as rhizodeposits, refers to a system that comprise two components: a membrane-bound sensor histidine protein kinase and a response regulator most often mediating differential gene expression (Faure et al., 2009). An example of such mechanism is chemotaxis. This is a primitive sensing mechanism activated by changes in pH, temperature, osmolarity, viscosity or environment chemical composition (Bacilio-jiménez et al., 2003). Chemotaxis is mediated by two-component regulatory system consisting of a sensor kinase -CheA and a response regulator -CheY. Chemoreceptors, methyl-accepting proteins (MCPs), localized in the cytoplasmic membrane, monitor the concentration of the chemicals in the environment. Through the methylation of the MCPs a signal transduction occurs and auto-phosphorylation of CheA takes place. Subsequently, P-CheA donates the phosphate group to CheY and P-CheY will interact with the flagellar motor. Whenever the signal drops below a certain threshold, CheY will be phosphorylated, and clockwise rotation will occur. In this case, bacteria will start to tumble to change swimming direction. If this signal rises above the threshold value, CheY will be dephosphorylated and counter clockwise rotation will occur, resulting in a run of the bacterial cell (De Weert et al., 2002). The capacity of the microorganisms to colonize the rhizosphere can be stimulated by the presence of root exudates (Bacilio-jiménez et al., 2003), but in the same time depends on the genetic traits of the host plant and the colonizing microorganisms (Chin-a-Woeng and Lugtenberg, 2004).

The most studied quorum-sensing (QS) signal among bacteria is the N-acyl-homoserine lactones (AHLs), which can be synthesized by a large number of Gram-negative bacteria. The AHL synthesis is dependent on synthases belonging to two classes: LuxI and AinS homologs. Signal perception relies on a sensor protein, a LuxR homolog, which is also a transcriptional regulator controlling the expression of quorum-sensing-regulated genes (Faure et al., 2009). Intra-specific communication in the Gram-positive bacteria is mediated by peptide-signaling molecules (Walker et al., 2003). This peptide is secreted via ATP-binding cassettes (ABC) transporters. Migration of microorganisms towards the roots is followed by movement along the roots, agglutinability by root exudates and adherence (Bacilio-jiménez et al., 2003). Because of the heterogeneous release of exudates along the root axis, bacterial distribution and colonization on the root is not uniform, the rich areas in organic compounds being preferred (Compant et al., 2010). Therefore, a high number of bacteria is founded at the root base 10^7-10^8 CFU/ cm (Macmillan, 2002) and rapidly decrease to 10^3-10^4 CFU/cm at the root tip. Furthermore, it is estimated that microorganisms can colonixe less than 10% of the root surfaces. For example, *Pseudomonas* cells on the tomato root are mainly present on junctions between epidermal cells and the deeper parts of the root epidermis and root hairs (Chin-a-Woeng and Lugtenberg, 2004).

5.1. Quorum Sensing

Many bacteria such as *Erwinia* spp., *Pseudomonas* spp. and *Agrobacterium* spp. possess quorum-sensing (QS) systems that control the production and secretion of virulence factors (Fray, 2002). Briefly, QS is a form of cell-to-cell communication between bacteria mediated by small diffusible signaling molecules (autoinducers, AI), which are responsible for the coordinated actions of bacteria reaching the "quorum." The most commonly reported types of AI signals are AHLs for Proteobateria and peptide-signaling molecules for Firmicutes (Waters and Bassler, 2005). Bacteria that use QS also have a receptor that can specifically detect signaling molecules, and when the inducer binds the receptor it activates transcription of target genes, including those required for inducer synthesis. The rhizosphere contains a higher proportion of AHL-producing bacteria as compared to the bulk soil, suggesting that they play a role in root colonization (Elasri et al., 2001).

Plants have evolved strategies to interfere with the bacterial AHL signaling system to avoid pathogenesis. Such interference includes the production of signal mimics, signal blockers, signal degrading enzymes, or compounds that block the activity of the AHL-producing enzymes (Fray, 2002; Rasmussen and Givskov, 2006). A few years ago, it was reported that root exudates of pea seedlings (*Pisum sativum* L.) had several separable activities that mimicked AHL signals in well-characterized bacterial reporter strains, stimulating AHL-regulated behaviors in some strains and inhibiting them in others (Teplitski et al., 2000). For example, the addition of root exudates from pea seedlings to a *Chromobacterium violaceum* culture inhibits the production of violacein and exoenzymes, which are under the control of QS and involving AHLs. The compounds responsible have not been identified, but they preferentially partition into polar solvents. Various species of higher plants, including pea (*P. sativum* L.), crown vetch (*Coronilla varia* L.), M. truncatula, rice (*Oryza sativa* L.), soybean (*Glycine max*) and tomato (*Lycopersicum lycopersicon* L.), were found to secrete compounds having AHL-mimicking activities (Teplitski et al., 2000; Rasmussen et al., 2005a). AHL signal-mimicking compounds seem to be important in determining the outcome of interactions between higher plants and a diverse range of microorganisms against stress.

As mentioned above, another QS interference strategy is inactivation or complete degradation of generated signal molecules. This can be achieved by different methods such as chemical and enzymatic degradation. A simple way to inactivate AHL signal molecules is to increase the pH to above 7, which causes AHL lactonolysis (ring opening) (Yates et al., 2002). A number of higher organisms use this strategy in defense against invading QS bacteria. Plants infected with the tissue-macerating plant pathogen *Erwinia carotovora* will, as a first response, actively increase the pH at the attack site (Yates et al., 2002).

6. Factors Affecting Microbe-Assisted Biodegradation of Organic Pollutants

6.1. Toxicity and Bio-Availability

Despite huge promise, biological remediation often suffers with a number of limitations in the field. Many soil-bound and insoluble compounds pose a particular problem in terms of their uptake by remediating microbes and plants. This lack of availability is much more pronounced in the case of organic contaminants, which sometimes puts severe restrictions on the remediating potential of plants. The uptake and translocation of organic compounds through the transpiration stream is optimal for slightly hydrophobic compounds but may be severely restricted for soil-bound and highly hydrophobic compounds in many higher plants (Briggs et al., 1982; Ryan et al., 1988; Bromilow and Chamberlain, 1995).

Further limitations to the growth of plants and microbes may arise from the severity of a contamination and the heterogeneous nature of soil. Plants and microbes also seem to have a limited ability to degrade certain organic compounds, in particular those containing chlorines and/or aromatic ring structures. The tolerance of a remediating species may also vary considerably towards different toxic chemicals, even towards some closely related ones. For example, chlorinated ethane has been reported to be more toxic to hybrid poplar trees than the corresponding chlorinated ethane (Dietz and Schnoor, 2001). Our observations on the root growth of alfalfa in kerosene-contaminated soils have shown that it occurred mainly in the plume, where contaminant levels were below the critical toxic concentration. The root growth in the first instance occurred at a distance of 45 and 25 cm from center of the plume, whilst rooting in the center only occurred almost a year after planting. However, whether this resulted from gradual decomposition of kerosene, or from adaptation of the roots remains to be established. The rates of removal and/or degradation of pollutants during biological remediation are often much slower than those that can be achieved by physical or chemical methods. This is particularly evident in remediation by plants, the growth of which is dependent on many environmental factors such as soil conditions, availability of nutrients, water etc.

6.2. Microbial Composition of the Rhizosphere

The exogenous metabolic activities of rhizosphere microorganisms may be controlled by plants in addition to the root exudates (Walton et al., 1994; El and Makhadmeh, 2001). This has led to speculation that such complementary interactions may be part of a plant's response to chemical stress in the soil (Walton et al., 1994). A strong indication for this has emerged from the consistent findings that microbial numbers in a rhizosphere are generally several orders of magnitude greater than those in a non-vegetated soil (Reynolds et al., 1999), a phenomenon that has been observed up to a few mm away from the soil-root interface (Badalucco et al., 1997). This interaction may, however, be restricted to a much narrower zone for hydrophobic organic compounds, due to their limited aqueous diffusion, different soil textures and organic matter content etc. Thus, whilst (Corgié et al., 2003) noted a root effect on phenanthrene degradation that extended >6 mm from the surface of ryegrass roots in pure fine sand, Joner and Leyval (Joner and Leyval, 2003) found that enhanced phenanthrene degradation extended only <0.6 mm from the root surface of clover and ryegrass in two industrially polluted soils.

6.3. Root Exudation

Indeed, as much as 20% of the assimilated carbon, may be exuded by plants into the rhizosphere (Guckert, 1992), which is readily utilized by rhizospheric microorganisms as a direct source of energy (Helal and Sauerbeck, 1989). The rate of exudation, however, changes with the age of a plant (Haller and Stolp, 1985), and can also be affected by the availability of mineral nutrients (Rovira et al., 1983) and presence of certain contaminants in soil. Thus, P deficiency is known to induce exudation of organic acids and certain enzymes, such as acid phosphatases (Wasaki et al., 2001). Similarly, iron or zinc deficiency in soil may induce the production of certain phytosiderophores (plant-exuded chelators of metal ions), whilst K deficiency has been reported to increase the exudation of organic acids and free amino acids in the rhizospheres of wheat and mung bean (Tu et al., 1996). The main proportion of root-exuded organic acids has been reported to exist in soil as anions, such as citrate, oxalate, malate, malonate, fumarate, and acetate (Ryan et al., 2001). These anions can chelate to metal ions to varying degrees, and thus decrease their toxicity to plants. The tricarboxylate anions (*e.g.*, citrate) are known to chelate metal cations more strongly than the dicarboxylate anions (*e.g.*, oxalate), which in turn chelate more strongly than the monocarboxylate anions (*e.g.*, acetate).

Phenolic compounds are another important class of compounds exuded by plant roots. Microorganisms capable of using phenolic compounds as a carbon source often have enzymes that can co-metabolize pollutants with similar structures. The levels of phenolic compounds and carbohydrates have been found to be higher in the rhizosphere soil of certain plants such as *Bouteloua gracilis* and *Artemisia frigida* (Bokhari et al., 1979). In particular, the levels released by roots of red mulberry (*Morus rubra* L.) towards the end of the growing season, have been considered sufficient to support the growth of PCB- and PAH degrading microorganisms (Fletcher and Hegde, 1995; Hegde and Fletcher, 1996). The late season accumulation of three flavones (morusin, morusinol, and kuwanon C) in the fine roots of mulberry (*Morus* sp.) has been shown to support the growth of a PCB-degrading bacterium *Burkholderia* sp. LB400 (Leigh et al., 2002). It has also been suggested that dead mulberry roots may serve as a source of substrate for PCB-degrading bacteria. Many other phenolic compounds such as naringin, myricetin, coumarin, phloridzin, catechin, maclurin, released by the roots of certain plants, have also been shown to support the growth of PCB-degrading bacteria in the rhizosphere (Donnelly et al., 1994). The exudation of other secondary metabolites, such as terpenes, also appears to play an important role in activating or transforming specific bacterial habitats (Schnoor et al., 1995; Singer et al., 2003). Addition of the terpenes to soil has also been shown to increase the population of PCB-degrading bacteria by up to 100-fold, as indicated by growth of the soil bacterium *P. pseudoalcaligenes* KF707 that was genetically tagged with a green fluorescent protein reporter gene (Oh et al., 2003). It is known from studies using synthetic artificially contaminated soils that *Quillaya* saponins can also enhance biological degradation of PCBs in soils (Fava and Di Gioia, 1998). Thus, plants that have a greater capacity to release secondary metabolites, especially those that also have a surfactant activity, may have an even greater potential for use in phytoremediation (Fava and Di Gioia, 1998). The deposition of specific phytochemicals in soil offers the potential for developing 'rhizosphere breeding' as a method to stimulate the molecular biological techniques. Soils amended with plant root exudates have also been shown to exert a similar stimulation of microbial activity. The addition of synthetic root exudates of wheat to a soil has been shown to stimulate growth of Actinomycetes, fungi, and many soil bacteria (Abdel-Nasser and Moawad, 1975). The presence of plant roots such as sunflower and pea has been shown to increase phenanthrene metabolism by more than 50% (Liste and Alexander, 1999). Amendment of a soil with root exudates from slender oat (*Avena barbata* L.) has been reported to increase biodegradation of phenanthrene (Miya and Firestone, 2001). The amended soil was reported to maintain greater populations of phenanthrene-degrading microorganisms than control soils. In a soil contaminated with heavy metals, Kozdroj and van Elsas (Kozdroj and van Elsas, 2000) showed a significant effect of root exudates on the development of bacterial populations, although flooding was also considered to stimulate bacterial activity. It was shown that the exudates reduced bacterial diversity towards domination by fast growing species, especially at higher contamination levels. The addition of root exudates from maize (*Zea mays* L.) to columns of soil contaminated with PAHs has also been shown to induce a significant increase in the mineralization of 14C-pyrene (Yoshitomi and Shann, 2001).

6.4. Specificity of Microbial Communities and Critical Success Factors

There is a substantial indirect evidence to suggest that the interactions that enable plants and microbes to survive soil contamination may have co-evolved in response to similar selective pressures (Walton et al., 1994). The mutual success of plants and rhizosphere microflora in degrading environmental pollutants, however, varies substantially between plant species and soil types. Some plant species appear to be much more successful in affecting the degradation processes than others. Despite our understanding of certain aspects, information on specific plant characteristics that promote microbial degradation of organic pollutants is still scarce. The factors responsible for a successful interaction between plants and microbes specialized in degrading organic pollutants are also not well understood. Introduction of O_2 into deeper soil layers is one such factor that can explain, at least partially, why some plant species such as *Salix* sp. and *Populus* sp. are so successful phytoremediators (Corseuil and Moreno, 2001, Thompson et al., 1998). These plants possess specialized root vessels (aerenchyma) that may comprise up to 60% of the intracellular volume, and which mediate O_2 diffusion under situations of low O_2 in soil (Grosse et al., 1992).

For a microbe-plant synergy to be effective, a soil microbial population must contain species that are capable of transforming soil-bound and recalcitrant organic compounds. An adaptation and selection within the microhabitats may take place and a population may change considerably with time in response to the type and degree of a contamination. For example, the composition of a microbial community has been shown to change in a soil contaminated with dibenzothiophene (DBT) with time favoring those species that were specialized in degrading DBT. A progressive change over time following treatment with DBT-containing petroleum, and selection of *Rhodococcus erythropolis* with specific capacity to desulfurise DBT, has been demonstrated by PCR, DGGE analysis (Duarte et al., 2001). Similarly, greater activities of denitrifies, *pseudomonads*, and BTEX-degrading microbes have been reported in the rhizosphere of poplar trees than in adjacent soils. In trichloroethylene (TCE)-contaminated soils, there was a greater microbial activity and a faster TCE degradation in the rhizosphere soil than in bulk soil (Walton and Anderson, 1990).

The presence of higher numbers of methanotrophic bacteria in rhizosphere soils, and on roots of *Lespedeza cuneata* and *Pinus taeda*, has been speculated to be the basis of a faster mineralization of 14C-TCE (Brigmon et al., 1999). Rhizosphere of certain desert and crop plants grown in oil polluted soils have been shown to contain more hydrocarbon utilizing bacteria, such as *Cellulomonas flavigena*, *Rhodococcus erythropolis* and *Arthrobacter sp.*, than in control soils (Radwan et al., 1998). Larger populations, and selective enrichment, of organic compound degrading microorganisms have been found in the rhizosphere of alfalfa and bluegrass than in non-contaminated soils (Nichols et al., 1997). The hydrocarbon degrading potential of rhizosphere soil has been shown to increase through changes in the composition of a microbial community towards those containing the degradative genes ndoB, alkB, and xylE (Siciliano et al., 2003).

As mentioned also before, the exudation of certain compounds by plant roots can enhance the growth of specific pollutant degrading microbes in a rhizosphere, e.g., PAH degraders (Günther et al., 1996; Joner et al., 2002). The nature of the compounds in root exudates, the quantity in which they enrich the rhizosphere with substrates for microbes, and the timing of exudation, all seem crucial for a remediation process. Obviously, the size of a root system and the release capacity of carbon compounds determine the sphere of influence of a plant species. Beyond the rooting zone, the microbial growth-stimulating effects of exudates seem to diminish rapidly (Newman and Watson, 1977; Badalucco et al., 1997). The changes in the composition of a rhizosphere microbial community, and ability to retain a particular degradation capability, are also affected by pre-exposure to a soil contaminant. It is well known that pre-exposure to a particular compound may enhance metabolizing potential of soil microorganisms upon re-exposure to the same compound. This is due to the growth of selected microbial populations that may retain a specific metabolic capability over long periods of time (a phenomenon known as 'soil memory'). Thus, mineralization of *p*-nitrophenol by microorganisms in rice rhizosphere has been reported to be faster in both planted and unplanted soil systems that were pre-exposed to the compound, compared to a soil without such pre-exposure (Reddy and Sethunathan, 1994). Other factors, such as presence of plant roots (Wetzel et al., 1997), and soil matrix as opposed to a hydroponic system (Siciliano and Germida, 1999), may also affect the pollutant-degrading potential of rhizospheric microorganisms.

7. Mycorrhizae Role in the Rhizospheric Phytoremediation

Plants associate with many soil microbial symbionts that improve their nutrition. The most widespread association is mycorrhizal symbiosis, involving soil fungi and plant roots (Smith and Read, 1997), a symbiosis supposed to be ancestral and to date back to the colonization of land ecosystems (Selosse and Tacon, 1998). In a mycorrhiza, the plant receives water and mineral nutrients collected in soil by the fungal partner (Smith and Read, 1997). However, since symbiosis can be defined as a reciprocal enhancement of fitness, a bilateral nutrient flux is not exclusively required and any protective effect of one partner toward the other, enhancing the latter's survival or reproduction, is sufficient to establish a symbiotic relationship. It has been claimed that, in temperate ecosystems, plants with highly branched root systems mainly benefit from protection by the fungus; whereas, the others, having more reduced root systems, may profit more from nutrition by the fungus. Mycorrhizal fungi can buffer stress reactions entailed by these compounds, e.g., heavy metal-induced oxidative stress (Schutzendubel and Polle, 2002).

7.1. Arbuscular Mycorrhizal Fungi

Arbuscular mycorrhizal (AM) fungi are soil fungi developing nonspecific symbiosis with most of the terrestrial plants. This association is beneficial for both symbionts as the host plant supplies the fungi with C in the exchange for water and nutrients. AM fungi are important components of the ecosystem because they can significantly increase ecosystem efficiency. The communications of signal molecules between the two symbionts is necessary for the onset of symbiosis. Fungal spores germinate and grow toward the roots and eventually form the extensive hyphal network inside the root cortical cells and the surrounding soil. Fungal hyphae also produce arbuscules and vesicles. AM fungi diversity can considerably influence ecosystem functioning through affecting the plant growth as well as its structural combination. Different AM species differ in their ability to affect plant growth and yield production. It has been indicated that AM fungi are also able to enhance plant growth under stress by their unique abilities. AM fungi have the following effects on plant growth and ecosystem efficiency: 1) increase plant water and nutrient uptake,

2) improve soil structure, 3) interact with other soil microbes, 4) control pathogens, and 5) alleviate soil stresses including heavy metals (Smith and Read, 2008).

7.2. AM Fungi and Heavy Metals

Heavy metals are a very important source of pollution resulted by anthropogenic activities (Schachtschabel et al., 1992). Plants growing on polluted soils can influence the behavior of heavy metals by absorbing heavy metals in their tissues affecting the quality of food chain. Use of plant cover on soil can avoid the distribution of heavy metals to other parts by water or wind erosion. The uptake of heavy metals by plant and their toxic effects on plant and soil is determined by soil pH; carbonate concentration, soil colloids including clay and organic matter, root exudates and soil microbes (Wenzel et al., 1999; Adriano, 2001). Soil microbes and organic matter can immobilize heavy metals such as Cd; however, root exudates including organic acids can enhance the mobility of heavy metals by forming soluble complexes (Janouskova et al., 2006).

Production of H^+ by plant roots can influence rhizosphere pH affecting the solubility and hence bioavailability of metals in soil (Hinsinger, 1998). Heavy metals such as Cd can adversely affect the activity of proton pump and hence production of H^+ by roots (Tu et al., 1989). Proton pump supplies energy for the uptake and movement of nutrients across the plasma membrane. Accordingly, as heavy metals alter the rhizosphere pH their solubility and hence their bioavailability is decreased with reducing soil pH (Xian, 1989). The exchangeable Cd and not total Cd concentration, determines Cd uptake and accumulation by plant (Wu and Zhang, 2002; Cheng et al., 2004). AM fungi are able to alleviate the stress of heavy metals on plant growth, by soil bioremediation (bioaugmentation). Janouskova et al. (2006) found that AM fungi are able to alleviate the unfavorable effects of cadmium (Cd) on plant growth by the process of phytostabilization. Compared with non-mycorrhizal plants, significantly lower amounts of Cd were found in the mycorrhizal plants because AM hyphae were able to accumulate 10–20 times higher rates of Cd relative to the plant roots.

As previously mentioned, the physiological properties of *T. praecox* allow it to act as a hyperaccumulator for heavy metals such as Zn, Cd and Pb. However, a very important aspect regarding such ability is how it can be improved. Although *T. praecox* is from the *Brassicaseae* family, it is able to develop symbiosis with AM fungi (Leyval et al., 1997; Pawlowska et al., 2000). This can be beneficial to improve the tolerating and hyperaccumulating ability of *T. praecox* under heavy metal stress. In their research work Vogel-Mikus et al. (2006) examined the effects of AM fungi on AM colonization and uptake of Zn, Cd and Pb by *T. praecox* in a polluted soil. *T. praecox* developed symbiosis with AM fungi during the stages of enhanced nutrient uptake such as the reproductive stage. There was positive correlation between root colonization by AM fungi and soil total Zn, Cd and Pb. AM colonization resulted in higher and lower nutrient and heavy metal uptake by plant roots, respectively. The decrease of heavy metal uptake by mycorrhizal *T. praecox* indicates that this plant can physiologically behave differently under heavy metal stress. This may however, enhance the ability of mycorrhizal plant to alleviate heavy metal stress more efficiently. Interestingly and similar to the results of other researchers (Miransari et al., 2007, 2008, 2010a, b) elevated levels of stress enhanced the effectiveness of AM fungi on stress alleviation, which was also evident from the increased root colonization (Vogel-Mikus et al., 2006).

The important point about treating polluted soil with mycorrhizal plants is the selection of appropriate AM species. Although experimental results indicate the effectiveness of AM species on the alleviation of heavy metals stress, the most efficient species are the ones, which are selected from areas polluted with heavy metals. Such species have attained the ability to survive under such conditions and hence can act more efficiently relative to the other AM species. Such species have developed mechanisms with time, which can make AM fungi resist or tolerate the heavy metal stress and they can accordingly develop more efficient symbiosis with their host plant under such conditions (Khan, 2005). Chen et al. (2004) indicated that in soils, not polluted with Zn, *Glomus caledonium* is able to colonize the roots at a rate of more than 70%, however, when the soil was polluted with Zn concentrations at 300 and 600 mg/kg the rate of colonization decreased to 50%. Similar results were also found by Duponnois et al. (2006) as the use of 560 mg/kg Cd decreased the rate of root colonization from 60 to 20%. They also mentioned that the interaction effects between AM fungi and florescent pseudomonas increased the colonization rate of host plant from 32 to 45%.

It has yet not been likely to differentiate between heavy metals allocation to the roots and fungal hyphae. However, the mechanisms, directing heavy metal movement to plant roots by AM fungi have been indicated. They include: 1) heavy metals may be deposited in the cellular wall or in the fungal vacuoles, 2) sequestration of heavy metals by siderophores may deposit heavy metals in root apoplasm or in soil, 3) metallothionein or phytochelatins may result in the deposition of heavy metals in fungal or plant cells and 4) the allocation of heavy metals from cytoplasm is performed by metal transporters located at the plasmalemma or tonoplast of both symbionts (Galli et al., 1994; Leyval et al., 1997; Schutzendubel and Polle, 2002). Heavy metal content in roots of mycorrhizal plants is highly altered indicating that the related genes are expressed at transcriptional and translation levels by AM fungi (Ouziad et al., 2005).

There are indications that AM fungi result in the expression of some genes in the host plant when subjected to heavy metal stress (Repetto et al., 2003; Rivera-Becerril et al., 2005). When subjected to Zn contamination the Zn transporter gene (GintZnT1) in the hyphae of *G. intraradices* was expressed indicating its role in the alleviation of Zn stress (Gonzalez-Guerrero et al., 2005). In their other experiment, Gonzalez-Guerrero et al. (2006) indicated that Cd and Cu result in the expression of the related transporters (GintABC1), which may detoxify their toxic effects in the fungal hyphae. Experiments have shown the expression of genes in the hyphae of *G. intraradices* inoculating *M. truncatula* in soils polluted with Zn. When expressed, the related four genes, which are important in the alleviation of Zn stress, result in the production of glutathione S transferase as a Zn transporter (Hildebrandt et al., 2006, 2007). This enzyme is the catalyzer for the combination of glutathione with some electron receivers resulting in the alleviation of oxidative stress (Moons, 2003; Smith et al., 2004).

Use of a fungicide resulted in the enhanced uptake of Pb by mycorrhizal corn (Burke et al., 2000). In soils with low or medium level of fertility, AM fungi can enhance plant nutrient uptake including micronutrients such as Fe, Zn, Mn, and Cu. However, with increasing the concentration of such micronutrients in a polluted soil, mycorrhizal plants are able to reduce the concentration of these heavy metals in their shoots. This indicates that in combination with plant roots AM fungi are able to use some strategies, which can control the uptake of heavy metals by plant under such conditions. One of these strategies is the increased growth of mycorrhizal plant, although other mechanisms may also be responsible.

8. Molecular Insights into the Rhizosphere

There are two different methods by which soil microbes can assist the plant to alleviate the stress: 1) enhanced metal mobility by microbial production of biosurfactants (Herman et al., 1995), siderophores (Dubbin and Louise Ander, 2003) and organic acids (Di Simine et al., 1998), and/or 2) increasing plant growth by PGPRs (Zhuang et al., 2007) and/or AM fungi (Khan, 2006). In the combined use of plant and soil microbes, plant supplies C source for microbes, which absorb, degrade or make the plant absorb the released elements. The energy source, used for such a process is usually up to 40% of plant photosynthates exudated by plant roots (Lynch and Moffat, 2005). As a general rule, an increase in the activities of indigenous microbial species in a rhizosphere is likely to enhance the degradation of soil contaminants due to co-metabolic processes or diauxic growth. Thus, provision of appropriate nutrients and growth conditions may have a significant impact on microbial numbers in the rhizosphere and consequently the remediation of contaminated soils.

Depending on the mediators involved, five different IAA synthesis ways using tryptophan as precursor were identified (Spaepen et al., 2007; Gamalero and Glick, 2011). The best characterized IAA synthesis pathway in bacteria is the indole-3-acetamide (IAM), in which tryptophan is converted to IAM by the enzyme tryptophan-2-monooxygenase (IaaM), encoded by the *iaaM* gene and IAM is converted to IAA by an IAM hydrolase (IaaH), encoded by *iaaH*. The genes that encode IaaM and IaaH was cloned and characterized in many bacterial species such as: *A. tumefaciens, P. syringae, Pantoea agglomerans, Rhizobium* and *Bradyrhizobium* (Spaepen et al., 2007). The indole-3-pyruvate (IPyA) pathway is considered to be a major way of IAA synthesis in plants. This pathway frequently occurs in beneficial bacteria such as *Bradyrhizobium, Azospirillum*, *Rhizobium* spp. and *Enterobacter cloacae*. In this pathway, tryptophan is converted to IPyA by an aminotransferase, decarboxylated to indole-3-acetaldehyde (IAAld) by indole-3-pyruvate decarboxylase (IPDC) and finally oxidized to IAA. The encoded gene for IPDC enzyme was isolated and characterized in *Azospirillum brasilense, Enterobacter cloacae*, *P. putida* and *Pantoaea agglomerans* (Spaepen et al., 2007). IAA synthesis through tryptophan side-chain oxidase (TSO) activity has only been characterized in *P. fluorescens* CHA0. In this pathway tryptophan is directly converted to IAAld, which can be oxidized to IAA (Spaepen et al., 2007). The last pathway dependent on tryptophan is indole-3-acetonitrile (IAN). For this route, found also in plants, only the last step is known: the conversion of IAN to IAA by a nitrilase. The identification of nitrilase in *Alcaligenes faecalis* was possible due to the specificity for indole-3-acetonitrile (Spaepen et al., 2007). The results showed that 90% of IAA is synthesized via the tryptophan-independent pathway, while 0.1% is produced via the IAM pathway. Moreover, it has been observed that the same bacteria can possess several ways of IAA production (Spaepen et al., 2007). Auxin synthesis is affected by either environmental factors as well as genetic factors. Therefore, medium pH, osmotic and matrix stress, carbon starvation and the composition of the root exudates influence the bacterial abilities to produce IAA (Gamalero and Glick, 2011). For example, in the case of *Azospirillum brasilense* strains it has been shown that the amount of IAA is increased and the expression of *ipdC* genes take place only in environments with limited carbon concentration, acid pH and only when the bacterial cells enter in the stationary phase (Ona et al., 2005). Therefore, when the genes are located on plasmids, the amount of IAA is significantly higher compared with the situations when are situated on the chromosome. This is happening because the plasmids are generally present in various copies in the bacterial cells, providing a higher number of IAA biosynthesis genes (Spaepen et al., 2007). Moreover, expression of IAA biosynthesis gene is influenced by two transcriptional regulators *RpoS* which regulates the transcription of genes in response to stress conditions and starvations and the two-component system GacS/GacA, which controls the expression of genes of which are induced during a late logarithmic growth phase and have a role in maintaining the competitiveness of the bacterium in the rhizosphere (Spaepen et al., 2007). The main mechanism of phosphate solubilization by rhizobacteria is the production of low molecular weight organic acids such as gluconic acid, citric acid, oxalic acid, lactic acid, acetic acid etc. These acids that are the result of the bacterial metabolism mostly by oxidative respiration or by fermentation of organic carbon sources (Sharma et al., 2013), bind phosphate with their hydroxyl and carboxyl groups thereby chelating cations and also inducing soil acidification, both resulting in the release of soluble phosphate (Gamalero and Glick, 2011). Regarding the organic phosphate solubilization (mineralization) an important role is played by a number of enzymes as phosphatases (also known as phosphohydrolase), phytases, phosphonates and C-P lyases (Hayat et al., 2010; Sharma et al., 2012).

CONCLUSION

Chemical pollution can be kept under control using microbial community associated with plant root rhizosphere. Efforts were made herein to discuss the major belowground processes associated with the plant rhizosphere. This chapter also covered the interactions between plants and microbes, their chemical communications, the biological processes they sustain, and the costs and benefits to plants associated with these interactions particularly in phytoremediation. Alienated into several distinct zones including endorhizosphere, the rhizoplane and the ectorhizosphere rhizosphere can be considered as a stage for the microbial activity. In addition to modulating plant growth, development and productivity changes in the relative diversity and functional characteristics of microbial communities at the rhizosphere can significantly influence plant's ability to cope with soil contaminations. The microbial activity in the immediate vicinity of the root seems to offer a favorable environment for co-metabolism of soil-bound and recalcitrant chemicals. Plants depend on the ability of roots to communicate with microbes. Many bacteria and fungi are dependent on associations with plants that are often regulated by root exudates, known to play roles in root–microbe interactions. The root exudates composition and concentration can vary with the plant species, cultivars, plant's age, soil properties, level of biotic and abiotic stress etc. Living root hairs, as well as actively growing primary and secondary roots release large quantities of root exudates. Root exudates also act as antimicrobials against rhizospheric microflora and provides the defense to plants. In addition, chemicals present in the root exudates metabolize, degrade or transform may chemical pollutants and minimize their load in environmental compartments.

REFERENCES

Abdel-Nasser, M., Moawad, H. (1975). Changes in numbers of micro-organisms during decomposition of root exudates in soil. *Zentralblatt fur Bakteriologie, Parasitenkunde, Infektionskrankheiten und Hygiene* 130, 738–744.

Adesemoye, A., Kloepper, J. (2009). Plant–microbes interactions in enhanced fertilizer-use efficiency. *Applied Microbiology and Biotechnology* 85, 1-12.

Adriano, D. C. (2001). Trace elements in terrestrial environments: biogeochemistry, bioavailability, and risks of metals. New York, Berlin, Heidelberg: Springer-Verlag.

Ahemad, M., Kibret, M. (2014). Mechanisms and applications of plant growth promoting rhizobacteria: Current perspective. *Journal of King Saud University - Science* 26, 1-20.

Albert, D. R. (1969). Plant root exudates. *Botanical Review* 35(1), 35-57.

Anderson, T. A., Coats, J. R. (1995). An overview of microbial degradation in the rhizosphere and its implications for bioremediation. *Bioremediation: Science and Applications*, American Society of Agronomy, Madison, 1995, USA, 135–143.

Anderson, T. A., Coats, J. R. (1994). Bioremediation through rhizosphere technology. In *ACS Symposium Series (USA)*. American Chemical Society, 1994, 82–92.

Bacilio-Jiménez, M., Aguilar-Flores, S., Ventura-Zapata, E., Pérez-Campos, E., Bouquelet, S., Zenteno, E. (2003). Chemical characterization of root exudates from rice (*Oryza sativa*) and their effects on the chemotactic response of endophytic bacteria. *Plant and Soil* 249, 271-277.

Badalucco, L., Kuikman, P. J., Nannipieri, P. (1997). Protease and deaminase activities in wheat rhizosphere and their relation to bacterial and protozoan populations. *Biology and Fertility of Soils* 23, 99–104.

Bais, H. P., Weir, T. L., Perry, L. G., Gilroy, S., Vivanco, J. M. (2006). The role of root exudates in rhizosphere interactions with plants and other organisms. *Annual Reviews of Plant Biology* 57, 233-266.

Bais, H. P., Broeckling, C. D., Vivanco, J. M. (2008). Root exudates modulate plant - microbe interactions in the rhizosphere. In *Secondary metabolites in soil ecology*, pp. 241-252. Springer Berlin Heidelberg.

Bakker, M., Manter, D., Sheflin, A., Weir, T., Vivanco, J. (2012). Harnessing the rhizosphere microbiome through plant breeding and agricultural management. *Plant and Soil* 360, 1-13.

Barea, J. M., Pozo, M. J., Azco´n, R., Azco´n-Aguilar, C. (2005). Microbial co-operation in the rhizosphere. *Journal of Experimental Botany* 56, 1761–1778.

Battey, N. H., Blackbourn, H. D. (1993). The control of exocitosis in plant cells. *New Phytologist* 125, 307–308.

Bécard, G., Loverine, P., Taylor, D. D. Douds, P. E. P., Landis W. D. Flavonoids are not necessary plant signal compounds in arbuscular mycorrhizal symbioses. (1995) *Molecular Plant Microbe Interactions.* 252-258.

Bending, G. D. (2003). Litter decomposition, ectomycorrhizal roots and the 'Gadgil' effect. *New Phytologist* 158, 228–229.

Berendsen, R. L., Pieterse, C. M. J., Bakker, P. A. H. M. The rhizosphere microbiome and plant health. (2012). *Trends in Plant Science*,. 17, 478-486.

Bhattacharyya, P. N., Jha, D. K. Plant growth-promoting rhizobacteria (PGPR): emergence in agriculture. (2012). *World Journal of Microbiology and Biotechnology*, 28, 1327-1350.

Blaylock, M. S. D. E., Dushenkv, S., Zakharova, O., Gussman, C., Kapulnik, Y., Ensley, B., Raskin, E. (1997). Enhanced accumulation of Pb in Indian mustard by soil-applied chelating agents. Environmental Science and Technology, 31, 860–865.

Bloemberg, G. V., Lugtenberg, B. J. J. (2001). Molecular basis of plant growth promotion and biocontrol by rhizobacteria. *Current Opinion in Plant Biology*, 4, 343-350.

Bokhari, U. G., Coleman, D. C., Rubink, A. (1979). Chemistry of root exudates and rhizosphere soils of prairie plants. *Canadian Journal of Botany*, 57, 1473–1477.

Bottini, R., Cassán, F., Piccoli, P. (2004). Gibberellin production by bacteria and its involvement in plant growth promotion and yield increase. *Applied Microbiology and Biotechnology*, 65, 497-503.

Brady, N. C., Weil, R. R. (1999). The Nature and Property of Soils. Prentice Hall, Upper Saddle Hall, NJ.

Briggs, G. G., Bromilow, R. H., Evans, A. A. (1982). Relationship between lipophilicity and root uptake and translocation of nonionized chemicals by barley. *Pesticide Science*, 13, 495–504.

Brigmon, R. L., Anderson, T. A., Fliermans, C. B. (1999). Methanotrophic bacteria in the rhizosphere of trichloroethylene-degrading plants. *International Journal of Phytoremediation*, 1, 241–253.

Brimecombe, M. J., De Leij Frans, A. A. M., Lynch, J. M. (2001). Nematode community structure as a sensitive indicator of microbial perturbations induced by a genetically modified *Pseudomonas fluorescens* strain. *Biology and Fertility of Soils*, 34, 270–275.

Bromilow, R. H., Chamberlain, K. (1995). Principles governing uptake and chemicals. In *Plant Contamination: Modelling and Simulation of Organic Chemical Processes*. Trapp, S., McFarlane, J. C. (Eds), Lewis Publishers, 37–68.

Burke, S. J., Angle, J. S., Chaney, R. L., Cunningham, S. D. (2000). Arbuscular mycorrhizae effects on heavy metal uptake by corn. *International Journal of Phytoremediation*, 2,23–9.

Carvalhais, L. C., Dennis, P. G., Fedoseyenko, D., Hajirezaei, M. R., Borriss, R., Von Wirén, N. (2011). Root exudation of sugars, amino acids, and organic acids by maize as affected by nitrogen, phosphorus, potassium, and iron deficiency. *Journal of Plant Nutrition and Soil Science*, 174, 3-11.

Chaiharn, M., Lumyong, S. (2011) Screening and optimization of indole-3-acetic acid production and phosphate solubilization from rhizobacteria aimed at improving plant growth. *Current Microbiology*, 62, 173 – 181.

Chaparro, J., Sheflin, A., Manter, D., Vivanco, J. (2012). Manipulating the soil microbiome to increase soil health and plant fertility. *Biology and Fertility of Soils*, 48, 489-499.

Chen, B., Shen, H., Li, X., Feng, G., Christie, P. (2004). Effects of EDTA application and arbuscular mycorrhizal colonization on growth and zinc uptake by maize (*Zea mays* L.) in soil experimentally contaminated with zinc. *Plant and Soil*, 261, 19–29.

Cheng, H., H. (1995). Characterization of the mechanisms of allelopathy: modeling and experimental approaches. *In Allelopathy: Organisms, Processes, and Applications*. Eds. Inderjit, K., Dakshini, M. M., Einhellig, F. A. American Chemical Society, Washington, DC, 132–141.

Cheng, W. D., Zhang, G. P., Yao, H. G., Dominy, P., Wu, W. F., Wang, R. Y. (2004). Possibility of predicting heavy metal contents in rice grains based on DTPA-extracted levels in soil. *Communication in Soil Science and Plant Analysis*, 35, 2731–45.

Chin-A-Woeng, T. C., Lugtenberg, B. J. (2004). Root Colonisation Following Seed Inoculation, In *Plant Surface Microbiology* (A. Varma *et al.*, eds.), Springer, Berlin, Heidelberg.

Chrispeels, M. (1991). Sorting of proteins in the secretory system. *Annual Reviews in Plant Physiology*, 42, 21.

Chrispeels, M., Raikhel, N. V. (1992). Short peptide domains target proteins to plant vacuoles. *Cell*, 68, 613–616.

Compant, S., Clément, C., Sessitsch, A. (2010). Plant growth-promoting bacteria in the rhizo- and endosphere of plants: Their role, colonization, mechanisms involved and prospects for utilization. *Soil Biology and Biochemistry*, 42, 669-678.

Corgié, S. C., Joner, E. J., Leyval, C. (2003). Rhizospheric degradation of phenanthrene is a function of proximity to roots. *Plant and Soil*, 257, 143–150.

Corseuil, H. X., Moreno, F. N. (2001). Phytoremediation potential of willow trees for aquifers contaminated with ethanol-blended gasoline. *Water Research*, 35, 3013–3017.

Criquet, S., Joner, E. J., Léglize, P., Leyval, C. (2000). Anthracene and mycorrhiza affect the activity of oxidoreductases in the roots and the rhizosphere of lucerne (*Medicago sativa* L.). *Biotechnology Letters*, 22, 1733–1737.

Curl, E. A., Truelove, B. (1986). The Rhizosphere. Springer, New York.

Dalton, B. R. (1999). The occurrence and behavior of phenolic acids in soil environment and their potential involvement in allelochemicals interference interactions: methodological limitations in establishing conclusive proof of allelopathy. In *Principles and Practices in Plant Ecology: Allelochemical Interaction*. Eds. Inderjit, K. M., Dakshini, M., Foy, C. L. CRC Press, Boca Raton, FL, 57–74.

De Weert, S., Vermeiren, H., Mulders, I. H. M., Kuiper, I., Hendrickx, N., Bloemberg, G. V., Vanderleyden, J., De Mot, R., Lugtenberg, B. J. J. (2002). Flagella-driven chemotaxis towards exudate components is an important trait for tomato root colonization by *Pseudomonas fluorescens*, *Molecular Plant-Microbe Interactions*, 15, 1173-1180.

Di Simine, C. D., Sayer, J. A., Gadd, G. M. (1998). Solubilization of zinc phosphate by a strain of *Pseudomonas fluorescens* isolated from a forest soil. *Biology and Fertility of Soils*, 28, 87–94.

Dietz, A. C., Schnoor, J. L. (2001). Advances in phytoremediation. *Environmental Health Perspectives*, 109, 163–168.

Dittmann, J., Heyser, W., Bucking, H. (2002). Biodegradation of aromatic compounds by white rot and ectomycorrhizal fungal species and the accumulation of chlorinated benzoic acid in ectomycorrhizal pine seedlings. *Chemosphere*, 49, 297–306.

Dixon, R. A. (2001). Natural products and plant disease resistance. *Nature*, 411, 843–847.

Donnelly, P. K., Fletcher, J. S. (1994). Potential use of mycorrhizal fungi as bioremediation agents. In: Anderson, T. A., Coats, J. R. (eds), *Bioremediation through rhizosphere technology*, ACS Symposium Series 563, American Chemical Society, Washington DC, 93–99.

Donnelly, P. K., Hegde, R. S., Fletcher, J. S. (1994). Growth of PCB degrading bacteria on compounds from photosynthetic plants. *Chemosphere*, 28, 981–988.

Duarte, G. F., Rosado, A. S., Seldin, L., de Araujo, W., van Elsas, J. D. (2001). Analysis of bacterial community structure in sulfurousoil-containing soils and detection of species carrying dibenzothiophene desulfurization (dsz) genes. *Applied Environmental Microbiology*, 67, 1052–1062.

Dubbin, W. E., Louise Ander, E. (2003). Influence of microbial hydroxamate siderophores on Pb (II) desorption from a-FeOOH. *Applied Geochemistry*, 18, 1751–6.

Duponnois, R., Kisa, M., Assigbetse, K., Prin, Y., Thioulouse, J., Issartel, M., Moulin, P., Lepage, M. Fluorescent pseudomonads occurring in *Macrotermes subhyalinus* mound structures decrease Cd toxicity and improve its accumulation in sorghum plants. *Science of Total Environment*, 370, 391–400.

Dutta, S., Rani, T. S., Podile, A. R. (2013). Root exudate-induced alterations in *Bacillus cereus* cell wall contribute to root colonization and plant growth promotion. *PloS One*, 8, e78369.

El Shatnawi, M. K. J., Makhadmeh, I. M. (2001). Ecophysiology of the plant-rhizosphere system. *Journal of Agronomy and Crop Science*, 187, 1–9.

Elasri, M., Delorme, S., Lemanceau, P., Stewart, G., Laue, B., Glickmann, E., Oger, P.M., Dessaux, Y. (2001). Acyl-homoserine lactone production is more common among plant-associated *Pseudomonas* spp. than among soil borne *Pseudomonas* spp. *Applied Environmental Microbiology*, 67, 1198-1209.

Fan, T. W. M., Lane, A. M., Crowley, D., Higashi, R. M. (1997). Comprehensive analysis of organic ligands in whole root exudate using nuclear magnetic resonance and gas chromatography-mass spectrometry. *Analytical Biochemistry*, 251, 57.

Faure, D., Vereecke, D., Leveau, J. J. (2009). Molecular communication in the rhizosphere *Plant and Soil*, 321, 279-303.

Fava, F., Di Gioia, D. (1998). Effects of triton X-100 and *Quillaya* saponin on the *ex situ* bioremediation of a chronically polychlorobiphenyl-contaminated soil. *Applied Microbiology and Biotechnology*, 50, 623-630.

Felici, C., Vettori, L., Giraldi, E., Forino, L. M. C., Toffanin, A., Tagliasacchi, A. M., Nuti, M. (2008). Single and co-inoculation of *Bacillus subtilis* and *Azospirillum brasilense* on *Lycopersicon esculentum:* Effects on plant growth and rhizosphere microbial community. *Applied Soil Ecology*, 40, 260-270.

Fletcher, J. S., Hegde, R. S. (1995). Release of phenols by perennial plant roots and their potential importance in bioremediation. *Chemosphere*, 31, 3009–3016.

Frans, A. A. M. D. L., James, M. L., Melissa, J. B. (2007). Rhizodeposition and Microbial Populations, In *The Rhizosphere: Biochemistry and Organic Substances at the Soil-plant interface* (R. Pinton *et al.*, eds.), CRC Press, New York.

Fray, R.G., (2002). Altering plant-microbe interaction through artificially manipulating bacterial quorum sensing. *Annals of Botany*, 89, 245-253.

Gafny, R., Okon, Y., Kapulnik, Y., Fischer, M. (1986). Adsorption of *Azospirillum brasilense* to corn roots. *Soil Biology and Biochemistry*, 18, 69-75.

Gagnon, H., Seguin, J., Bleichert, E., Tahara, S. Ibrahim, R. K. (1992). Biosynthesis of white lupin isoflavonoids from [U-14C] Lphenylalanine and their release into the culture medium. *Plant Physiology*, 100, 76–79.

Galli, U., Schüepp, H., Brunold, C. Heavy metal binding by mycorrhizal fungi. *Physiologia Plantarum*, 92, 34-368.

Gamalero, E., Glick, B. R. (2011). Mechanisms Used by Plant Growth-Promoting Bacteria, In *Bacteria in Agrobiology: Plant Nutrient Management* (Maheshwari, D. K. eds.), Springer, Berlin, Heidelberg.

Glick, B. R. (2010). Using soil bacteria to facilitate phytoremediation. *Biotechnology Advances*, 28, 367–74.

Gonzalez-Guerrero, M., Azcon-Aguilar, C., Ferrol, N. (2006). *GintABC1* and *GintMT1* are involved in Cu and Cd homeostasis in *Glomus intraradices. Abstracts of the 5th International Conference on Mycorrhiza*, Granada, Spain.

Grayston, S. J., Vaughan, D., Jones, D. (1996). Rhizosphere carbon flow in trees, in comparison with annual plants: the importance of root exudation and its impact on microbial activity and nutrient availability. *Applied Soil Ecology*, 5, 29–56.

Grosse, W, Frye, J., Lattermann, S. (1992). Root aeration in wetland trees by pressurized gas transport. *Tree Physiology*, 10, 285–295.

Guckert, A. (1992). Significance of roots and their exudates as sources of organic soil substances. [German]. *Berichte uber Landwirtschaft, Sonderheft,* 97–113.

Guern, J., Renaudin, J. P. Brown, S. C. (1987). The compartmentation of secondary metabolites in plant cell cultures. *In Cell Culture and Somatic Cell Genetics of Plants*. Eds. Constabel, F., Vasil, I., K. Academic Press, San Diego, CA, 43.

Günther, T., Dornberger, U., Fritsche, W. (1996). Effects of ryegrass on biodegradation of hydrocarbons in soil. *Chemosphere,* 33, 203–215.

Hale, M. G., Moore, L. D. Griffin, G. J. (1978). Root exudate and exudation. *In Interactions between Non-pathogenic Soil Microorganisms and Plants.* Eds. Domergues, V. R., Krupa, S. V. Elsevier, Amsterdam, 163.

Hale, M. G., Moore, L. D. (1979). Factors affecting root exudation. *Advance in Agronomy*, 31, 93–124.

Haller, T., Stolp, H. (1985). Quantitative estimation of root exudation of maize plants. *Plant and Soil,* 86, 207–216.

Hanzlikova, A., Sotolova, I., Jandera, A. (1988). Chitinase in the rhizosphere and on plant roots. *In Interrelationships between microorganisms and plants in soil*, Elsevier Science Publishers, Amsterdam, Netherlands, 293–299.

Hayat, R., Ali, S., Amara, U., Khalid, R., Ahmed, I. (2010). Soil beneficial bacteria and their role in plant growth promotion: a review. *Annals of Microbiology*, 60, 579-598.

Hees van, P. A. W., Vinogradoff, S. I., Edwards, A. C., Godbold, D. L., Jones, D. L. (2003). Low molecular weight organic acid adsorption in forest soils: effects on soil solution concentrations and biodegradation rates. *Soil Biology and Biochemistry*, 35, 1015–1026.

Hegde, R. S., Fletcher, J. S. (1996). Influence of plant growth stage and season on the release of root phenolics by mulberry as related to development of phytoremediation technology. *Chemosphere*, 32, 2471–2479.

Heinonsalo, J., Jorgensen, K. S., Haahtela, K., Sen, R. (2000). Effects of *Pinus sylvestris* root growth and mycorrhizosphere development on bacterial carbon source utilization and hydrocarbon oxidation in forest and petroleum-contaminated soils. Canadian Journal of Microbiology, 46, 451–464.

Heinrich, D., Hess, D. (1985). Chemotactic attraction of *Azospirillum lipoferum* by wheat roots and characterization of some attractants. *Canadian Journal of Microbiology*, 31, 26-31.

Helal, H. M., Sauerbeck, D. (1989). Carbon turnover in the rhizosphere. *Z Pflanzenernahr Bodenk* 152, 211–216.

Herman, D., Artiola, J., Miller, R. (1995). Removal of cadmium, lead, and zinc from soil by a rhamnolipid biosurfactant. Environmental Science and Technology, 29, 2280–5.

Hildebrandt, U., Ouziad, F., Marner, F. J., Bothe, H. (2006). The bacterium *Paenibacillus validus* stimulates growth of the arbuscular mycorrhizal fungus *Glomus intraradices* up to the formation of fertile spores. *FEMS Microbiology Letters*, 254, 258–67.

Hildebrandt, R. M, Bothe, H. (2007). Arbuscular mycorrhiza and heavy metal tolerance. *Phytochemistry*, 68, 139–46.

Hiltner, L. (1904). Uber neue erfahrungen und probleme auf dem gebiete der bodenbakteriologie. *Arbeiten der Deutschen Landwirtschaft Gesellschaft,* 98, 59–78.

Hinsinger, P. (1998). How do plant roots acquire mineral nutrients? Chemical processes involved in the rhizosphere. *Advances in Agronomy*, 64, 225–65.

Hoagland, R. E., Williams, R. D. (1985). The influence of secondary plant compounds on the associations of soil microorganisms and plant roots. In *The chemistry of allelopathy. Biochemical interactions among plants*. American Chemical Society, Washington DC, USA, 301–325.

Huang, P. M., Wang, M. C., Wang, M. K. (1999). Catalytic transformation of phenolic compounds in the soil. *In Principles and Practices in Plant Ecology: Allelochemical Interactions*. Eds. Inderjit, K. M., Dakshini, M., Foy, C. L. CRC Press, Boca Raton, FL, 287-306.

Husain, S. S., Mckeen, W. E. (1963). Interactions between strawberry roots and *Rhizoctonia fragariae*. *Phytopathology*. 53, 541-545.

Hussain, A., Hasnain, S. (2009). Cytokinin production by some bacteria: its impact on cell division in cucumber cotyledons. *African Journal of Microbiology Research*, 3, 704-712.

Inderjit, Dakshini, K. M. M. (1999). Bioassays for allelopathy: interactions of soil organic and inorganic constituents. *In Principles and Practices in Plant Ecology: Allelochemicals Interactions*. CRC Press LLC, Boca Raton, FL, 35-44.

Janouskova, M., Pavlıkova, D., Vosatka, M. (2006). Potential contribution of arbuscular mycorrhiza to cadmium immobilisation in soil. *Chemosphere*, 65, 59–65.

Joner, E. J., Corgié, S., Amellal, N., Leyval, C. (2002) Nutritional constraints to PAH degradation in a rhizosphere model. *Soil Biology and Biochemistry*, 34, 859–864.

Joner, E. J., Leyval, C. (2003). Rhizosphere gradients of polycyclic aromatic hydrocarbon (PAH) dissipation in two industrial soils, and the impact of arbuscular mycorrhiza. *Environmental Science and Technology*, 37, 2371–2375.

Joo, G. J., Kim, Y. M., Kim, J. T., Rhee, I. K., Kim, J. H., Lee, I. J. (2005). Gibberellins-Producing Rhizobacteria Increase Endogenous Gibberellins Content and Promote Growth of Red Peppers. *The Journal of Microbiology*, 43, *6*, 510-515.

Khan, A., Kuek, C., Chaudhry, T., Khoo, C., Hayes, W. (2000). Role of plants, mycorrhizae and phytochelators in heavy metal contaminated land remediation. *Chemosphere,* 41, 197–207.

Khan, A. (2005). Role of soil microbes in the rhizospheres of plants growing on trace metal contaminated soils in phytoremediation. *Journal of Trace Elements in Medicine and Biology*, 18, 355–64.

Khan, A. A., Jilani, G., Akhtar, M. S., Naqvi, S. M. S., Rasheed, M. (2009). Phosphorus solubilizing bacteria: occurrence, mechanisms and their role in crop production. Journal of Agricultural and Biological Sciences, 1, 48-58.

Khan, A. G. (2006). Mycorrhizoremediation an enhanced form of phytoremediation. Journal of Zhejiang University Science B, 7, 03–14.

Knauff, U., Schulz, M., Scherer, H. W. (2003). Arylsulfatase activity in the rhizosphere and roots of different crop species. European Journal of Agronomy, 19, 215–223.

Kozdroj, J., van Elsas, J. D. (2000). Response of the bacterial community to root exudates in soil polluted with heavy metals assessed by molecular and cultural approaches. Soil Biology and Biochemistry, 32, 1405–1417.

Kravchenko, L. V., Azarova, T. S., Makarova, N. M., Tikhonovich, I. A. (2004). The Effect of Tryptophan Present in Plant Root Exudates on the Phytostimulating Activity of Rhizobacteria. *Microbiology*, 73, 156-158.

Krol, M. J. (1999). Production of exoenzymes from microorganisms of barley rhizosphere. [Polish]. *Folia Universitatis Agriculturae Stetinensis, Agricultura*, 117–127.

Kuiper, I., Bloemberg, G. V., Lugtenberg, B. J. J. (2001). Selection of a plant-bacterium pair as a novel tool for rhizostimulation of polycyclic aromatic hydrocarbon-degrading bacteria. *Molecular Plant Microbe Interaction,* 14,1197–205.

Kuiper, I., Lagendijk, E. L., Bloemberg, G. V., Lugtenberg, B. J. J. (2004). Rhizoremediation: a beneficial plantmicrobe interaction. *Molecular Plant-Microbe Interaction,* 17, 6–15.

Kuiper, I., Kravchenko, L. V., Bloemberg, G. V., Lugtenberg, B. J. J. (2002). *Pseudomonas putida* strain PCL1444, selected for efficient root colonization and naphthalene degradation, effectively utilizes root exudate components. *Molecular Plant-Microbe Interactions*, 15, 734-741.

Kuzyakov, Y., Raskatov, A., Kaupenjohann, M. (2003). Turnover and distribution of root exudates of *Zea mays*. *Plant Soil*, 254, 317–327.

Lehmann, R. G., Cheng, H. H., Harsh, J. B. (1987). Oxidation of phenolic acids by iron and manganese oxides. *Soil Science Society of America Journal*, 51, 352–356.

Leigh, M. B., Fletcher, J. S., Fu, X., Schmitz, F. J. (2002). Root turnover: An important source of microbial substrates in rhizosphere remediation of recalcitrant contaminants. *Environmental Science Technology*, 36, 1579–1583.

Leyval, C., Binet, P. (1998). Effect of polyaromatic hydrocarbons in soil on arbuscular mycorrhizal plants. *Journal of Environmental Quality*, 27, 402–407.

Leyval, C., Turnau, K., Haselwandter, K. (1997). Effect of heavy metal pollution on mycorrhizal colonisation and function, physiological, ecological and applied aspects. *Mycorrhiza,* 7, 139–53.

Linderman, R. G. (1988). Mycorrhizal interactions with the rhizosphere microflora – the mycorrhizosphere effect. *Phytopathology,* 78, 366–371.

Liste, H. H., Alexander, M. (1999). Rapid screening of plants promoting phenanthrene degradation. *Journal of Environmental Quality*, 28, 1376–1377.

Liu, Y., Zhang, N., Qiu, M., Feng, H., Vivanco, J. M., Shen, Q., Zhang, R. (2014). Enhanced rhizosphere colonization of beneficial *Bacillus amyloliquefaciens* SQR9 by pathogen infection. *FEMS Microbiology Letters*, 353, 49-56.

Lugtenberg, B., Kamilova, F. (2009). Plant-Growth-Promoting Rhizobacteria. *Annual Review of Microbiology*, 63, 541-556.

Lugtenberg, B. J., Chin-A-Woeng, T. C., Bloemberg, G. (2002). Microbe–plant interactions: principles and mechanisms. *Antonie van Leeuwenhoek*, 81, 373-383.

Lynch, J., Moffat, A. Bioremediation — prospects for the future application of innovative applied biological research. *Annals of Applied Biology*, 146, 17–21.

Lynch, J. M. (ed.) (1987). The rhizosphere. Chichester: Wiley Interscience.

Macmillan, J., (2002). Occurrence of gibberellins in vascular plants, fungi, and bacteria. *Journal of Plant Growth Regulation*, 20, 387–442.

Marschner, H. (1995). Mineral Nutrition of Higher Plants. *Academic Press*, London.

Marschner, P., Yang, C. H., Lieberei, R., Crowley, D. E. (2001). Soil and plant specific effects on bacterial community composition in the rhizosphere. *Soil Biology and Biochemistry,* 33, 1437–1445.

Marschner, P., Crowley, D., Yang, C. (2004). Development of specific rhizosphere bacterial communities in relation to plant species, nutrition and soil type. *Plant and Soil*, 261, 199-208.

Meharg, A. A., Cairney, J. W. G. (2000). Ectomycorrhizas – extending the capacities of rhizosphere remediation? *Soil Biology and Biochemistry*, 32, 1475–1484.

Meharg, A. A., Dennis, G. R., Cairney, J. W. G. (1997). Biotransformation of 2, 4, 6-trinitrotoluene (TNT): by ectomycorrhizal basidiomycetes. *Chemosphere,* 35, 513–521.

Miransari, M., Bahrami, H. A., Rejali, F., Malakouti, M. J., Torabi, H. (2007). Using arbuscular mycorrhiza to reduce the stressful effects of soil compaction on corn (*Zea mays* L.) growth. *Soil Biology and Biochemistry*, 39, 14–26.

Miransari, M., Bahrami, H. A., Rejali, F., Malakouti, M. J. (2008). Using arbuscular mycorrhiza to reduce the stressful effects of soil compaction on wheat (*Triticum aestivum* L.) growth. *Soil Biology and Biochemistry*, 40, 197–206.

Miransari, M. (2010). Arbuscular mycorrhiza and soil microbes. In: *Thangadurai D, Busso CA, Hijri M, editors. Mycorrhizal Biotechnology*. USA: Science, CRC and Taylor and Francis Publishers, 226.

Miransari, M. (2010). Contribution of arbuscular mycorrhizal symbiosis to plant growth under different types of soil stresses. *Plant Biology*, 12, 3–9.

Mithofer, A. (2002). Suppression of plant defense in rhizobia–legume symbiosis. *Trends in Plant Sciences*. 7, 440–444.

Miya, R. K., Firestone, M. K. (2001). Enhanced phenanthrene biodegradation in soil by slender oat root exudates and root debris. *Journal of Environmental Quality*, 30, 1911–1918.

Moons, A. (2003). *Osgstu3* and *osgstu4*, encoding *tau* class glutathione S-transferases, are heavy metal- and hypoxic stress-induced and differentially salt stress-responsive in rice roots. *FEBS Letters*, 553, 27–32.

Morgan, J. A. W., Bending, G. D., White, P. J. (2005). Biological costs and benefits to plant–microbe interactions in the rhizosphere. *Journal of Experimental Botany*, 56, 1729-1739.

Naseby, D. C., Lynch, J. M. (2002). Enzymes and microorganisms in the rhizosphere, *In Enzymes in the environment: activity, ecology, and applications*. Marcel Dekker Inc., New York, USA, 109–123.

Neumann, G., Romheld, V. (2000). The release of root exudates as affected by the plant's physiological status. *In The Rhizosphere– Biochemistry and Organic Substances at the Soil-Plant Interface*. Eds. Pinton, R., Varanini, Z., Nannipieri, P. Marcel Dekker, Inc, New York, 41–93.

Newman, E. I., Watson, A. (1977). Microbial abundance in the rhizosphere: a computer model. *Plant and Soil,* 48, 17–56.

Newsham, K. K., Fitter, A. H., Watkinson, A. R. (1995). Multi-functionality and biodiversity in arbuscular mycorrhizas, *Trends in Ecology and Evolution*, 10, 407–411.

Nicholas, C. U. (2007). Types, Amounts, and Possible Functions of Compounds Released into the Rhizosphere by Soil-Grown Plants, In *The Rhizosphere: Biochemistry and Organic Substances at the Soil-Plant Interface* (R. Pinton *et al.*, eds.), CRC Press, New York.

Nichols, T. D., Wolf, D. C., Rogers, H. B., Beyrouty, C. A., Reynolds, C. M. (1997): Rhizosphere microbial populations in contaminated soils. *Water, Air and Soil Pollution,* 95, 165–178.

Nihorimbere, V., Ongena, M., Smargiassi, M., Thonart, P. (2011). Beneficial effect of the rhizosphere microbial community for plant growth and health. *Biotechnology, Agronomy, Society and Environment*, 15, *2*, 327-337.

Nishizawa, N. K., Mori, S. (1987). The particular vesicles aooearing in barley root cells and its relation to mugeneic acid secretion. *Journal of Plant Nutrition*, 10, 1013–1020.

Norman, A. G. (1961). Microbial products affecting root development. Trans. 7th Congr. Wiscons. *International Soil Science Society*, 2, 531-536.

Oh, E. T., Koh, S. C., Kim, E., Ahn, Y. H. So, J. S. (2003). Plant terpenes enhance survivability of polychlorinated biphenyl (PCB) degrading *Pseudomonas pseudoalcaligenes* KF707 labeled with *gfp* in microcosms contaminated with PCB. *Journal of Microbiology and Biotechnology*, 13, 463–468.

Ona, O., Van Impe, J., Prinsen, E., Vanderleyden, J. (2005). Growth and indole-3-acetic acid biosynthesis of *Azospirillum brasilense* Sp245 is environmentally controlled. *FEMS Microbiology Letters*, 246, 125-132.

Ouziad, F., Hildebrandta, U., Schmelzerb, E., Bothe, H. (2005). Differential gene expressions in arbuscular mycorrhizal-colonized tomato grown under heavy metal stress. *Journal of Plant Physiology*, 162, 34–49.

Park, S. W., Stevens, N. M., Vivanco, J. M. (2002). Enzymatic specificity of three ribosomeinactivating proteins against fungal ribosomes, and correlation with antifungal activity. *Planta*, 216, 227–234.

Patten, C. L., Glick, B. R. (2002). Regulation of indoleacetic acid production in *Pseudomonas putida* GR12-2 by tryptophan and the stationary-phase sigma factor RpoS. *Canadian Journal of Microbiology*, 48, 635-642.

Pawlowska, T. E., Chaney, R. L., Chin, M., Charavat, I. (2000). Effects of metal phytoextraction practices on the indigenous community of arbuscular mycorrhizal fungi at a metal contaminated landfill. *Applied Environmental Microbiology*, 66, 26–30.

Phillips, D. A. (2000) Biosynthesis and release of rhizobial nodulation gene inducers by legumes. *In Prokaryotic Nitrogen Fixation: A Model System for the Analysis of a Biological Process* (Triplett, E.W., ed.), Horizon Scientific, 349–364.

Pinton, R., Varanini, Z., Nannipieri, P. (2001). The rhizosphere: biochemistry and organic substances at the soil-plant interface. New York: Marcel Dekker.

Podile, A., Kishore, G. K. (2006). Plant growth-promoting rhizobacteria, *In Plant-Associated Bacteria* (S. Gnanamanickam, eds.), Springer, Berlin, Heidelberg.

Raaijmakers, J., Paulitz, T., Steinberg, C., Alabouvette, C., Moënne-Loccoz, Y., (2009). The rhizosphere: a playground and battlefield for soil borne pathogens and beneficial microorganisms. *Plant and Soil*, 321, 341-361.

Radwan, S. S., Al-Awadhi, H., Sorkhoh, N. A., El-Nemr, I. M. (1998). Rhizospheric hydrocarbon-utilizing microorganisms as potential contributors to phytoremediation for the oily Kuwaiti desert. *Microbiology Research*, 153, 247–251.

Rangarajan, S., Loganathan, P., Saleena, M., Nair, S. (2001). Diversity of *Pseudomonads* isolated from three di variant plant rhizospheres. *Journal of Applied Microbiology*, 91, 742–749.

Rasmussen, T. B., Bjarnsholt, T., Skindersoe, M. E., Hentzer, M., Kristoffersen, P., Kote, M., Nielsen, J., Eberl, L., Givskov, M., (2005a). Screening for quorum sensing inhibitors (QSI) by use of a novel genetic system, the QSI selector. *Journal of Bacteriology*, 187, 1799-1814.

Rasmussen, T. B., Givskov, M., (2006). Quorum-sensing inhibitors as anti-pathogenic drugs. *International Journal of Medical Microbiology*. 296, 149-161.

Reddy, B. R., Sethunathan, N. (1994). Mineralization of p-nitrophenol in the rhizosphere of rice. *Agriculture, Ecosystem and Environment*, 47, 313–317.

Repetto, O., Bestel-Corre, G., Dumas-Gaudot, E., Berta, G., Gianinazzi-Pearson, V., Gianinazzi, S. (2003). Targeted proteomics to identify cadmium-induced protein modifications in *Glomus mosseae*-inoculated pea roots. *New Phytologist*, 157, 555–67.

Reynolds, C. M., Wolf, D. C., Gentry, T. J., Perry, L. B., Pidgeon, C. S., Koenen, B. A., Rogers, H. B., Beyrouty, C. A. (1999). Plant enhancement of indigenous soil micro-organisms: A low-cost treatment of contaminated soils. *Polar Record,* 35, 33–40.

Richardson, A., Barea, J. M., Mcneill, A., Prigent-Combaret, C. (2009). Acquisition of phosphorus and nitrogen in the rhizosphere and plant growth promotion by microorganisms. *Plant and Soil*, 321, 305-339.

Rivera-Becerril, F., van Tuinen, D., Martin-Laurent, F., Metwally, A., Dietz, K. J., Gianinazzi, S., Gianinazzi-Pearson, V. (2005). Molecular changes in *Pisum sativum* L. roots during arbuscular mycorrhizal buffering of cadmium stress. *Mycorrhiza,* 16, 51–60.

Rodriguez, H., Fraga, R. (1999). Phosphate solubilizing bacteria and their role in plant growth promotion. *Biotechnology Advances*, 17, 319-339.

Ross, I. L., Alami, Y., Harvey, P. R., Achouak. W., Ryder, M. H. (2000). Genetic diversity and biological control activity of novel species of closely related *Pseudomonads* isolated from wheat weld soils in South Australia. *Applied Environmental Microbiology*, 66, 1609–1616.

Rothballer, M., Schmid, M., Fekete, A., Hartmann, A. (2005). Comparative *in situ* analysis of *ipdC-gfpmut3* promoter fusions of *Azospirillum brasilense* strains Sp7 and Sp245. *Environmental Microbiology*, 7, 1839-1846.

Rougier, M. (1981). Secretory activity of the root cap. *In Plant Carbohydrates II, Extracellular Carbohydrates: Encyclopedia of Plant Physiology*. Eds. W Tanner and F A Leowus. Springer, Berlin.

Rovira, A. D. (1969). Plant root exudates. Botanical Review, 35, 35.

Rovira, A. D. (1959). Plant root excretions in relation to the rhizosphere effect. I. The nature of root exudate from oats and peas. *Plant and Soil*, 7, 178-194.

Rovira, A. D., Bowen, G. D., Foster, R. C. (1983). The significance of rhizosphere microflora and mycorrhizas in plant nutrition. *Encyclopedia of plant physiology.* Volume 15A. Inorganic plant nutrition, Springer-Verlag, Berlin, Germany, 61–93.

Rudrappa, T., Czymmek, K. J., Paré, P. W., Bais, H. P. (2008). Root-secreted malic acid recruits beneficial soil bacteria. *Plant Physiology*, 148, 1547-1556.

Ryan, J. A., Bell, R. M., Davidson, J. M., O'Conner, G. A. (1988). Plant uptake of non-ionic chemicals from soil. *Chemosphere* 17, 2299–2423.

Ryan, P. R., Delhaize, E., Jones, D. L. (2001): Function and mechanism of organic anion exudation from plant roots. *Annual Review of Plant Physiology and Plant Molecular Biology*, 52, 527–560.

Ryu, R. J., Patten, C. L. (2008). Aromatic amino acid-dependent expression of indole-3- pyruvate decarboxylase is regulated by *TyrR* in *Enterobacter cloacae* UW5. *Journal of Bacteriology*, 190, 7200-7208.

Sarand, I., Timonen, S., Koivula, T., Peltola, R., Haahtela, K., Sen, R., Romantschuk, M. (1999). Tolerance and biodegradation of mtoluate by Scots pine, a mycorrhizal fungus and fluorescent pseudomonads individually and under associative conditions. *Journal of Applied Microbiology*, 86, 817–826.

Sauerbeck, D., Nonnen, S. Allard, J. L. (1981). Consumption and turnover of photosynthesis in the rhizosphere depending on plant species and growth conditions. *Landw. Forschung Sonderheft,* 37, 207–216.

Schachtschabel, P., Blume, H. P., Brummer, G., Hartge, K. H., Schwertmann, U. (1992). *Lehrbuch der Bodenkunde. Stuttgart: Ferdinand Enke Verlag.*

Schnoor, J. L., Light, L. A., McCutcheon, S. C., Wolfe, N. L., Carreira, L. H. (1995). Phytoremediation of organic and nutrient contaminants. *Environmental Science and Technology*, 29, 318–323.

Schutzendubel, A., Polle, A. (2002). Plant responses to abiotic stresses: heavy metal-induced oxidative stress and protection by mycorrhization. *Journal of Experimental Botany*, 53, 51–65.

Schutzendubel, A., Polle, A. (2002). Plant responses to abiotic stresses: heavy metal-induced oxidative stress and protection by mycorrhization. *Journal of Experimental Botany*, 53, 1351–1365.

Selosse, M. A., Le Tacon, F. (1998). The land flora: a phototroph fungus partnership? *Trends in Ecology and Evolution*,. 13, 15–20.

Shann, J. R. (1995). The role of plants and plant/microbial systems in the reduction of exposure. *Environmental Health Perspective,* 103, 13–15.

Sharma, B. C., Subba, R., Saha, A. (2012). *In vitro* solubilization of tricalcium phosphate and production of IAA by phosphate solubilizing bacteria isolated from tea rhizosphere of *Darjeeling Himalaya. Plant Sciences Feed*, 2, 96-99.

Sharma, S., Sayyed, R., Trivedi, M., Gobi, T. (2013) Phosphate solubilizing microbes: sustainable approach for managing phosphorus deficiency in agricultural soils, *SpringerPlus*, 2, 587.

Shetty, K. G., Hetrick, B. A. D., Figge, D. A. H., Schwab, A. P. (1994). Effects of mycorrhiza and other soil microbes on revegetation of heavy metal contaminated mine spoil. *Environment Pollution*. 86, 181–188.

Siciliano, S. D., Germida, J. J. (1997). Bacterial inoculants of forage grasses that enhance degradation of 2-chlorobenzoic acid in soil. *Environmental Toxicology Chemistry*, 16, 1098–1104.

Siciliano, S. D., Germida, J. J. (1999). Enhanced phytoremediation of chlorobenzoates in rhizosphere soil. *Soil Biology and Biochemistry*, 31, 299–305.

Siciliano, S. D., Germida, J. J, Banks, K., Greer, C. W. (2003). Changes in microbial community composition and function during a polyaromatic hydrocarbon phytoremediation field trial. *Applied Environmental Microbiology*, 69, 483–489.

Singer, A. C., Crowley, D. E., Thompson, I. P. (2003). Secondary plant metabolites in phytoremediation and biotransformation. *Trends in Biotechnology*, 21, 123–130.

Smalla, K., Wieland, G., Buchner, A., Zock, A., Parzy, J., Kaiser, S., Roskot, N., Heuer, H., Berg, G. (2001). Bulk and rhizosphere soil bacterial communities studied by denaturing gradient gel electrophoresis: plant-dependent enrichment and seasonal shifts revealed. *Applied and Environmental Microbiology*, 67, 4742–4751.

Smith, A. P., DeRidder, B. P., Guo, W. J., Seeley, E. H., Regnier, F. E., Goldsbrough, P.B. (2004). Proteomic analysis of Arabidopsis glutathione S-transferases from benoxacor- and coppertreated seedlings. *Journal of Biological Chemistry*, 279, 26098–104.

Smith, S. E., Read, D. J. (2008). Mycorrhizal Symbiosis. *San Diego: Academic Press*.

Smith, S. E., Read, D. J. (1997). Mycorrhizal symbiosis. *Academic Press*, New York, (1997).

Sood, S.G. (2003). Chemotactic response of plant-growth-promoting bacteria towards roots of vesicular-arbuscular mycorrhizal tomato plants. *FEMS Microbiology Ecology*, 45, 219-227.

Spaepen, S., Vanderleyden, J., Remans, R. (2007). Indole-3-acetic acid in microbial and microorganism-plant signaling. *FEMS Microbiology Reviews*, 31, 425-448.

Stefan, M. (2008). Biologia microorganismelor rizosferice. *Aplicaţii biotehnologice*, Technopress, Iaşi.

Tan, S., Yang, C., Mei, X., Shen, S., Raza, W., Shen, Q., Xu, Y. (2013). The effect of organic acids from tomato root exudates on rhizosphere colonization of *Bacillus amyloliquefaciens* T-5. *Applied Soil Ecology*, 64, 15-22.

Teplitski, M., Robinson, J. B., Bauer, W. D. (2000). Plants secrete substances that mimic bacterial N-acyl homoserine lactone signal activities and affect population density-dependent behaviors in associated bacteria. *Molecular Plant Microbe Interaction*, 13, 637-648.

Thompson, P. L., Ramer, L. A., Schnoor, J. L. (1998). Uptake and transformation of TNT by hybrid poplar trees. *Environmental Science and Technology,* 32, 975–980.

Torsvik, V., Sorheim, R., Goksoyr, J. (1996). Total bacterial diversity in soil and sediment communities – a review. *Journal of Indian Microbiology*, 17, 170– 178.

Trieu, A. T., Van Buuren, M. L., Harrison, M. J. (1997). Gene expression in mycorrhizal roots of *Medicago truncatula. In Radical Biology: Advances and Perspectives on the Function of Plant Roots* (Flores, H.E. et al., eds), American Society of Plant Physiologists, 498–500.

Tsavkelova, E. A., Klimova, S. Y., Cherdyntseva, T. A., Netrusov, A. I. (2006). Microbial producers of plant growth stimulators and their practical use: A review. *Applied Biochemistry and Microbiology*, 42, 117-126.

Tu, C., Yuan, L., Wei, S., Qiong, C. (1996). Effects of soil K status on rhizospheric exudates of plants. [Chinese]. *Journal of Southwest Agriculture University*, 18, 276–280.

Tu, S. I., Nungesser, E., Brauer, D. (1989). Characterization of the effects of divalent cations on the coupled activities of the H+–ATPase in tonoplast vesicles. *Plant Physiology*. 10, 36–43.

Uren, N. C. (2000). Types, amounts, and possible functions of compounds released into the rhizosphere by soil-grown plants. *In The Rhizosphere: Biochemistry and Organic Substances at the Soil- Plant Interface*. Eds. R Pinton, Z Varanini and P Nannipieri. Marcel Dekker, Inc, New York, 19–40.

Van Loon, L. C. (2007). Plant responses to plant growth-promoting rhizobacteria. *European Journal of Plant Pathology*. 119, 243-254.

Vance, C. P. (2001). Symbiotic Nitrogen Fixation and Phosphorus Acquisition. Plant Nutrition in a World of Declining Renewable Resources. *Plant Physiology*, 127, 390-397.

Vancura, V. (1964). Root exudates of plants and analysis of root exudates of barley and wheat in their initial phases of growth. *Plant and Soil*, 21, 231-248.

Vancura, V., Hovadik, A. (1965). Composition of root exudates in the course of plant development. In *Proceedings Symposium Relationships between Soil Microorganisms and Plant Roots*, Prague, Czechoslovak Academy of Sciences, 21-25.

Vessey, J. K (2003). Plant growth promoting rhizobacteria as biofertilizers. *Plant and Soil*, 255, 571-586.

Vogel-Mikus, K., Pongrac, P., Kump, P., Necemer, M., Regvar, M. (2006). Colonisation of a Zn, Cd and Pb hyperaccumulator *Thlaspi praecox* Wulfen with indigenous arbuscular mycorrhizal fungal mixture induces changes in heavy metal and nutrient uptake. *Environmental Pollution*, 139, 362–71.

Von Bodman, S. B., Bauer, W. D., Coplin, D. L. (2003). Quorum sensing in plant-pathogenic bacteria. *Annual Review of Phytopathology*, 41, *1*, 455-482.

Walker, T. S., Bais, H. P., Grotewold, E., Vivanco, J. M. (2003). Root Exudation and Rhizosphere Biology. *Plant Physiology*, 132, 44-51.

Walton, B. T., Anderson, T. A. (1990). Microbial degradation of trichloroethylene in the rhizosphere: potential application to biological remediation of waste sites. *Applied Environmental Microbiology*, 56, 1012–1016.

Walton, B. T., Hoylman, A. M., Perez, M. M., Anderson, T. A., Johnson, T. R., Guthrie, E. A., Christman, R. F. (1994). Rhizosphere microbial communities as a plant defense against toxic substances in soil. In: *Anderson TA, Coats JR (eds), Bioremediation through rhizosphere technology, ACS Symposium Series 563*, American Chemical Society, Washington DC, 82–9.

Wasaki, J., Dateki, H., Shinano, T., Osaki, M., Tadano, T. (2001). Characterization of secretory acid phosphatase gene of lupin roots and the transformation of the gene into tobacco plants. In *Plant nutrition: food security and sustainability of agro-ecosystems through basic and applied research*, Kluwer Academic Publishers, Dordrecht, Netherlands, 36–37.

Waters, C. M., Bassler, B. L., (2005). Quorum sensing: cell-to-cell communication in bacteria. *Annual Reviews in Cell Development and Biology*, 21, 319-346.

Wenzel, W. W., Lombi, E., Adriano, D. C. (1999). Biogeochemical processes in the rhizosphere: role in phytoremediation of metal-polluted soils. In: *Prasad MNV, Hagemeyer J, editors. Heavy Metal Stress in Plants*. Berlin, Heidelberg: Springer-Verlag, 273–303.

Wetzel, S. C., Banks, M. K., Schwab, A. P. (1997). Rhizosphere effects on the degradation of pyrene and anthracene in soil. *Phytoremediation of soil and water contaminants*, American Chemical Society, Washington USA, 254–262.

Whipps, J. M. (1990). Carbon economy. *In The Rhizosphere.* Ed. J M Lynch. J Wiley and Son, Chichester, UK, 59.

Wu, F. B., Zhang, G. P. (2002). Genotypic variation in kernel heavy metal concentrations in barley and as affected by soil factors. *Journal of Plant Nutrition*, 25, 1163–73.

Xian, X. (1989). Effect of chemical form of cadmium, zinc, and lead in polluted soil on their uptake by cabbage plants. *Plant and Soil*, 113, 57–64.

Xiaoe, Y. F., Zhenli, H., Stoffellab, P. J. (2005). Molecular mechanisms of heavy metal hyperaccumulation and phytoremediation. *Journal of Trace Elements in Medicine Biology,* 18, 339-353.

Yang, C. H., Crowley. D. E., Menge. J. A. (2001). 16S rDNA fingerprinting of rhizosphere bacterial communities associated with healthy and Phytophthora infected avocado roots. *FEMS Microbiology Ecology*, 35, 129–136.

Yang, C. H., Crowley, D. E. (2000). Rhizosphere microbial community structure in relation to root location and plant iron nutritional status. *Applied and Environmental Microbiology,* 66, 345–351.

Yaryura, P. M., Leon, M., Correa, O. S., Kerber, N. L., Pucheu, N. L., García, A. F. (2008). Assessment of the role of chemotaxis and biofilm formation as requirements for colonization of roots and seeds of soybean plants by *Bacillus amyloliquefaciens* BNM339. *Current Microbiology*, 56, 625-632.

Yates, E. A., Philipp, B., Buckley, C., Atkinson, S., Chhabra, S. R., Sockett, R. E., Goldner, M. (2002). N Acyl homoserine lactones undergo lactonolysis in a pH, temperature-, and acyl chain length-dependent manner during growth of Yersinia pseudotuberculosis and *Pseudomonas aeruginosa. Infection and Immunity*, 70, 5635-5646.

Yoshitomi, K. J., Shann, J. R. (2001). Corn (*Zea mays* L.): root exudates and their impact on 14C-pyrene mineralization. *Soil Biology and Biochemistry,* 33, 1769–1776.

Zablotowicz, R. M., Hoagland, R. E., Locke, M. A. (1994). Glutathione Stransferase activity in rhizosphere bacteria and the potential for herbicide detoxification. In: *Anderson TA, Coats JR (eds), Bioremediation through rhizosphere technology*, ACS Symposium Series 563, American Chemical Society, Washington DC, 184–198.

Zablotowicz, R. M., Hoagland, R. E., Locke, M. A., Hickey, W. J. (1995). Glutathione-S-transferase activity and metabolism of glutathione conjugates by rhizosphere bacteria. *Applied Environmental Microbiology*, 61, 1054–1060.

Zamioudis, C., Pieterse, C. M. J. (2011). Modulation of Host Immunity by Beneficial Microbes. *Molecular Plant-Microbe Interactions*, 25, 139-150.

Zheng, X. Y., Sinclair, J. B. (1996). Chemotactic response of *Bacillus megaterium* strain B153-2-2 to soybean root and seed exudates. *Physiological and Molecular Plant Pathology*, 48, 21–35.

In: Chemical Pollution Control …
Editor: Naser A. Anjum
ISBN: 978-1-53611-034-0

Chapter 6

PLANT–BACTERIA PARTNERSHIPS: A MAJOR POLLUTANT-REMEDIATION APPROACH

***C. Karthik*[1]*, M. Oves*[2,]**, M. S. Khan*[3]*, R. K. Sharma*[1]*, P. I. Arulselvi*[1] *and H. A. Qari*[2]**

[1]Department of Biotechnology, Periyar University,
Periyar Palkalai Nagar, Salem, Tamil Nadu, India
[2]Center of Excellence in Environmental Studies (CEES),
King Abdulaziz University, Jeddah, Kingdom of Saudi Arabia
[3]Department of Biochemistry,
Protein Research Chair College of Science, King Saud University,
Riyadh, Kingdom of Saudi Arabia

ABSTRACT

The boom in the development of industries and modernization as a whole has no doubt made our life easier. On the contrary, the environment has been hit hard due to unchecked and uncontrolled release of toxic wastes into the environment. This leads to the accumulation of detrimental wastes to the soil, thereby affecting not only agronomically

* Corresponding author: M. Oves, Center of Excellence in Environmental Studies (CEES), King Abdulaziz University, Jeddah 21589, KSA; E-mail: owais.micro@gmail.com; maaldin@kau.edu.sa.

but also the health of the animals and human beings. Contamination of soil is one of the major environmental issues, which needs to be addressed by the environmentalists and scientists at the foremost level. The issue can be solved by employing physicochemical remediation treatment methods and bioremediation. Physicochemical remediation of the contaminated soil adopts one of these methods viz physical dumping, incineration, in situ acid treatment, electrokinetic reclamation and soil washing. However, being expensive and disruptive in nature, their effectiveness is questioned economically and impact wise. Bioremediation in contrast is a low cost, environmental friendly and practical technology utilized in order to reclaim contaminated soil. Rhizospheric dwelling PGPR plays significant role in providing a base for contaminants in soil and plant to interact during the process of phytoremediation. Plant Growth Promoting Rhizobacteria (PGPR) release chelating and bioleaching agents in the rhizospheric region during phytoremediation process. They cause acidification, enzymatic reduction and redox changes enhancing the mobility and availability of contaminants to the plant. There arises a necessity to understand and study the mechanism of removal and mobilization of contaminants by microbes especially PGPR for the restoration and reclamation on the road to sustainability of contaminant free environment.

Keywords: heavy metals, pollutants, bioremediation, bacteria, plants, environment

1. Introduction

Industrial revolution has modernized the world in a significant way, but the environment and society at large are being affected with the resulting pollution, which has increased exponentially. Most of the contaminants originate from the industrial waste effluents, incineration activities, smelting, agrochemicals and municipal sewage wastes. These activities further accelerate the spread and migration of contaminants in the form of vapors through wind and leachate through aqueous activities in natural ecosystems (Khan, 2005). Pollutants, thus magnified in areas are difficult to degrade, as they are less soluble. The added characteristics of high carcinogenicity and mutagenicity, these pollutants possess, affect both the environment and living beings critically (Diels et al., 2002). Inorganic and organic compounds, toxic and explosive gases, combustible and putrescible substances, hazardous wastes and explosives are the main material causing pollution (Cheng, 2003). Among them the widely dispersed inorganic pollutants released as wastes from industries cause serious problems as far as their disposal technique is concerned. With large amounts of constituents of heavy metals such as arsenic, lead, mercury, cadmium, cobalt, nickel, zinc, chromium, manganese, copper, etc. they pollute the environment either through sewage, industrial waste or mine washings. In addition to inorganic compounds, another category of wastes consisting of organic compound, are polluting the soil system. These organic contaminants can be chlorinated solvents like, trichloroethylene, explosives such as trinitrotoluene (TNT), 1, 3, 5-trinitro-1, 3, 5-hexahydrotriazine (RDX), petroleum hydrocarbons including benzene, toluene and xylene (BTX), polyaromatic hydrocarbons (PAHs) and pesticides. There is no doubt about the capability of soil micro flora to reduce the pollutants, nevertheless the process of degradation slow and sometimes inefficient due to the low occurrence of degradative microorganisms (Glick, 2010; Karamalidis et al., 2010). Soils polluted with single or mixture of pollutants are of grave concern for the public and thus receiving increasing attention from the governmental bodies, particularly in developing countries (Yanez et al., 2002). The remediation of the site contaminated with heavy metals with the main objective of reclamation and restoration of sites is a challenging task and a major environmental concern (Khan et al., 2009; Oves et al., 2012). Heavy metal-contaminated areas are large and unfit for agricultural and other human use. Therefore, an urgent need is to develop suitable *in situ* remediation technologies so that the challenges of *ex situ* remediation activities can be reduced.

2. Hazardous Pollutants Sources in Soil

Soil is a very intricate mixture of organic and inorganic materials that covers the Earth's surface. The uppermost top part is 'humus' dark in color. It is composed of degenerated organic remains of plants and animals. The inorganic portion is made up of minerals and rock fragments. Fertile and productive soils are necessary for effective agrarian activities.

Soil pollution is defined as "an unwanted change in the natural, physical, chemical or biological components of the soil". Soil is dumping place for most of the wastes generated from industrial, agricultural and anthropogenic sources. A huge quantity of hazardous chemicals and several thousand tons of industrial wastes are dumping on fertile earth surface without proper treatment, it is an alarmingly rate. The pollutants get leached during rain and other sewage seeps inside the ground, resulting in ground water pollution enhancement day by day. The situation is further worsened when human use agrochemicals such as pesticides, fungicides, insecticides, biocides and chemical fertilizer for pest management and higher yield of crops. Air and water pollution allows the dispersion of the pollutants thereby making it less concentrated. However, pollution of soil is different since the pollutants are dumped and mobility is very less, finally resulting in exponential magnification of these toxic pollutants for relatively long periods.

2.1. Sources of Soil Pollutants

Soil continuously gets polluted by various natural and artificial sources. Major aspects related with natural and artificial sources of soil pollutants are summarized in Figure 6.1 and briefly discussed hereunder.

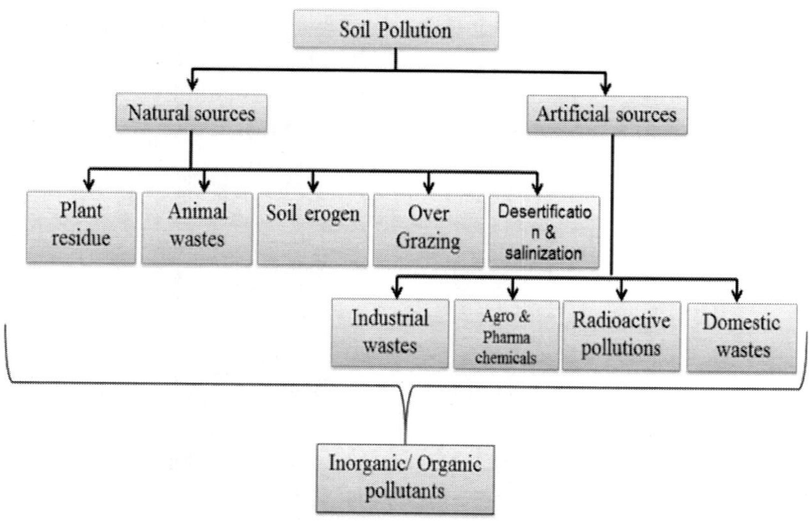

Figure 6.1. Major sources of soil pollutants.

2.1.1. Natural Sources

Natural pollutants mostly are of plant and animal origin. Another major reason continuous grazing and nutrient lost by same crop farming, soil erosion by natural climatic condition like flood and high-speed wind damage the soil fertility and originality.

2.1.1a. Plant Residues

Normally plants, on death and decay, contribute organic matter to the soil and thereby increase soil fertility. Sometimes, naturally occurring or man-made forest fires and continuous crops add unwanted organic matter to the soil and destroy soil micro flora, which create unfavorable conditions leading to soil pollution.

2.1.1b. Animal Wastes

Animal wastes such as fecal matter, urine, blood, slaughterhouse waste in the form of liquid or particulate matter, bodies of dead animals etc., are all indiscriminately dumped into the soil. These create unhealthy conditions detrimental to the growth of plants and soil organisms. Excessive organic contents are harmful to root growth as they create hypertonic conditions in the soil causing wilting or stunted growth of the plants.

2.1.1c. Over Grazing

Continuous long time grazing on agriculture land by heavy livestock has a major impact on soil structure and composition. It is also divided in two-part (a) physical damage (b) chemical change. (a) Physical damage by livestock grazing on fertile land is destroying surface primary vegetation and soil structure by poaching. Cattle grazing can be lost up to 5-10 kg dry matter/ha/d where 50% of area treading, it recovery takes time few month (Sheath et al., 1998). After grazing, upper soil layer becomes stronger that prevents the new emerging shoots and buds. Further, soil loses the absorbing capability when a moderate rainfall becomes ponds form. Grazing area surface soil particles are very susceptible for erosion (Warren et al., 1986). Furthermore, waterlogged soil become anaerobic zone and it will encourage denitrification ultimately soil nitrogen lost. (b) Grazing animal same time excretes urine and faces in land which also harmful in other point of view. Grazing cattle, if release 2-liter urine in green land approximately spread 0.4 m^2 (Addiscott et al., 1991). It instantaneous affect the surrounded area and burn vegetation due to toxic for plant roots and takes up to 12 month in recovery. Urine urea in soil is quickly hydrolyzed the grass root nitrogen and atmosphere also became bad smell. In another report substantial amount of soil calcium and magnesium are also lost due to livestock urine patches on soil (Early et al., 1998).

2.1.1d. Soil Erosion

Topsoil fertile layer lost due to high-speed wind and water is widely known as erosion. It is basically the mobilization of the surface soil nutrient by climatic reason due heavy rainfall and cyclone. Eroded soil particle are sifted to other location and deposited as sediment. It is natural process that removes fertile soil from the crop land (Toy et al., 2002). Soil erosion we can decrease by the vegetation but recently it is increasing due to anthropogenic activities like deforestation for urbanization, industrialization and overgrazing. In USA in the year 1930, >30 million hectares soil was damaged due to dust bowl. Recently, an increasing soil erosion rate was supposed replace over 1/3 of the world's croplands. Annually, USA spends about $125B on the eroded soil to maintain their fertility (World Resource, 1992-1999; Mackenzie and Mackenzie, 1995; Humberto and Rattan, 2010; Julien and Pierre, 2010).

2.1.1e. Desertification

Desertification is increasing and the non-desert areas are becoming desert mainly as a result of unwanted lost of water and declining rate of rainfall (United Nation, 1977). In the last half century, the Southern Sahara edge has became true desert. Similarly, many other African countries are under the major threat of desertification. However, the soil-desertification can be controlled by proper management of fertile soil and forestation in a sustainable way (World Resource, 1992-1999).

2.1.1f. Salinization and Nutrient Loss

Many people have observed that the soil-salinization can develop due to irrigation of seawater and other high-level water containing salts and also due to salt rock contamination and nearest salt surface exposure. Soil-salinization can damage important soil properties. Notably, salinization has become one of the most common problems in dry lands because these areas require high irrigation, where occur high evaporation and salt concentrations. High salt concentration in soil has been shown to decrees the water uptake by plants roots. Hence, further vegetation could not sustain for long-time. Furthermore, due to high salt concentrations soil nutrient levels get declined. Nutrient loss problem is more common problem mainly in African countries, where occur the low input agriculture facilities and continuous farming of same type crop.

2.1.2. Artificial Sources

Anthropogenic activities based pollutants can be classified into industrial wastes, mining, agrochemicals, radioactive pollutants, and urban and domestic wastes. These pollutants are briefly discussed hereunder.

2.1.2a. Industrial Wastes

Industrial wastes are the major cause of soil pollution. These wastes are discharged from various industrial sources such as leather tanning, paper, textile, petrochemical, food processing, soap and detergent, dairy, metal processing, cement and engineering industries. With the advancement of new technology, new types of industrial wastes are produced and discharged on to the land. The furnaces of industries like thermal and electric power plants produced fly ash severely pollute air, water and soil. Industrial sludges are more dangerous than industrial solid wastes. These pollutants not only pollute soil but also water bodies. The pollution of soil affects and alters the physio-biochemical properties of the soil. The entry point for these pollutants in the biological system is mainly through the food chain from the soil. Once intercalated in the food chain, they start affecting the biochemical processes and seriously affected the living organisms. The most common industrial waste chemicals involved in causing soil pollution are toxic heavy metals, petroleum related hydrocarbons and industrial solvents. Industrial wastes can be categorized into two major groups based on the nature of pollutants such as inorganic/elemental or heavy metal wastes and organic wastes.

2.1.2b. Inorganic/Elemental or Heavy Metal Wastes

Metals with atomic numbers greater than 23 (with few exceptions) can be grouped as heavy metals. They are hazardous and cause severe toxic consequences in the biological systems of all the life forms. They are notoriously difficult to degrade once entered in the biological systems and get biomagnified, causing slow but severe disorders. In general, arsenic, lead, mercury, cadmium, cobalt, nickel, zinc, chromium, manganese and copper are widely dispersed in the environment and they possess certain toxic attributes which impact human health on a serious note (Agency of toxic substances and disease registry, 2007) (Table 6.1). These metals are cheaply available and ease of their usage due to unique physio-chemical and toxic properties of have promoted their wide use in industries and as biocides (fungicide and herbicide) for agronomic activities. As a result of these uncontrolled usages, heavy metal accumulates in the environment making their concentration manifold. The health of the public and ecology become the main victims of these heavy metals. Removal of these metals is therefore a challenge to environmental management.

Table 6.1. Sources and toxicity of major heavy metals pollutants

Pollutants	Source	Effect
Arsenic	Herbicide, fungicide, wood preservative-Agro chemicals (70%), industrial chemicals – paints, bullets (20%), glass and glass wares (5%).	Accumulate in hair, nail, skin lesions, act as oxidative uncoupler, and cause damage to kidney, respiratory tract and nervous disorders.
Lead	Automobile exhaust of leaded petrol (50%), Batteries, Pipes, Soldiering.	Serious mutagenic effects on microorganisms, blackening in fish and paralysis in man.
Mercury	Methyl mercury fungicides, electrical and electronic industries, PVC, plastics and paints.	Irreversible neurological damage in man, Minamoto disease.
Cadmium	Pigment and stabilizer for PVC, plastics, tires, rechargeable cells, electroplating, coal, oil and phosphate rocks.	Bones become brittle–Itai itai disease in Japan, gastro enteric distress and pain.
Cobalt	Ore smelting process, ore deposits, fossil fuels, phosphate rocks, coal-fired power plants, vehicular exhaust and agricultural activities.	Asthma, pneumonia, vomiting, thyroid damage and hear problems.
Nickel	Power plants and trash incinerators	Lung and nasal cancer, decrease the lung function, asthma, skin rash and itching.
Zinc	Mining, purifying of zinc, steel production, coal burning, burning of wastes and fertilizers.	High concentration of Zn cause anemia, stomach cramps, skin irritation, vomiting, nausea, anemia nausea and decrease levels of high-density lipoprotein (HDL) cholesterol. Stomach cramps,
Chromium	Tanneries, electroplating and metal finishing processes, Khaki dyeing in textiles.	Toxic to aquatic organisms, absorbed through intestinal tract in man.
Manganese	metal processing facilities	Behavioral changes, affect the nervous system, slowed hand movements, lungs irritation, loss of sex drive and sperm damage.
Copper	Mining waste, domestic waste, combustion of fossil fuels and wastes, wood production, chemical fertilizers and natural sources.	Nose, mouth and eyes irritation, headaches, dizziness, nausea, diarrhea, vomiting, stomach cramps, liver and kidney damage.

2.1.2c. Organic Pollutants

Organic pollutants especially referred to persistent organic pollutants (POP) are compounds with gaseous, liquid or solid states having the content of carbon molecule. They are persistent and resistant to biochemical and photolytic degradation. The United Nations Environment Programme (UNEP) has investigated and recognized twelve harmful organic compounds, which needed immediate attention from global countries for their elimination under the Stockholm convention on persistent organic pollutants. They are aldrin, chlordane, dichlorodiphenyltrichloroethane (DDT), dieldrin, endrin, mirex, heptachlor, hexachloro-benzene, and toxaphene. These are used as insecticides, fungicides and pesticides. In addition to these polychlorinated biphenyls (PCB), dioxin and polychlorinated dibenzofurans generated from various industrial process. They are carcinogenic, endocrine, central nervous system, immune and reproductive disruptors, causing serious health hazards. The use of these POPs worldwide has been generally phased. However, the main problem lies in its removal techniques, which is not very clear because of their toxicity and persistence byproducts and they continue to pose a threat to biological living forms.

2.1.2d. Agrochemicals

The agrochemicals like pesticides (insecticides, fungicides, herbicides etc.) and fertilizers are commonly used for improvement of agricultural crops.

Pesticide Pollution

With the advent of modernization and the green revolution, the utilization of agrochemicals like pesticides, herbicides, insecticides, and fungicides has become a common practice in order to improve the yield of the crops and prevent them from various pests. The entry source of the pesticides in the environment comes during the application during seed treatment or foliar application. Their residual accumulation in the biosphere contaminates the soil, water and ultimately becomes part of the food chain, causing ecological stress. Moreover, the extensive pesticide usage also affects beneficial insects as they have a broad range of action. The pesticides get accumulated in the food and tissues of all living forms, giving direct or long-term health effects. Even their minute residual exposure for a longer duration of time is carcinogenic because of their high mutation coefficient. Their degraded metabolites affect not only the higher living forms, but also the beneficial microbes in the soil are also hampered. For instance, seed treatment with mercury fungicides inhibits nodulation and nitrogen fixation activity in *Rhizobium* and nitrification activity in *Nitrosomonas and Nitrobacter*.

Fertilizer Pollution

Fertilizers have become a part and parcel of cultivation these days since the farmers, in the urge to get more out of the crops, use more fertilizers for better yield. Long-term application of fertilizers in the field deteriorates the natural characteristics of soil by changing its chemical and physical properties. Application of certain fertilizers such as ammonium sulphate, ammonium chloride and urea makes the soil pH acidic. Adoption of organic manures, biofertilizers, and integrated nutrient managements are necessary to sustain the productivity of our soils and a way to replace the chemical fertilizers.

2.1.2e. Radioactive Pollutants

The radioactive pollution is defined as the distribution of radioactive materials in soil, air and water. Certain materials possess radioactivity when their nuclei have the ability to emit high penetrating and powerful electrons, proton and gamma rays. Sources of such pollutants are nuclear explosions, nuclear testing, nuclear reactors, etc. Radioactive dust particles produced after an atomic explosion slowly settle down on the soil surface and gradually permeate the soil strata. When absorbed by the plants, these enter the human body via the food chain and cause serious damages. The great problem with radioactive elements is that they cannot be destroyed once they have been created and it is only the passage of time, which can reduce their harmful properties. The major sources of radioactive pollutants in soil are: radio isotopes and accidents in nuclear reactors.

Radioisotope

These are used in research, industry and medicine and are also generated by the operation of nuclear reactors for the generation of power and the nuclear fuel processing plants. Wastes from nuclear reactor, which ultimately dumped in the soil consists of Ruthenium-106, Iodine-131, Barium-140, Lanthenium-140, Caesium-144, Promethium-444, Strontium-90 and Caesium-137.

Accidents in Nuclear Reactors

These are rare; however, if taken place, they become the major cause of emission of radioisotopes into the soil and land. Recently, Fukushima Daiichi nuclear disaster was a nuclear reactor at Fukushima (Japan), it happened due to by the tsunami of the Tōhoku earthquake 11 March 2011. Tsunami hit the some part of nuclear reactor and created equipment-control failures that finally caused the release of radioactive materials. It was the second largest nuclear disaster after the Chernobyl nuclear disaster of 1986 and it classified as a Level 7 event according International Nuclear Event Scale. According to Kyodo News (Japan) radioactive material dispersion distance from Fukushima was 1/10 of Chernobyls nuclear disaster. According to Wire service report, Fukushima Daiichi plant disaster material dispersion distance was approximately 33 km; whereas, Chernobyls nuclear disaster material spread up to 250 Km. During nuclear disaster, frozen soil materials were used as a barrier to control exposure of running water and meltdown nuclear fuel. Nuclear plant release malted down the release radioactive material. Exposed water significantly contaminated the soil, where the pollutants might be persist for long time. Additionally, along the costal areas, these pollutants can be continuously transported in to sea and thereby can create severe problems.

2.1.2f. Urban and Domestic Wastes

The urban and domestic wastes are jointly referred as refuse. This comprises of garbage and waste materials like plastics, glass, metallic containers, fibers, papers and a number of discarded house made wastes. The urban domestic wastes are more harmful since most of these cannot be easily degraded. Certain items like (paint, varnishes, oil, etc.) create more soil pollution problems (Abdullahi et al., 2014). The leachates from the dumping sites and sewage along with industrial effluent are extremely harmful to biota. These leachates contain poisonous gases along with partially decomposed materials from food remains, vegetables, toxic chemicals and pathogenic bacteria that can be very harmful and may cause various diseases. These are responsible not only for soil but also for water bodies (FAO, 2013).

3. Plant-Microbes Interaction and Bioremediations

Remediation of contaminated soil is a challenging task. Most of the remediation technologies such as dumping, incineration, physico-chemical treatment and electrokinetic reclamation are not suitable for long term and practical applications. The challenge lies in the fact that they are costly, low efficiency, high specificity of the site with concerned contaminants and long term with vast area of soil characteristic disruption. The expensiveness of conventional restoration program can be scaled by a prediction of the United States endeavor to restore all contaminated sites, which will have a huge cost of approximately $1.7 trillion. In addition, the conventional techniques are not always sufficient (Dixon, 1996). During conventional methods of remediation the destruction of wastes process itself give way to more secondary pollution like air pollution from incineration, ground water and well pollution from leaching of landfill and toxic gas emissions from the excavation of soil. Conventional approaches employed for the remediation of metal from contaminated sites are illustrated in Figure 6.2. Therefore, the exploration of alternative methods, which are less expensive, less labor intensive, safe and eco- friendly way to restore polluted sites, is essential. An alternative method that can substitute the above-mentioned conventional practices is bioremediation, a process where environmental pollutants are degraded by the actions of microbes or other biological systems (Dua et al., 2002; khan et al., 2012).

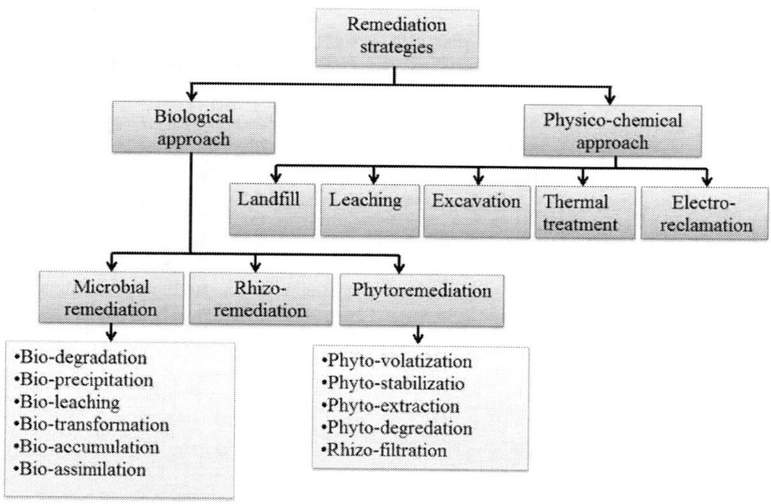

Figure 6.2. Overall classification of remediation strategies for heavy metals.

One form of bioremediation is called phytoextraction. It is emerging as a promising and cost effective approach to remediate metal contaminated sites and has advantages over the alternative chemical and physical remediation technology approaches (Zhuang et al., 2007). The effectiveness of successful phytoremediation at any contaminated site depends on the magnitude of contamination, the availability of the metal contaminant for the plants involved and the capacity of the plant to entrap metals in their biomass. But the higher capacity the plant possesses for metal accumulation in high metal concentrated soil, the slower they grow with low biomass yield (Tangahu et al., 2011). However, the success of phytoremediation can be maximized by utilizing soil microbes inhabiting the rhizosphere are called plant growth promoting rhizobacteria (PGPR). When PGPRs are introduced to a contaminated site, they increased the potential for plants that grow there to sequester heavy metals and recycle nutrients, maintain soil structure, detoxify chemicals and control diseases and pests (Karthik et al., 2016; Rashid et al., 2016). Moreover, PGPR also decrease the toxicity of metals by various mechanisms such biotransformation, biosorption, bioaccumulation, biominerlization, bioleaching and biodegradation. In this multicomponent system, the plants in turn provide the microorganisms with various essential root exudates such as free amino acids, proteins, carbohydrates, alcohols, vitamins and hormones, which are important sources of their nutrition (Bacilio-Jimenez et al., 2003). The rhizosphere has high concentrations of root-exuded nutrients and attracts more bacteria than does bulk soil (Han et al., 2005; Babalola, 2010). These bacteria of the rhizosphere, therefore, facilitate plant growth.

3.1. Organic Pollutants Removal

Previously, before understanding the full potential of PGPR, it was used as a beneficial microbe for plant growth promotion and as a bio control agent against plant pathogens. However, with new profound knowledge it has got much attention for its bioremediation capability of organic pollutants. In contrast to the bioremediation of inorganic compounds, microorganisms can degrade and mineralize organic compounds more effectively in association with plants (Saleh et al., 2004; Glick, 2010). Studies on the discovery of effective pathways for degradation and mineralization of organic compounds have to be emphasized as it plays an important role in the future. So far, a number of toxic organic compounds in soil have been successfully bioremediate using rhizospheric bacteria (Table 6.2). Scientists and researchers have devised numerous effective methods to improve the degradation efficiency and the tolerance of bacteria to contaminants in polluted soils. According to Huang et al. (2004), PGPR stimulates the plants to grow faster and accumulate more root biomass thereby helps in removal of organic pollutant (polycyclic aromatic hydrocarbon and creosote) probably by enhancing plant germination and survival in soils. Moreover, Kuiper et al. (2001 and 2004) reported that a multi component system of grass species with naphthalene degrading *Pseudomonas* species worked together as the bacteria protected the grass seed from the toxic effects of naphthalene and the growing roots assisted the naphthalene degrading bacteria penetrate deep into the soil that would have been impossible in the absence of roots.

Table 6.2. Reported cases of successful multi component bioremediation of organic pollutants by plant and microbial partnership

Organic pollutants	Plants	Microbes	References
PCBs, TCP	Wheat (*Triticum* sp.)	*Herbaspirillum* sp.	Mannisto et al. (2001)
Chlorobenzoic acid	Wild rye (*Elymus dauricus*)	*P. aeruginosa* *P. savastanoi*	Siciliano et al. (1998)
Pesticide 2,4-D	Pea (*Pisum sativa*)	*P. putida*	Germaine et al. (2006)
Toluene	Poplar (*Populus*)	*Burkholderia cepacia*	Barac et al. (2004)
MTB, BTEX, TCE	Poplar (*Populus* cv, *Hazendans* and *Hoogvorst*)	*Pseudomonas* sp.	Germaine et al. (2004) Porteus-Moore et al. (2006)
HCs Naphthalene	Pea (*Pisum sativa*)	*P. putida*	Germaine et al. (2009)
Explosives TNT, RDX, HMX	Polar tissues (*Populus deltoidesnigra* DN34)	*Methylobacterium populi*	Van Aken et al. (2004a)
Total petroleum hydrocarbon (TPH)	*Festuca arundinacea*	*Klebsiella* sp.	Liu et al. (2014)

3.2. Inorganic or Elemental Pollutants Removal

The model plants for phytoremediation should have the attributes for rapid growth, high biomass production and accumulation and tolerance to high concentrations of metals in tissues. Hyper accumulator plants have some exceptional ability to accumulate heavy metal but most of these have a slow growth rate consequently produced lesser biomass during very high metal concentration (Khan et al., 2008; Oves et al., 2013). In order to negate the disadvantages a plants faced, an alternative is to provide them an associated plant growth promoting rhizobacteria, which helps in reducing the stress level a plant encounter during phytoremediation process (Glick, 2003). The unique feature of rhizospheric region is that large microbial populations with high metabolic activity exist when compared with bulk soil (Anderson et al., 1993). Microbial populations helps in heavy metals mobility and availability to the plants through the release of chelating agents, phosphate solubilization and redox change (Abou-Shanab et al., 2003; Zaidi et al., 2009, 2010). As mentioned earlier, PGPR can influence the plant growth by various plant growth promoting mechanisms such as transformation of nutrient elements (phosphate solubilization) and production of phytohormones, ammonia and siderophores (Kloepper et al., 1989; Glick, 1995; Oves et al., 2009). The use of PGPR in combination with plants is expected to provide high efficiency for phytoremediation (Whiting et al., 2001; Abou-Shanab et al., 2003a). The potential and the exact mechanisms of rhizobacteria to enhance phytoremediation of soil heavy metals pollution have been reviewed extensively and also have received major attention. Successful case studies of heavy metals enhanced by PGP bacteria are listed in Table 6.3.

Table 6.3. Successful case studies on rhizobacterial enhanced phytoremediation of heavy metals

PGPR	Associated plants	Heavy metals	References
Pseudomonas fluorescens G10, *Microbacterium* sp. G16	*Brassica napus*	Pb	Sheng et al. (2008)
Enterobactor cloacae CAL2	*B. napus*	As	Nie et al. (2002)
Cupriavidus taiwanensis	*Mimosa pudica*	Pb, Cu, Cd	Chen et al. (2008)
Azotobactor chroococum, Rhizobium leguminosarum	*Zea mays L.*	Pb	Hadi and Bano (2010)
Bacillus thuringiensis GDB-1	*Alnus firma*	As	Babu et al., (2013)
B. thuringiensis GDB-1	*A. firma*	Cu	
B. pumilus E2S2, *Bacillus* sp. E1S2	*Sedum plumbizincicola*	Cd, Zn	Ma et al. (2015)
P. tolaasii ACC23, P. *Fluorescens* ACC9	*B. napus*	Cd	Dell'Amico et al. (2008)
Pseudomonas sp. LK9	*Solanum nigrum*	Cd, Zn, Cu	Chen et al. (2014)

Table 6.3. (Continued)

PGPR	Associated plants	Heavy metals	References
Pseudomonas sp. PsM6, *P. Jessenii* PjM15	*Ricinus communis*	Zn	Rajkumar and Freitas (2008)
PGPR	Associated plants	Heavy metals	References
Rahnella sp. JN6	*B. napus*	Cd, Pb, Zn	He et al. (2013a,b)
P. veronii	*S. alfredii*	Zn	Long et al. (2013)
Burkholderia sp. J62	*Lycopersicum esculentum, Z. mays*	Pb, Cd	Jiang et al. (2008)
Staphylococcus arlettae NBRIEAG-6	*Brassica juncea*	As	Srivastava et al. (2013)
Achromobacter xylosoxidans Ax10	*B. juncea*	Cu	Ma et al. (2009a)
Micrococcus sp. MU1, *Klebsiella* sp. BAM1	*Helianthus annus*	Cd	Prapagdee et al. (2013)
Psychrobacter sp.SRA1 and SRA2, *B. cereus* SRA10	*B. juncea*	Ni	Ma et al. (2009b)
Pseudomonas sp. SRI2, *Psychrobacter* sp. SRS8, *Bacillus* sp. SN9	*B. juncea*	Ni	Ma et al. (2009c)
Pseudomonas sp. SRI2, *Psychrobacter* sp. SRS8, *Bacillus* sp. SN9	*Brassica oxyrrhina*	Ni	Ma et al. (2011)
B. subtilis, B. cereus, B. megaterium, P. aeruginosa	*Orychophragmus violaceus*	Cd	Liang et al. (2014)
Paenibacillus macerans NBRFT5, *B. endophyticus* NBRFT4, *B.pumilus* NBRFT9	*B. juncea*	Cu, Ni, Zn	Tiwari et al. (2012)
R. leguminozarum	*B. juncea*	Zn	Adediran et al. (2015)
Azotobacter chroococcum HKN-5	*B. juncea*	Pb	Wu et al. (2006a)
B. megaterium HKP-1, *B. mucilaginosun* HKK-1	*B. juncea*	Zn, Cu, Cd	
Pantoea agglomerans Jp3-3, *P. thivervalensis* Y1-3- 9	*B. napus*	Cu	Zhang et al. (2011a,b)
Enterobacter sp. JYX7, *Klebsiella* sp. JYX10	*Polygonum pubescens*	Cd, Pb, Zn	Jing et al. (2014)

PGPR	Associated plants	Heavy metals	References
Rahnella sp.	*Amarant hushypochondria cus,* *A. Mangostanus,* *S. nigrum*	Cd	Yuan et al. (2013)
B. subtilis SJ-101	*B. juncea*	Ni	Zaidi et al. (2006)
PGPR	Associated plants	Heavy metals	References
Sinorhizobium sp. Pb002	*B. juncea*	Pb	Di Gregorio et al. (2006)
Xanthomonas sp. RJ3, *Azomonas* sp. RJ4, *Pseudomonas* sp. RJ10, *Bacillus* sp. RJ31	*B. napus*	Cd	Sheng and Xia (2006)
P. fluorescens Avm, *R. leguminosarum bv phaseoli* CPMex46	Alfalfa	Cu	Carrillo-Castaneda et al. (2003)
Bacillus thuringiensis	Mustard	Cd, Cr, Cu, Ni	Oves et al. (2013)
Pseudomonas aeruginsa	Chick pea	Cr	Oves et al. (2011, 2013)

4. Mechanisms of PGPRs-Assisted Enhanced Phytoremediation

The concept of using green plants to extract metals is not new. The original concept has, however, been expanded to include the use of an interdisciplinary approach. Microbes, in close association with plants are employed in the interdisciplinary approach to enhance the removal, immobilization or degradation of certain metals from polluted soils. Such enhanced action results from the presence of PGPRs. PGPRs comprise several different genera including *Pseudomonads*, *Bacillus, Acinetobacter, Microbacterium* and *Flavobacterium*. In fact, PGPRs can enhance the phytoremediation abilities of non-hyper accumulating plants by increasing their growth and biomass (Lippmann et al., 1995). Some plant growth promoting bacteria can also significantly increase the growth of plants in the presence of heavy metals including nickel, lead and zinc (Burd et al., 1998; 2000). Thus, PGPRs allow plants to develop long roots and get better established during the early stages of growth (Glick et al., 1998). The rhizospheric microbial populations enhance the phytoremediation of pollutants either with their direct or indirect involvement and associated mechanisms. The direct mechanism involved in the solubilization, bioavailability and accumulation of pollutants by plants is known as direct process. Indirect mechanism, rhizobacterial population improves the plant growth and prevents phytopathogens from inhibiting plant growth, which facilitates the accumulation of pollutants. The overall mechanism, including the direct and indirect processes is presented in Figure 6.3.

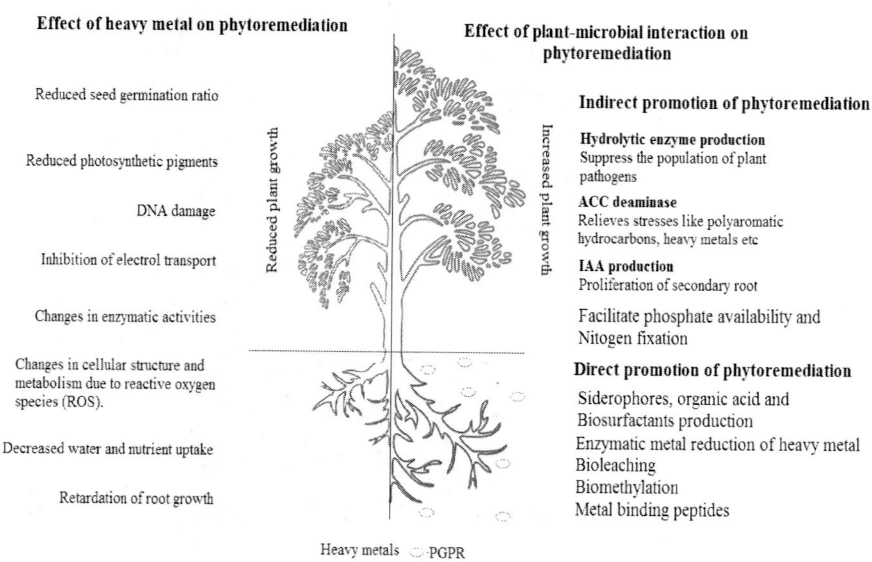

Figure 6.3. Overall mechanism of phytoremediation and rhizoremediation of contaminated soil.

4.1. Direct Mechanism of Phytoremediation

Rhizospheric microbes assist in different reactions and metabolic processes that occur in the soil structure maintenance, biogeochemical cycling of nutrients and detoxification of pollutants (Khan et al., 2010). PGPRs have been reported to produce metal chelating agents termed siderophores and they have a wide range of chemical structure and properties. Siderophores are able to bind with a variety of metals and thus enhance their bioavailability in the rhizosphere through a complexation reaction (Yan-de et al., 2007; Gadd, 2010; Rajkumar et al., 2010). Siderophore-producing bacterial strains significantly increase the bioavailability of Pb and Cr in the rhizosphere and made them available for uptake by maize (Braud et al., 2009). Moreover, PGPR produce biosurfactants and low molecular weight organic acids such as gluconi, oxalic and citric acids, which are playing a key role in the mobilization and solubility of the pollutants (Gadd, 2010; Rajkumar et al., 2012 and Ullah et al., 2015). For instance, Saravanan et al. (2007) demonstrated the potential of *Gluconacetobacter diazotrophicus* to produce a gluconic acid derivative (5-ketogluconic acid), which aids in the solubilization of Zn compounds.

In addition, the reduction/oxidation mechanism of rhizobacteria has potential to mobilize metals, metalloids and attack organicmetallic compounds (Bolan et al., 2014). Such as, rhizobacteria can enzymatically reduce Hg^{2+} to metallic mercury (Gadd, 2010). Similarly, Fe(III) and Mn(IV) solubility is increased by reduction to Fe(II) and Mn(III), respectively. Moreover, PGPR enhance Cu mobilization in contaminated soil and subsequently Cu uptake in plant tissues (Shi et al., 2012). Bioleaching is an important process for mobilization of metals from mineral resources. Gadd (2010) demonstrated that HCN can produce bacteria that in turn can mobilize Ni and Cu as various cyanide complexes and compounds from solid materials such as copper containing ores and electronic scarp. Microorganisms can mobilize metals through biomethylation, which can result in volatilization. Various heavy metals such as Pb, Hg, Se, As, Tn and Sn can be remediated by bacterial methylation. These bacteria can transfer a methyl group to the metals and result in the formation of methylated metal compounds, which may differ in volatility, solubility and toxicity (Gadd, 2004; Bolan et al., 2014). Moreover, some fungi and plants enzymatically synthesize phytochelatins (PCs), which are responsible for metal stress tolerance (Gadd, 2010). Some engineered bacteria can also produce phytochelatein (PC) synthase, which is activated by heavy metal stress. These enzymes bind to heavy metals, especially Cd with high affinity through thiolate complexes, which lead to mobility and availability of heavy metals (Kang et al., 2007).

4.2. Indirect Promotion of Phytoremediation

In indirect mechanism, PGPRs have the potential to confer plant metal tolerance, enhance biomass production and control the plant pathogens from infecting plants, which facilitates the removal of pollutants by plants (Luo et al., 2012). Heavy metals seriously affect the nutrient uptake, which retards plant growth and development (Ouzounidou et al., 2006). Under such nutrient limitation, PGPRs have the ability to enhance the plant growth by providing sufficient nutrients to plants. In which, many of PGPR bacteria are capable to producing phytohormones such as IAA, cytokinins and gibberellins even under stress conditions and therefore affect plant hormone levels and their response to stress (Glick, 2012). In addition, PGPRs can fix atmospheric nitrogen and ammonia even under metal stress. For instant, *Rhizobium leguminosarum* bv *trifolii* able to fix nitrogen in the presence of heavy metals (Nonnoi et al., 2012). Ammonia and nitrogen based elements are essential nutrients for all forms of life and is a basic requisite for nucleic acid, protein and other nitrogenous compounds (Das et al., 2013). The ammonia produced from the bacteria supplies nitrogen source to their host plants and promotes root and shoot elongation, consequently increasing plant biomass (Marques et al., 2010).

Soil is a major storehouse of several forms of phosphates, including inorganic and organic phosphate. In general, 20-80% of phosphates are in the organic form. But plants may poorly/not possess an innate ability to acquire phosphorus directly from soil. Soil-inhabiting microorganisms exhibit their capacity to mineralize the most organic phosphorous compounds. In addition, PGPRs release organic acids that can solubilize phosphorus (Chen et al., 2006; Rodriguez et al., 2006). Phosphate solubilizing bacteria play a key role in plant growth promoting and root proliferation by making the availability of solubilized phosphate minerals in the metal amended soil. Moreover, PGPR solubilize minerals (Fe^{3+}) by siderophores, which can be taken up by plant roots as an iron nutrient (Rajkumar et al., 2010; Ma et al., 2011). Additionally, PGPRs can also produce hydrolytic enzymes, which can act as a antimicrobial agent in the close vicinity of the rhizospheric plant pathogens (Pandey et al., 2013). In nature, the cell wall of fungal pathogens contains fibrillar materials bound together by sugar, proteins, lipid and a variety of polysaccharides like β-1,3-glucan and chitin, which are essential for fungal growth and pathogenicity (Lynd et al., 2002; Macagnan et al., 2008). Lysis of the cell wall by hydrolytic enzyme producing microbes leads to serious effects on pathogens and inhibits their population in rhizospheric environment. Previously, Malleswari and Bagyanarayana (2013) reported that, production of hydrolytic enzymes from plant growth promoting bacteria would be the most prominent traits of antagonism.

PGPRs have also been reported to stabilize the plant physiology and biochemistry. To this end, ethylene was evidenced to play a key role in various physiological changes in plants, when the concentration of this hormone exceeds inhibits the plant growth. Arguably, the PGPR trait that is key in facilitating plant growth is the possession of the enzyme ACC deaminase. This enzyme is responsible for the cleavage of the plant ACC, the biosynthetic precursor for ethylene in plants, into ammonia and ketobutyrate (Honma and Shimomura, 1978). By decreasing the levels of ACC in plants, ACC deaminase producing PGPRs decrease plant ethylene levels, which when present in high concentrations can lead to plant growth inhibition or even death. Collectively, these PGPRs improve plant growth in polluted soil, which in turn facilitates the phytoremediation process.

CONCLUSION AND FUTURE PROSPECTS

The plant-microbial interaction for environmental cleanup is a recent outshoot of bioremediation technology. The major attraction of this technology lies in fact that it is sustainable and inexpensive and therefore offers a feasible alternative to conventional remediation methods. Hence, we have no other option but to develop suitable strategies like those using rhizosphere biology to cope with the challenges of contaminated soil. In addition, the technology helps to improve the soil fertility for future agronomic application, which is the main benefit for sustainable development of soil agroecosystem. There are few reasons that plant bacterial interactions are advantageous when applied to bioremediation. The plant associated rhizobacteria population would be more potential than the native soil bacterial population because plant exudates provide sufficient nutrients. In rhizospheric environment, plants would act as natural pumps that draw contaminated soil pore water towards the plant associated remediating bacteria. The combined plant and bacterial mediated bioremediation is an ideal strategy for cleanup of mixed contaminants. Further research on selection and improving the expression of catabolic genes in the rhizosphere and in the selection of potential plant and bacterial combinations will have to be translated into field studies that can demonstrate the usefulness of this approach. Optimizing the process with potential fast growing plants with better metal uptake ability and high biomass along with suitable microbial diversity in rhizosphere will provide new insights.

REFERENCES

Abdullahi, I., Ajibike, M.A., Man-ugwueje, A.P., Ndububa O.I. (2014). Environmental impact of indiscriminate waste disposal - a case study of Nigerian air force base Kaduna. *International Journal of Engineering and Applied Sciences* 1 (1), 25-33.

Abou-Shanab, R. A., Angle, J. S., Delorme, T. A., Chaney, R. L., van Berkum, P., Moawad, H., Ghanem, K., Ghozlan, H. A. (2003a). Rhizobacterial effects on nickel extraction from soil and uptake by *Alyssum murale*. *New Phytologist* 158(1), 219-224.

Addiscott, T. M., Whitmore, A. P., Powlson, D. S. (1991). Farming, Fertilizers and the Nitrate Problem. C.A.B. International, Wallingford, UK. pp. 170.

Adediran, G. A., Ngwenya, B. T., Mosselmans, J. F. W., Heal, K. V., Harvie, B. A. (2015). Mechanism behind bacteria induced plant grwoth promotion and Zn accumulation in *Brassica juncea*. *Journal of Hazardous Materials* 283, 490–499.

Anderson, T. A., Guthrie, E. A., Walton, B. T. (1993). Bioremediation in the rhizosphere: plant roots and associated microbes clean contaminated soil. *Environmental Science and Technology* 27(13), 2630-2636.

Babalola, O. O. (2010). Beneficial bacteria of agricultural importance. *Biotechnology Letters* 32, 1559–1570.

Babu, A. G., Kim, J. D., Oh, B. T. (2013). Enhancement of heavy metal phytoremediation by *Alnus firma* with endophytic *Bacillus thuringiensis* GDB-1. *Journal of Hazardous Materials* 250, 477–483.

Bacilio-Jimenez, M., Aguilar-Flores, S., Ventura-Zapata, E., Perez-Campos, E., Bouquelet, S., Zenteno, E. (2003). Chemical characterization of root exudates from rice (Oryza sativa) and their effects on the chemotactic response of endophytic bacteria. *Plant and Soil* 249, 271-277.

Barac, T., Taghavi, S., Borremans, B., Provoost, A., Oeyen, L., Colpaert, J. V., Vangronsveld, J., van der Lelie, D. (2004), Engineered endophytic bacteria improve phytoremediation of water-soluble, volatile, organic pollutants. *Nature Biotechnology* 22, 583-588.

Bolan, N., Kunhikrishnan, A., Thangarajan, R., Kumpiene, J., Park, J., Makino, T., Kirkham, M. B., Scheckel, K. (2014). Remediation of heavy metal (loid)s contaminated soils-to mobilize or to immobilize? *Journal of Hazardous Materials* 266, 141–166.

Braud, A., Jezequel, K., Bazot, S., Lebeau, T. (2009). Enhanced phytoextraction of an agricultural Cr, Hg and Pb-contaminated soil by bio-augmentation with siderophore producing bacteria. *Chemosphere* 74, 280–286.

Braud, A., Jezequel, K., Vieille, E., Tritter, A., Lebeau, T. (2006), Changes in extractability of Cr and Pb in a polycontaminated soil after bio-augmentation with microbial producers of biosurfactants, organic acids and siderophores. *Water, Air and Soil Pollution* 6, 261–279.

Carrillo-Castaneda, G., Munoz, J. J., Peralta-Videa, J. R., Gomez, E., Gardea-Torresdey, J. L. (2003). Plant growth-promoting bacteria promote copper and iron translocation from root to shoot in alfalfa seedlings. *Journal of Plant Nutrition* 26, 1801–1814.

Chen, L., Luo, S., Li, X., Wan, Y., Chen, J., Liu, C. (2014). Interaction of Cd hyperaccumulator *Solanum nigrum* L. and functional endophyte *Pseudomonas* sp. Lk9 on soil heavy metals uptake. *Soil Biology and Biochemistry* 68, 300–308.

Chen, W. M., Wu, C. H., James, E. K., Chang, J. S. (2008). Metal biosorption capability of *Cupriavidus taiwanensis* and its effects on heavy metal removal by nodulated *Mimosa pudica*. *Journal of Hazardous Materials* 151, 364–371.

Cheng, S. (2003). Heavy metal pollution in China: origin, pattern and control. *Environmental Science and Pollution Research* 10, 192–198.

Das, A. J., Kumar, M., Kumar, R. (2013). Plant growth promoting rhizobacteria (pgpr): an alternative of chemical fertilizer for sustainable, environment friendly agriculture. *Research Journal of Agriculture and Forestry Sciences* 1, 21-23.

Dell'Amico, E., Cavalca, L., Andreoni, V. (2008). Improvement of *Brassica napus* growth under cadmium stress by cadmium-resistant rhizobacteria. *Soil Biology and Biochemistry* 40, 74–84.

Di Gregorio, S., Barbafieri, M., Lampis, S., Sanangelantoni, A. M., Tassi, E., Vallini, G. (2006). Combined application of Triton X-100 and *Sinorhizobium* sp. Pb002 inoculum for the improvement of lead phytoextraction by *Brassica juncea* in EDTA amended soil. *Chemosphere* 63, 293–299.

Diels, L., van der Lelie, N., Bastiaens, L. (2002). New development in treatment of heavy metal contaminated soils. *Reviews in Environmental Science and Biotechnology* 1, 75–82.

Dixon, B. (1996). Bioremediation is here to stay. *ASM News* 62, 527-528.

Dua, M., Sethunathan, N., Johri, A. K. (2002). Biotechnology and bioremediation: successes and limitations. *Applied Microbiolied and Biotechnology* 59, 143-152.

Early, M. S. B., Cameron, K. C., Fraser, P. M. (1998). The fate of potassium, calcium and magnesium in simulated urine patches on irrigated dairy pasture soil. *New Zealand Journal of Agricultural Research* 41, 117-124.

FAO (2013). Guidelines to control water pollution from agriculture in China: Decoupling water pollution from agricultural production FAO WATER REPORTS 40, Rome.

Gadd, G. M. (2010). Metals: minerals and microbes: geomicrobiology and bioremediation. *Microbiology* 156, 609–643.

Germaine, K., Keogh, E., Garcia-Cabellos, G., Borremans, B., van der Lelie, D., Barac, T., Oeyen, L., Vangronsveld, J., Moore, F. P., Moore, E. R. B., Campbell, C. B., Ryan, D., Dowling, D. N. (2004). Colonisation of poplar trees by *gfp* expressing bacterial endophytes. *FEMS Microbiology and Ecology* 48, 109-118.

Germaine, K. J., Keogh, E., Ryan, D., Dowling, D. N. (2009). Bacterial endophyte-mediated naphthalene phytoprotection and phytoremediation. *FEMS Microbiology Letters* 296, 226-234.

Germaine, K. J., Liu, X., Cabellos, G. G., Hogan, J. P., Ryan, D., Dowling, D. N. (2006). Bacterial endophyte-enhanced phytoremediation of the organochlorine herbicide 2,4-dichlorophenoxyacetic acid. *FEMS Microbiology and Ecology* 57, 302-310.

Glick, B. R. (1995). The enhancement of plant growth by free-living bacteria. *Canadian Journal of Microbiology* 41, 109-117.

Glick, B. R. (2003). Phytoremediation: synergistic use of plants and bacteria to clean up the environment. *Biotechnology Advances* 21(5), 383-393.

Glick, B. R. (2010). Using soil bacteria to facilitate phytoremediation. *Biotechnology Advances* 28, 367–374.

Glick, B. R., Penrose, D. M., Li, J. (1998). A model for the lowering of plant ethylene concentrations by plant growth promoting bacteria. *Journal of Theoretical Biology* 190, 63–68.

Hadi, F., Bano, A. (2010). Effect of diazotrophs (*Rhizobium* and *Azobactor*) on growth of maize (*Zea mays* L.) and accumulation of Lead (Pb) in different plant parts. *Pakistan Journal of Botany* 42, 4363–4370.

Han, J., Sun, L., Dong, X., Cai, Z., Sun, X., Yang, H., Wang, Y., Song, W. (2005). Characterization of a novel plant growth-promoting bacteria strain *Delftia tsuruhatensis* HR4 both as a diazotroph and a potential biocontrol agent against various plant pathogens. *Systematic and Applied Microbiology* 28, 66–76.

He, H., Ye, Z., Yang, D., Yan, J., Xiao, L., Zhong, T., Yuan, M., Cai, X., Fang, Z., Jing, Y. (2013a). Characterization of endophytic *Rahnella* sp. JN6 from *Polygonum pubescens* and its potential in promoting growth and Cd Pb, Zn uptake by *Brassica napus*. *Chemosphere* 90, 1960–1965.

He, J., Ma, C., Ma, Y., Li, H., Kang Liu, T., Polle, A., Peng, C., Luo, Z. (2013b). Cadmium tolerance in six poplar species. *Environmental Science and Pollution Reseach* 20, 163–174.

Honma, M., Shimomura, T. (1978). Metabolism of 1-aminocyclopropane-1-carboxylic acid. *Agricultural and Biological Chemistry* 42, 1825–1831.

Huang, X. D., El-Alawi, Y., Penrose, D. M., Glick, B. R., Greenberg, B. M. (2004). A multi-process phytoremediation system for removal of polyaromatic hydrocarbons from contaminated soils. *Environmental Pollution* 130, 465-476.

Humberto, B., Rattan L. (2010). "Soil and water conservation". Principles of Soil Conservation and Management. Springer. p. 2. ISBN 978-90-481-8529-0.

Jiang, C., Sheng, X., Qian, M., Wang, Q. (2008). Isolation and characterization of a heavy metal-resistant *Burkholderia* sp. from heavy metal-contaminated paddy field soil and its potential in promoting plant growth and heavy metal accumulation in metal-polluted soil. *Chemosphere* 72, 157–164.

Jing, Y. X., Yan, J. L., He, H. D., Yang, D. J., Xiao, L., Zhong, T., Yuan, M., Cai, X. D., Li, S. B. (2014). Characterization of bacteria in the rhizosphere soils of polygonum pubescens and their potential in promoting growth and Cd Pb, Zn uptake by *Brassica napus*. *International Journal of Phytoremediation* 16, 321–333.

Julien, Pierre Y. (2010). Erosion and Sedimentation. Cambridge University Press. p. 1. ISBN 978-0-521-53737-7.

Karamalidis, A. K., Evangelou, A. C., Karabika, E., Koukkou, A. I., Drainas, C., Voudrias, E. A. (2010). Laboratory scale bioremediation of petroleum-contaminated soil by indigenous microorganisms and added *Pseudomonas aeruginosa* strain Spet. *Bioresource Technology* 101, 6545–6552.

Karthik, C., Oves, M., Sathya, K., Ramkumar, V.S., Arulselvi, P.I. (2016). Isolation and characterization of multi-potential Rhizobium strain ND2 and its plant growth promoting activities under Cr (VI) stress. *Archives of Agronomy and Soil Science* doi:10.1080/03650340.2016.1261116.

Khan, A. G. (2005). Role of soil microbes in the rhizospheres of plants growing on trace metal contaminated soils in phytoremediation. *Journal of Trace Elements in Medicine and Biology* 18, 355–364.

Khan, M. S., Zaidi, A., Oves, M., Wani, P. A. (2008). Heavy metal toxicity to legumes Heavy metal pollution. Nova Science, Hauppauge, pp 197-225.

Khan, M. S., Zaidi, A., Wani, P. A., Ahemad, M., Oves, M. (2013). Functional diversity among plant growth-promoting rhizobacteria: Current Status. Microbial Strategies for Crop Improvement, pp 105-132.

Khan, M. S., Zaidi, A., Wani, P. A., Oves, M. (2009). Role of plant growth promoting rhizobacteria in the remediation of metal contaminated soils. *Environmental Chemistry Letters* 7, 1-19.

Khan, S., Hesham, A. E. L., Qiao, M., Rehman, S., He, J. Z. (2010). Effects of Cd and Pb on soil microbial community structure and activities. *Environmental Science and Pollution Reseach* 17, 288–296.

Kloepper, J. W., Lifshitz, R., Zablotowicz, R. M. (1989). Free-living bacterial inocula for enhancing crop productivity. *Trends in Biotechnology* 7(2), 39-44.

Kuiper, I., Bloemberg, G. V., Lugtenberg, B. (2001). Selection of a plant-bacterium pair as a novel tool for rhizostimulation of polycyclic aromatic hydrocarbon-degrading bacteria. *Molecular Plant-Microbe Interactions Journal* 14, 1197-1205.

Kuiper, I., Lagendijk, E. L., Bloemberg, G. V., Lugtenberg, B. J. J. (2004). Rhizoremediation, a beneficial plant-microbe interaction. *Molecular Plant-Microbe Interactions Journal* 17, 6-15.

Liang, X., He, C. Q., Ni, G., Tang, G. E., Chen, X. P., Lei, Y. R. (2014). Growth and Cd accumulation of *Orychophragmus violaceus* as affected by inoculation of Cd tolerant bacterial strains. *Pedosphere* 24, 322–329.

Lippmann, B., Leinhos, V., Bergmann, H. (1995). Influence of auxin producing rhizobacteria on root morphology and nutrient accumulation of crops. 1. Changes in root morphology and nutrient accumulation in maize (*Zea mays* L.) caused by inoculation with indole-3-acetic acid (IAA) producing *Pseudomonas* and *Acinetobacter* strains or IAA applied exogenously. *Angewandte Botanik* 69, 31–36.

Liu, W., Jinyu, H., Qingling, W., Linlin, D., Yongming, L. (2014). Isolation and characterization of plant growth-promoting rhizobacteria and their effects on phytoremediation of petroleum-contaminated saline-alkali soil. *Chemosphere* 117, 303–308.

Long, X. X., Chen, X. M., Wong, J. W. C., Wei, Z. B., Wu, Q. T. (2013). Feasibility of enhanced phytoextraction of Zn contaminated soil with Zn mobilizing and plant growth promoting endophytic bacteria. *Transactions of Nonferrous Metals Society of China* 23, 2389–2396.

Lynd, L. R., Weimer, P. J., Van, Z. W. H., Pretorius, I. S. (2002). Microbial cellulose utilization: fundamentals and biotechnology. *Microbiology and Molecular Biology Reviews* 66, 506–577.

Ma, Y., Oliviera, R. S., Nai, F., Rajkumar, M., Luo, Y., Rocha, I., Freitas, H. (2015). The hyperaccumulator *Sedum plumbizincicola* harbors metal-resistant endophytic bacteria that improve its phytoextraction capacity in multi-metal contaminated soil. *Journal of Environmental Management* 156, 62–69.

Ma, Y., Rajkumar, M., Freitas, H. (2009a). Inoculation of plant growth promoting bacterium *Achromobacter xylosoxidans* strain Ax10 for the improvement of copper phytoextraction by *Brassica juncea*. *Journal of Environmental Management* 90, 831–837.

Ma, Y., Rajkumar, M., Freitas, H. (2009b). Improvement of plant growth and nickel uptake by nickel resistant-plant growth promoting bacteria. *Journal of Hazardous Materials* 166, 1154–1161.

Ma, Y., Rajkumar, M., Freitas, H. (2009c) Isolation and characterization of Ni mobilizing PGPB from serpentine soils and their potential in promoting plant growth and Ni accumulation by *Brassica* spp. *Chemosphere* 75, 719–725.

Macagnan, D., Romeiro, R. D. A., Pomella, A. M. V., DeSouza, J. T. (2008). Production of lytic enzymes and siderophores, and inhibition of germination of basidiospores of *Moniliophthora (ex Crinipellis) perniciosa* by phylloplane actinomycetes. *Biological Control* 47, 309–314.

Mackenzie, F.T., Mackenzie, J.A. (1995). Our Changing Earth: An Introduction to Earth System Science and Global Environmental Change, Prentice Hall, 1995.

Malleswari, D., Bagyanarayana, G. (2013). Plant growth-promoting activities and molecular characterization of rhizobacterial strains isolated from medicinal and aromatic plants. *International Journal of Pharmacy and Biological Sciences* 6, 30-37.

Mannisto, M. K., Tiirola, M. A., Puhakka, J. A. (2001). Degradation of 2,3,4,6-tetraclorophenol at low temperature and low dioxygen concentrations by phylogenetically different groundwater and bioreactor bacteria. *Biodegradation* 12, 291-301.

Marques, A. P. G. C., Pires, C., Moreira, H., Rgyange, A. O. S. S. et al. (2010) Assessment of the plantgrowth promotion abilities of six bacterial isolates using *Zea mays* as indicator plant. *Soil Biology and Biochemistry* 42, 1229–1235.

Nie, L., Shah, S., Burd, G. I., Dixon, D. G., Glick, B. R. (2002). Phytoremediation of arsenate contaminated soil by transgenic canola and the plant growth promoting bacterium *Enterobacter cloacae* CAL2. *Plant Physiology and Biochemistry* 40, 355–361.

Oves, M., Khan, M. S., Ahmad, E., Zaidi, A. (2011). Isolation and 16S R RNA gene sequence characterization of chromium reducing strain OS-4. *Journal of Natural Science, Biology and Medicine* 2, 146.

Oves, M., Khan, M. S., Zaidi, A. (2013). Biosorption of heavy metals by Bacillus thuringiensis strain OSM29 originating from industrial effluent contaminated north Indian soil. *Saudi Joural of Biological Science* 20, 121-129.

Oves, M., Zaidi, A., Khan, M. S. (2010). Role of metal tolerant microbes in legume improvement, Microbes for Legume Improvement, Springer, Vienna, pp 337-352.

Pandey, R., Chavan, P. N., Walokar, N. M., Sharma, N. (2013). *Pseudomonas stutzeri* RP1: A versatile plant growth promoting endorhizospheric bacteria inhabiting sunflower (*Helianthus annus*). *Research Journal of Biotechnology* 8, 48-55.

Porteus-Moore, F., Barac, T., Borremans, B., Oeyen, L., Vangronsveld, J., van der Lelie, D., Campbell, D., Moore, E. R. B. (2006). Endophytic bacterial diversity in poplar trees growing on a BTEX-contaminated site: the characterization of isolates with potential to enhance phytoremediation. *Systematic and Applied Microbiology* 29, 539-556.

Prapagdee, B., Chanprasert, M., Mongkolsuk, S. (2013). Bioaugmentation with cadmium-resistant plant growth-promoting rhizobacteria to assist cadmium phytoextraction by *Helianthus annuus*. *Chemosphere* 92, 659–666.

Rajkumar, M., Ae, N., Prasad, M. N. V., Freitas, H. (2010). Potential of siderophore producing bacteria for improving heavy metal phytoextraction. *Trends in Biotechnology* 28, 142–149.

Rajkumar, M., Freitas, H. (2008). Influence of metal resistant-plant growth promoting bacteria on the growth of *Ricinus communis* in soil contaminated with heavy metals. *Chemosphere* 71, 834–842.

Rajkumar, M., Sandhya, S., Prasad, M. N. V., Freitas, H. (2012). Perspectives of plant associated microbes in heavy metal phytoremediation. *Biotechnology Advances* 30, 1562–1574.

Rashid, M.I., Mujawar, L.H., Shahzad, T., Almeelbi, T., Ismail, I.M.I., Oves, M. (2016). Bacteria and fungi can contribute to nutrients bioavailability and aggregate formation in degraded soils. *Microbiological research* 183, 26-41.

Saleh, S., Huang, X. D., Greenberg, B. M., Glick, B. R. (2004). Phytoremediation of persistent organic contaminants in the environment. In: Singh, A., Ward, O. (eds). Applied bioremediation and phytoremediation, vol. 1, Soil biology. Springer, Berlin, pp 115–134.

Saravanan, V. S., Madhaiyan, M., Thangaraju, M. (2007). Solubilization of zinc compounds by the diazotrophic: plant growth promoting bacterium *Gluconacetobacter diazotrophicus*. *Chemosphere* 66, 1794–1798.

Sheatch, G. W., Carlson, W. T. (1998). Impact of cattle treading on hill land. 1. Soil damage patterns and pasture status. *New Zealand Journal of Agricultural Research* 41, 271-278.

Sheng, X. F., Xia, J. J. (2006). Improvement of rape (*Brassica napus*) plant growth and cadmium uptake by cadmium resistant bacteria. *Chemosphere* 64, 1036–1042.

Sheng, X. F., Xia, J. J., Jiang, C. Y., He, L. Y., Qian, M. (2008). Characterization of heavy metal-resistant endophytic bacteria from rape *Brassica napus* roots and their potential in promoting the growth and lead accumulation of rape. *Environmental Pollution* 156, 1164–1170.

Shi, R., Weber, G., Köster, J., Reza-Hajirezaei, M., Zou, C., Zhang, F., von Wirén, N. (2012). Senescence-induced iron mobilization in source leaves of barley (*Hordeum vulgare*) plants. *New Phytologist* 195, 372–383.

Siciliano, S. D., Goldie, H., Germida, J. J. (1998). Enzymatic activity in root exudates of Dahurian wild rye (*Elymus dauricus*) that degrades 2-chlorobenzoic acid. *Journal of Agricultural and Food Chemistry* 46, 5-7.

Srivastava, S., Verma, P. C., Chaudhary, V., Singh, N., Abhilash, P. C., Kumar, K. V., Sharma, N., Singh, N. (2013). Inoculation of arsenic-resistant *Staphylococcus arlettae* on growth and arsenic uptake in *Brassica juncea* (L.) Czern. Var. R-46. *Journal of Hazardous Materials* 262, 1039–1047.

Stockholm Convention on Persistent Organic Pollutants (POPs). Available online: http://chm.pops.int/Home/tabid/2121/Default.aspx#convtext (accessed November 27, 2016).

Tangahu, B.V., Abdullah, S.R.S., Basri, H., Idris, M., Anuar, N., Mukhlisin M. (2011). A Review on Heavy Metals (As, Pb, and Hg) Uptake by Plants through Phytoremediation. *International Journal of Chemical Engineering,* Article ID 939161, 31.

Tiwari, S., Singh, S. N., Garg, S. K. (2012). Stimulated phytoextraction of metals from fly ash by microbial interventions. *Environmental Technology 33, 2405–2413.*

Toy, T.J., Foster, G.R., Renard, K.G. (2002). Soil Erosion: Processes, Prediction, Measurement, and Control. John Wiley & Sons. p. 1. ISBN 978-0-471-38369-7.

Ullah, A., Mushtaq, H., Ali, H., Munis, M. F. H., Javed, M. T., Chaudhary, H. J. (2015). Diazotrophs-assisted phytoremediation of heavy metals: a novel approach. *Environmental Science and Pollution Research* 22, 2505–2514.

Van Aken, B., Peres, C., Doty, S., Yoon, J., Schnoor, J. (2004a). *Methylobacterium populi* sp. nov., a novel aerobic, pink-pigmented, facultative methylotrophic, methane-utilising bacterium isolated from poplar trees (*Populus deltoids x nigra* DN34). *International Journal of Systematic and Evolutionary Microbiology* 54, 1191-1196.

Warren, S. D., Thurow, T. L., Blackburn, W. H., Garza, N. E. (1986). The influence of livestock trampling under intensive rotation grazing on soil hydrologic characteristics. *Journal of Range Management* 39, 491-495.

Whiting, S. N., deSouza, M. P., Terry, N. (2001). Rhizosphere bacteria mobilize Zn for hyperaccumulation by *Thlaspi caerulescens. Environmental Science and Technology* 35(15), 3144-3150.

Wu, S., Cheung, K., Luo, Y., Wong, M. (2006a). Effects of inoculation of plant growth promoting rhizobacteria on metal uptake by *Brassica juncea. Environmental Pollution* 140, 124–135.

Yan-de, J., Zhen-li, H., Xiao-e, Y. (2007). Role of soil rhizobacteria in phytoremediation of heavy metal contaminated soils. *Journal of Zhejiang University SCIENCE B* 8, 192–207.

Yanez, L., Ortiz, D., Calderon, J., Batres, L., Carrizales, L., Mejia, J. (2002). Overview of human health and chemical mixtures: problems facing developing countries. *Environmental Health Perspectives* 10, 901–909.

Yuan, M., He, H., Xiao, L., Zhong, T., Liu, H., Li, S., Deng, P., Ye, Z., Jing, Y. (2013). Enhancement of Cd phytoextraction by two Amaranthus species with endophytic *Rahnella* sp. JN27. *Chemosphere* 103, 99–104.

Zaidi, A., Khan, M. S., Ahemad, M., Oves, M. (2009). Plant growth promotion by phosphate solubilizing bacteria. *Acta Microbiologica et Immunologica Hungarica* 56, 263-284.

Zaidi, A., Khan, M. S., Ahemad, M., Oves, M., Wani, P. A. (2009). Recent advances in plant growth promotion by phosphate-solubilizing microbes. In: Microbial Strategies for Crop Improvement (ed. Khan et al.) Springer Berlin Heidelberg, pp 23-50.

Zaidi, S., Usmani, S., Singh, B. R., Musarrat, J. (2006). Significance of *Bacillus subtilis* strain SJ 101 as a bioinoculant for concurrent plant growth promotion and nickel accumulation in *Brassica juncea*. *Chemosphere* 64, 991–997.

Zhang, X., Xia, H., Li, Z. A., Zhuang, P., Gao, B. (2011a). Identification of a new potential Cd-hyperaccumulator *Solanum photeinocarpum* by soil seed bank-metal concentration gradient method. *Journal of Hazardous Materials* 189, 414–419.

Zhang, Y., He, L., Chen, Z., Wang, Q., Qian, M., Sheng, X. (2011b). Characterization of ACC deaminase-producing endophytic bacteria isolated from copper-tolerant plants and their potential in promoting the growth and copper accumulation of *Brassica napus*. *Chemosphere* 83, 57–62.

Zhuang, P., Yang, Q., Wang, H., Shu, W. (2007). Phytoextraction of heavy metals by eight plant species in the field. *Water, Air and Soil Pollution* 184, 235–242.

In: Chemical Pollution Control … ISBN: 978-1-53611-034-0
Editor: Naser A. Anjum

Chapter 7

Remediation of Organic and Inorganic Pollutants from Soil: The Role of Plant-Bacteria Partnership

***Iftikhar Ahmad*[1,*], *Muhammad Imran*[2], *Muhammad Baqir Hussain*[3] and *Sabir Hussain*[4]**

[1]Department of Environmental Sciences, COMSATS Institute of Information Technology, Vehari, 61100, Pakistan

[2]Soil Science Division, Nuclear Institute for Agriculture and Biology, Faisalabad 38040, Pakistan

[3]Department of Soil Science, Muhammad Nawaz Shareef University of Agriculture, Multan, 61000, Pakistan

[4]Department of Environmental Sciences and Engineering, Government College University, Faisalabad 38040, Pakistan

Abstract

Environmental pollution resulting from organic and inorganic pollutants is a widespread problem for soil and water resources. Several of these pollutants are highly persistent and accumulated in soil beyond

* Corresponding author: Iftikhar Ahmad, Department of Environmental Sciences, COMSATS Institute of Information Technology, Vehari, Pakistan; E-mail: iftikharahmad@ciitvehari.edu.pk; iffises@yahoo.com.

the permissible limits. They are extremely toxic to soil biology and may be accumulated to higher trophic levels through food chain contamination. Thus, removal of such contaminants from the soil environment is not only mandatory but also a topic of interest nowadays. During the recent years, important role of bioremediation using either plants or bacteria has been elucidated as a primary method to decontaminate such polluted soils. However, a number of studies also indicate that phytoremediation (bioremediation by plants) can be partnered by rhizospheric or endophytic bacteria. The partnership of bacteria with plants not only improves plant health but also accelerates the remediation process of environmental pollutants. Sometimes bacteria perform this function by improving the bioavailability of pollutants to plants (hyper accumulation of metals) and sometimes by degrading the pollutants into plant nutrients or useable forms. The present book chapter presents an overview of the studies, which have been conducted to evaluate the potential of synergistic plant-bacteria partnerships for remediation of different organic and inorganic pollutants including polycyclic aromatic hydrocarbons, polychlorinated biphenyls, pesticides, explosives, heavy metals, metalloids and radionuclides from the soil environment. Furthermore, this book chapter also reports different factors, which may influence the performance of bacteria-plant partnership in remediation of contaminated sites.

Keywords: plant-bacteria-interaction, bio/phytoremediation, metal(loid)s, radionuclides, root colonization

1. Introduction

World soil and water resources are being polluted badly with different organic and inorganic pollutants, which has caused deterioration of both soil and food quality. The pollutants including polycyclic aromatic hydrocarbons (PAHs), pesticides, polychlorinated biphenyles (PCB), explosives, metals, metalloids and radionuclides are well reported in soils (Zhu and Shaw, 2000; Kumar et al., 2012; Testiati et al., 2013; Vane et al., 2014). These pollutants from soil are added into underground and surface water by leaching and runoff, respectively, thus polluting water bodies. Their presence in the soil and water resources is undesirable due to their toxic nature and anomalies they create. For instance, the presence of hydrocarbons and metals in the soils adversely affect seed germination, plant growth (Smith et al., 2006; Ahmad et al., 2012, 2013, 2014), soil microbial population and activities (Guo et al., 2012; Alrumman et al., 2015), the ability of plants and microorganisms to absorb water and nutrients from the soil (Nie et al., 2011). Pesticides have been reported as toxic to humans (Anderson et al., 2014). Many problems in humans can also be related with the exposure of pesticides (Isoda et al., 2005). Similarly, they have damaging effect to other forms of life including arthopodes (Desneux et al., 2007), plants (Siddiqui and Ahmed, 2006), butterflies (Longley et al., 1997), and soil microbes (Ahemad and Khan, 2012) etc. Metal and metalloids are phytotoxic (Kopittke et al., 2010; Ahmad et al., 2012) and dangerous for human health (Duruibe et al., 2007). Likewise, all other pollutants are toxic to all forms of life. Therefore, removal of these toxic substances from the soil is the hot topics in the field of environmental sciences.

Phytoremediation is an environment friendly technology, effective and inexpensive mean of removing the pollutants from soil (Gerhardt et al., 2009). Phytoremediation uses resistant and hyper-accumulator plants to degrade, transform, translocate, extract, sequester, and/or detoxify the organic and inorganic pollutants present in soil (Chigbo and Batty, 2014; Samardjieva et al., 2015), and thus remediate or restore the contaminated sites. This approach has been successfully employed in the bioremediation of various pollutant including PAHs (Sun et al., 2011; Meng et al., 2011), pesticides (Castelo-Grande et al., 2010), PCB (Passatore et al., 2014), metals, metalloids (Favas et al., 2014; Adiloglu et al., 2015) and radionuclides (Gupta et al., 2014). The most important challenge to improve effectiveness of this approach, is the limited access of pollutants to plants because of their reaction with soil constituents. However, phytoremediation process can be assisted by means of various chemicals which are added to contaminated sites to enhance the pollutant solubility and access to plants, thus resulting in high remediation rate (Yang et al., 2011). For instance, EDTA (ethylenediaminetetraacetic acid) and Tween 80 have been used to improve the solubility of metals, pesticides and PAHs (Yang et al., 2011; Garba et al., 2012; Farid et al., 2013; Jelusic and Lestan, 2014). Unfortunately, use of such chemicals is not an eco-friendly as these substances are not naturally synthesized, persistent in soil, toxic to biological systems and pollutes our soils. In addition, their use also increases the cost of treatment. Alternately, a natural approach is required that enhances plant access to these pollutants more effectively without deteriorating the soil health.

Table 7.1. Various bio/phytoremediation techniques are being used to treat contaminants

Techniques	Elucidation
Biostimulation	Nutrient elements like nitrogen and phosphorus incorporated to stimulate indigenous bacteria
Bioaugmentation	Isolation and reinocoulation of enriched bacteria capable to mobilize or immobilize pollutants
Biofilteration	Biofilm producing bacteria used to filter contaminants from polluted water
Bioventing	Insertion of some oxidants (oxygen, sulfate, nitrate, etc.) into the wells to stimulate resident bacteria usually sulfate oxidizing/reducing bacteria
Bioreactors	Use of bacteria in a special containers having contaminants in a slurry phase
Phytoextraction	Acquisition of higher amount of contaminants in harvestable part of plant
Phytostabilization	Immobilization of contaminants in soil by the complexation of compounds released by the plants
Phytodegradation	The conversion of contaminants into a simpler compounds due to the action of enzymes released by the host plant/microbe
Phytovolatilization	The accumulated pollutants converted into volatile compounds within plants and discharged to the atmosphere

The bacteria inhabiting rhizosphere and endosphere of plants are capable to improve growth and root development of plants growing in stressed conditions resulting in a greater efficiency of plants combated with resident pollutants. This association is studied in a broader spectrum under the terms: bioremediation (use of microbial technology to remediate contaminated soil for example biostimulation, bioaugmentation, biofilteration, bioventing, bioreactors) and phytoremediation (use of green crops and trees to decontaminate pollutants for example phytoextraction, phytostabilization, phytodegradation, phytovolatilization) (Table 7.1 above). Microbes and plants have been employed simultaneously to remediate contaminated sites (Zhang et al., 2011). Microbial-assisted phytoremediation is emerging as one of the most effective means by which plants and their associated rhizosphere/endophytic microbial community remediated the soils. Plants provide residency and nutrients to their associated rhizosphere and endophytic bacteria. In return, the bacteria support plant growth by the degradation and detoxification of pollutants. Bacteria have been reported to enhance the solubility of pesticides, metals, PAHs etc. in soil (Kuyukina et al., 2005; Kim and Vipulanandan, 2006; Manickam et al., 2012; Ahmad et al., 2016). Moreover, they improve plant growth and health due to their innate plant growth-promoting mechanisms. This book chapter provides a critical review of recent results of bacterial association of plants with special reference to remediation of organic and inorganic pollutants from soil. Moreover, how this association modulates in response to bacterial inoculation method and bacterial colonization to host plant has also been elucidated.

2. Plant-Bacteria Partnership and Pollutants' Remediation

The bacteria and plants have been extensively used for devising the strategies for cleanup of polluted environments. Despite their extensive use, both the bacteria and the plants have always been found to suffer from some limitations regarding their individual abilities in removal and/or breakdown of several pollutant compounds. One of the possible ways to deal with these problems and limitations is the synergistic use of plants and bacteria, which may provide a useful basis for enhancing remediation of contaminated environments by these organisms (Chaudhry et al., 2005). Sometimes, this partnership of plants and bacteria for removal of environmental pollutants is facilitated by plant growth-promoting bacteria, which have the ability to promote the growth and activity of plants through a number of different mechanisms (Glick, 2010). Similarly, some of the bacteria, often known as endophytes, may also facilitate this partnership by helping the host plants to adapt to the polluted environments either by enhancing the phytoremediation through mobilization/immobilization or degradation of the pollutants in the soil or through promotion of plant growth and decreasing the phytotoxicity (Germaine et al., 2009; Chen et al., 2010; Weyens et al., 2010; Li et al., 2012). They do it by several ways including production of phytohormones (Li et al., 2012), indirectly improving the plant growth through ACC deaminase production to adjust the ethylene levels in plants (Zhang et al., 2011; Li et al., 2012), nitrogen fixation for plants (Li et al., 2012) and increasing tolerance of plants to stresses (Ryan et al., 2008; Li et al., 2012). Hence, there is an increasing trend for exploiting the plants and their associated bacteria in a partnership for enhancing the remediation of the polluted environmental components. This partnership is also of considerable importance due to the mutuality of plants and bacteria as a result of co-existence and association in the rhizosphere zone (Badri et al., 2009). A detailed overview of the plant-bacteria partnership for removal/remediation of different organic and inorganic environmental pollutants has been presented below.

2.1. Organic Pollutants

During the past few decades, extensive use of the synthetic organic compounds has resulted into pervasive and long-term environmental problems. In particular, the widespread use of pesticides in agriculture and public health sectors around the world has resulted in numerous instances of contamination of foodstuffs, farm soils, and surface and ground water. Even several years after their application, residues of some of the pesticides are still reported to exist in the environment (Chaudhry et al., 2001, 2005). Similar problems have been found to be associated with other organic compounds including polychlorinated biphenyls (PCBs), polycyclic aromatic hydrocarbons (PAHs), BTEX (benzene, toluene, ethylbenzene, and xylenes) and explosives (Chaudhry et al., 2005). However, very high costs often associated with remediation of these organic compounds are considered as a major hurdle in devising the strategies for their cleanup solutions.

During the recent years several bacterial strains belonging to different genera have been reported to degrade a variety of toxic organic compounds including pesticides, synthetic dyes, PCBs, PAHs etc. under laboratory conditions (Chaudhry et al., 2005; Glick, 2010; Hussain et al., 2011; Hussain et al., 2013; Imran et al., 2015). Despite that several bacterial strains have been tested for their potential to degrade a variety of organic compounds under laboratory conditions, such transformations have often been found difficult to be achieved in the environment (Glick, 2010). Similar is the case with phytoremediation in which the growth of the phytoremediating plants on the contaminated sites is significantly reduced resulting into less biomass production required for efficient removal of contaminants within a reasonable time frame (Glick, 2010). In order to overcome some of these problems, recently a number of researchers have focused on using the biodegrading bacteria along with the phytoremediating plants for successful removal of organic contaminants from the soil (Glick, 2010). Such plant-bacteria partnerships have been reported for a number of organic pollutants including polycyclic and other aromatic hydrocarbons (Sheng et al., 2008a; Sheng et al.,

2009; Germaine et al., 2009; Glick, 2010), halogenated compounds (Leigh et al., 2006; Germaine et al., 2006; Uhlik et al., 2009) and petroleum products (Lin et al., 2008; Al-Awadhi et al., 2009; Glick, 2010). For example, Weyens et al. (2010) indicated that some engineered endophytes had the potential of increasing the root biomass and decreasing the activity of antioxidative defense enzymes by decreasing the trichloroethylene (TCE) phytotoxicity. Some of the plant and the bacterial strains working in partnership with each for bioremediation of different organic compounds have been presented in Table 7.2 (above).

One of the most important regions for plant-microbe interactions is the rhizosphere which is an area densely populated with soil borne microorganisms interacting with plant roots for enhanced biotransformation of several organic compounds (Liste and Alexander, 2000; Ryan et al., 2001; Dams et al., 2007). The plants release the enzymes and exudates from the roots resulting into the stimulation of microbial communities and ultimately the degradation of organic compounds (Dams et al., 2007). Such rhizoremediation through plants either occurs naturally or it can also be triggered by inoculating specific pollutant-degrading or plant-growth-promoting microorganisms (Gerhardt et al., 2009; Megharaj et al., 2011). This has been reported to be successfully done for removal of several compounds including PAHs, PCBs and pesticides from the polluted environments (Chekol et al., 2004; Kuiper et al., 2004; Germaine et al., 2006).

A detailed description of plant-bacteria partnership for bioremediation of PAHs, PCBs, pesticides and explosives has been presented below.

2.1.1. Polycyclic Aromatic Hydrocarbons

Polycyclic aromatic hydrocarbons (PAHs) are a group of hydrophobic compounds, which contain two or more aromatic benzene rings in their molecule and are considered among the most recalcitrant petroleum compounds. In order to cope with the organic compounds belonging to this group, several bacteria have been isolated for biodegradation of PAHs (Aitken et al., 1998; Bouchez et al., 1999; Chaudhry et al., 2005). However, it has been observed that several bacterial strains are able to metabolize the PAHs having only low and intermediate molecular weight but require some readily available carbon source for biodegradation of high molecular weight PAHs (Aitken et al., 1998; Bouchez et al., 1999; Chaudhry et al., 2005). This reason serves as a platform for use of plants in order to accelerate the microbial biodegradation of PAHs because of release of natural readily available carbon source in root exudates (Chaudhry et al., 2005). Hence, relatively faster dissipation of PAHs has been observed in the vegetated soils as compared to non-vegetated soils (Daane et al., 2001; Chaudhry et al., 2005). Similarly, Joner and Leyval (2003) observed relatively faster degradation of phenanthrene in the rhizosphere of clover and ryegrass in two industrially polluted soils. Ma et al. (2010) conducted a meta-analysis and suggested that the activity of PAH degrading populations in the soil is enhanced in response to the root exudation activities.

On the other hand, bacteria may also assist the plants by degrading the PAHs in the plant rhizosphere and hence reducing their toxic effects. For example, Kuiper et al. (2002) isolated a PAHs degrading bacterial strain *Pseudomonas putida* PCL1444. This strain had the potential to utilize the root exudates and lower the toxicity of the PAHs for the plants through their degradation. Similarly, Daane et al. (2001) isolated the bacteria having the potential for PAH degradation from the rhizosphere soil of a salt marsh plants and found that the use of bacteria associated with plants had the better potential for removal of PAHs from the contaminated sediments. In a pot study to examine the phytoremediation of PAHs by alfalfa inoculated with *Rhizobium meliloti,* Teng et al. (2011) found that alfalfa inoculated with *R. meliloti* had the more potential to significantly lower the PAH concentrations in the soil as compared to un-inoculated control. Recently, Xun et al. (2015) found that application of plant growth promoting (PGP) rhizobacteria along with arbuscular mycorrhizal fungi not only improved the soil quality by enhancing urease, sucrase, and dehydrogenase activities but also by degrading the petroleum hydrocarbon in the soil and hence supporting the plants. Some of the researchers have also shown a significant role of endophytes in improving the potential of their host plants for remediation of PAHs. For example, Siciliano et al. (2001) found that some plants when grown in the petroleum-contaminated soils had the potential for selectively increasing the level of the bacterial strains involved in biodegradation of petroleum compounds. Moreover, such increase in a bacterial population was found to depend on the type of pollutant, the plant species and the substrate on which the plant grows. Similarly, the endophytes isolated from the willow and poplar trees were found to tolerate the presence of different PAHs including naphthalene, phenanthrene and pyrene (Khan and Doty, 2011).

Table 7.2. Plant-bacteria partnership for bioremediation of different organic pollutants

Category	Target pollutant(s)	Plant(s) used	Bacterial strain(s)	Reference(s)
Pesticides	Lindane	*Zea mays*	*Streptomyces* sp.	Álvarez et al. (2015)
	Phenanthrene	*Salix sp.*	*Pseudomonas putida*	Khan et al. (2014)
	Crude oil	*Tectona grandis, Gmelina arborea, Azadirachta indica Michelia champaca*	*Pseudomonas aeruginosa*	Yenn et al. (2014)
	Hexachlorocyclohexane (HCH)	*Cytisus striatus*	*Rhodococcus erythropolis*	Becerra-Castro et al. (2013)
	Chlorpyrifos	*Lolium perenne*	*Bacillus pumilus*	Ahmad et al. (2012a)
	2,4-D	*Populus trichocarpa*	*Pseudomonas putida*	Germaine et al. (2009)
	Pentachlorophenol		*Sphingobium chlorophenolicum*	Dams et al. (2007)
	2,4-D	*Pisum sativum*	**Pseudomonas putida*	Germaine et al. (2006)
	Pentachlorophenol	*Lolium perenne*	Indigenous degraders	He et al. (2005)
	2,4-D	*Trifolium pratense*	Indigenous degraders	Shaw and Burns (2004)
	2-chlorobiphenyl	*Arabidopsis* sp.	*Pseudomonas putida*	Narasimhan et al. (2003)
	Dibenzofuran	*Macroptilium atropurpureum*	*Rhizobium tropici*	Saiki et al. (2003)
	2,4-D	*Hordeum vulgare*	*Burkholderia cepacia*	Jacobsen (1997)
	2,4-D	*Triticum aestivum*	*Pseudomonas putida*	Kingsley et al. (1994)
PCBs	3,3′,4,4′-TCB	*Astragalus sinicus*	*Mesorhizobium* sp.	Teng et al. (2015)
	PCB	*Zea mays*	Plant-growth-promoting bacterium	Luo et al. (2015)
	Weathered PCBs	*Medicago sativa*	*Rhizobium meliloti*	Teng et al. (2010)

Table 7.2. (Continued)

Category	Target pollutant(s)	Plant(s) used	Bacterial strain(s)	Reference(s)
	Biphenyl	*Morus* sp.	*Burkholderia* sp.	Leigh et al. (2002)
	2,3,4,6-tetraclorophenol	*Triticum* sp.	**Herbaspirillum* sp.	Mannisto et al. (2001)
	4-chlorobiphenyl	*Beta vulgaris*	*Pseudomonas fluorescens*	Brazil et al. (1995)
PAHs	Dibenzothiophene, Phenanthrene, naphthalene, and Fluorene	*Populus deltoides* × *Populus nigra (hybrid popular)*	*Burkholderia fungorum*	Andreolli et al. (2013)
	Different PAH compounds	*Medicago sativa*	*Rhizobium meliloti*	Teng et al. (2011)
	Phenanthrene and Pyrene	*Lolium multiflorum*	*Acinetobacter* sp.	Yu et al. (2011)
	Fluoranthene, Pyrene and Benzo[*a*] pyrene	*Medicago sativa*	*Flavobacterium* sp. *Bacillus* sp.	Zhang et al. (2010)
	Pyrene	*Allium macrostemon*	**Enterobacter* sp.	Sheng et al. (2008a)
	Different PAH compounds	*Medicago sativa*	**Pseudomonas* spp	Phillips et al. (2008)
	Naphthalene	Tall fescue grass	*Azospirillum brasilense*	Huang et al. (2004)
	Phenanthracene	*Hordeum sativum*	*Pseudomonas* sp.	Ankohina et al. (2004)
	Chrysene	*Trifolium repens*	*Rhizobium leguminosarum*	Johnson et al. (2004)
Explosives	TNT, RDX, HMX	*Populus deltoidesnigra*	*Methylobacterium populi*	Van Aken et al. (2004)
	Toluene	*Lupinus luteus*	*Burkholderia cepacia*	Barac et al. (2004)
	TNT	*Bromus erectus*	*Pseudomonas* sp.	Siciliano and Greer (2000)

*Endophytes

2.1.2. Polychlorinated Biphenyles

The plants have the ability to uptake a number of pollutants from the rhizospheric soil. Normally it has been observed that the mass and length of the plant roots determine their ability to take up these pollutant compounds. Hence improving the mass and length of the roots might serve as an important tool to improve the phytoremediating potential of the plants. In this regard, some of the bacterial strains have been found to assist the plants in phytoremediation of organic compounds including polychlorinated biphenyles (PCBs) through improvement in root length and structure (Chaudhry et al., 2005). For example, a naturally occurring soil bacterium *Agrobacterium tumefaciens* has been reported not only to induce root proliferation but also to improve the ability of the plant to absorb greater amounts of PCBs and other pollutants from soil and groundwater along with an increase in uptake of nutrients (Stomp et al., 1994; Chaudhry et al., 2005). This is possibly due to an increased rate of cell division and greater root mass, which leads to a higher uptake of soil chemicals. Such root proliferations may also expand the effectiveness of associated rhizospheric microorganisms for degrading organic pollutants in soils. Plant breeding techniques may thus also help 'rhizosphere breeding' by developing root systems with beneficial properties for degradation of organic pollutants through improvements in root length, root architecture and exudation processes (Chaudhry et al., 2005). In the similar way, sometimes plants also enhance the breakdown of the pollutants by promoting the growth and activity of the functional microbial communities through their root exudates. For example, high levels of root exudates released by red mulberry (*Morus rubra*) have been found to support the growth and activity of the bacterial populations involved in the biodegradation of PCBs and PAHs (Fletcher and Hegde, 1995; Hegde and Fletcher, 1996; Chaudhry et al., 2005). Similarly, the growth of a PCB-degrading bacterium *Burkholderia* sp. LB400 has also been found to be supported by the accumulation of the flavones including morusin, morusinol and kuwanon C in the fine roots of mulberry (*Morus* sp.) (Leigh et al., 2002; Chaudhry et al., 2005). Like flavons, some other compounds including coumarin, myricetin, naringin, maclurin, phloridzin and catechin, released through root exudation by some plants have also been found to assist the PCB degrading bacteria in the rhizosphere by enhancing their growth (Donnelly et al., 1994). Based on the studies conducted on synthetic artificially-contaminated soils, *Quillaya saponins* has also been found to enhance the biological degradation of PCBs in soils (Fava and Di, 1998; Chaudhry et al., 2005).

2.1.3. Pesticides

Both plants (Chaudhry et al., 2001, 2005; Atterby et al., 2002) and microorganisms (Chaudhry et al., 2005; Hussain et al., 2011) have metabolic pathways that can degrade most of pesticide compounds that are released into the environment. The long-term persistence of some pesticides, however, indicates certain limitations in the ability of microbes and plants to degrade them. In particular, organochlorine compounds are not readily metabolized by plants or marine phytoplankton (Chaudhry et al., 2001, 2005). The hydrophobic nature of many persistent pesticides prevents their uptake and translocation by plants. This is where microbial transformations in soil may play an important role in making them more available for uptake and degradation by plants. Thus, different chlorinated pesticides can be more degraded in a vegetated soil as compared to a non-vegetated soil, with a similar increase observed in the latter when amended with root-derived materials (Shann, 1995). Significant decreases in the concentration of weathered p,p'-DDE have been reported in the rhizospheres of lucerne, ryegrass, zucchini, pumpkin and spinach either in the near-root zone or in the rhizosphere as compared to that in bulk soil (Hodgson et al., 1995; White, 2001; Chaudhry et al., 2005). The higher microbial numbers and a faster rate of degradation of lindane, atrazine, metolachlor, trifluralin, butachlor and aldicarb has been observed in the rhizospheric soil of some herbicide tolerant plants compared to a bare soil (Singh, 2003; Yu et al., 2003; Sun et al., 2004; Chaudhry et al., 2005) In another study, the combined use of a chemical safener (benoxacor), and inoculation of maize rhizosphere with a herbicide-detoxifying strain of *Pseudomonas fluorescens* has been suggested for enhancing phytoremediation of chloroacetamide herbicides (Hoaglands et al., 1997). Similarly, the ability of a crested wheatgrass to tolerate PCP was profoundly enhanced after inoculation with a microbial consortium isolated from a contaminated site (Miller et al., 1998). A faster removal of dibenzofuran by the rhizosphere of *Macroptilium atropurpureum* has been reported after inoculation with a recombinant *Rhizobium tropici* strain (PBK3-IS) that constitutively expressed the oxygenase component of carbazole 1,9a-dioxygenase from *Sphingomonas* sp. (Saiki et al., 2003). Dams et al. (2007) found that PCP was more rapidly degraded when a plant occupied soil was inoculated with a bacterial strain *Sphingobium chlorophenolicum*. In this study, *S. chlorophenolicum* was found to decrease the phytotoxicity of PCP through its degradation. Shaw and Burns (2004) indicated that an increase in the size of 2,4-D degrading bacterial population resulted into an enhanced mineralization of 2,4-D in the rhizosphere soil.

Some of the researchers have also highlighted the role of endophytes in assisting the phytoremediation of different pesticides. For example, Germaine et al. (2006) found that colonization of the roots of a phytoremediating plant with some endophytes resulted into a decrease in phytotoxicity and an increase in phytoremediation of 2,4-D. In another study, three endophytic *Pseudomonas* strains originating from endophytes of hybrid cottonwood were found to have the ability for biodegradation of 2,4-D (Germaine et al., 2004) Moreover, these endophytes were also found to harbor the characteristics of plant growth promotion. Germaine et al. (2006) found that 2,4-D removing ability and biomass of the pea plants was increased after successful colonization of its roots with a poplar tree endophyte, *Pseudomonas putida*, having the potential for degradation of this compound. Hence, based on the above findings, it might be suggested that plant–endophyte partnership could serve as an important tool not only to enhance the phytoremediation of pesticide-contaminated sites but also to reduce the concentrations of toxic herbicide residues in crop plants.

2.1.4. Explosives

Some of the researchers have also exploited the potential of some endophytes for biodegradation of different explosives. For example, an endophyte *Methylobacterium populum* sp. Nov., strain BJ001 isolated from a poplar tree has the potential for biodegradation of TNT and RDX (Van Aken et al., 2004; Khan and Doty, 2011). Based on high potential to mineralize RDX, this strain was hypothesized to support the poplar plant in RDX contaminated soil. In another study, Siciliano and Greer (2000) inoculated the *Anthoxanthum odoratum*, *Lolium perenne* and *Bromus erectus* grasses with a TNT transforming bacterial strain *Pseudomonas* sp. strain I4 and they found that TNT levels in the soil were reduced due to this plant bacterial application. Inoculation of yellow lupine seeds with an engineered toluene-degrading plasmid (pTOM) harboring bacterial strain *Burkholderia cepacia* strain L.S.2.4 was found not only to protect the plants against the phytotoxic effects of toluene but also to decrease the emissions from the transpiration stream of the host plant (Barac et al., 2004: Megharaj et al., 2011).

One of the possible ways to remediate the phytotoxic nitroaromatics is to improve the tolerance and degrading ability of the plants by expressing the bacterial genes involved in degradation of the nitroaromatics plants (Doty, 2008). For example, French et al. (1999) found that the tolerance of transgenic tobacco (*Nicotiana tabacum*) to trinitroglycerin and TNT could be enhanced by introducing the bacterial pentaerythritol tetranitrate (PETN) reductase into these plants. Similarly, Hannink et al. (2001) found that phytoextraction, tolerance and metabolism of TNT in transgenic tobacco plants were increased when bacterial nitroreductase was expressed in this plant. In the same way, expression of bacterial flavodoxin reductase (*xplB*) and a fused flavodoxin cytochrome P450 (*xplA*) in a transgenic plant resulted into an enhanced removal of an explosive RDX in a military training area (Rylott et al., 2006).

Sometimes, the use of transgenic plants has also been found to be beneficial for the rhizospheric microbial communities. For example, Travis et al. (2007) found that the concentration of TNT was decreased in the root surroundings of an NR transgenic tobacco resulting into more survival of the microbial community as compared to the wild-type plants which resulted into a reduction in microbial density and diversity at higher TNT concentrations. Makris et al. (2009) indicated that rhizoremediation and genetic modification of plant/tree species might serve as an important tool for an efficient bioremediation of TNT. Moreover, they highlighted that this potential of rhizoremediation and phytoremediation of the explosives could further be improved by using some chemical amendments.

2.2. Inorganic Pollutants

The number of synthetic chemicals intentionally or unintentionally is being released into soil or water bodies every day. Some of them are essential having a specific purpose while most of them are challenging owing to their long persistence and resistance to degradation. This group mainly comprised of heavy metals, metalloids and radionuclides. These chemicals are associated with these processes (mining, fossil fuel burning, municipal solid waste, industrial effluent, and phosphate fertilizers application) through which they find their way into the food chain. In addition to this, these chemicals are known to suppress plant growth, although some plant species acclimatize and modify growth pattern to tolerate adverse conditions. The exploitation of tolerant plant species to cleanup these pollutants either through phytoextraction or phytostabilization is one way of managing metal contaminated soils, however, plant efficiency to accumulate/stabilize inorganic pollutants could be increased by the association of indigenous or augmented PGP bacteria.

The association of PGP bacteria to plants facilitates degradation, mobilization, immobilization, volatilization, and accumulation of organic and inorganic pollutants is called bacterial assisted phytoremediation. This method of inorganic pollutants remediation has gained considerable attention because metals, metalloids and radionuclides could not be degraded, however, these may be converted into one form to other with the help of metal tolerant bacteria and plants. Plant-bacteria interaction is useful as *in situ* remediation technique because bacteria either tolerate high concentrations of metals leading to their bioaccumulation or improved plant growth biomass in contaminated soil or facilitate metal uptake (Gadd, 2010; Glick et al., 2010, 2014; Ahmad et al., 2014, 2016).

Table 7.3. Plant-bacteria partnership for bioremediation of different inorganic pollutants

Target pollutant(s)	Plant(s) used	Bacterial strain(s)	Reference(s)
Heavy metals			
Cd	*Zea mays*	*Bacillus* sp. *Enterobacter* sp.	Ahmad et al. (2016)
Ni, Zn, Fe	*Brassica juncea* *Ricinus communis*	*Psychrobacter* sp. **Pseudomonas* sp.	Ma et al. (2015)
Cu	*Elsholtzia splendens*	*Pseudomonas putida*	Xu et al. (2015)
Cd, Zn, Pb	*Salix viminalis*	**Rahnella* sp.	Janssen et al. (2015)
Cd	*Triticum aestivum* *Zea mays*	*Klebsiella* sp. *Bacillus* sp.	Ahmad et al. (2014)
Cd	*Zea mays*	*Ralstonia eutropha* *Chryseobacterium humi*	Moreira et al. (2014)
Cd, Zn, Pb	*Brassica napus*	*Klebsiella* sp. *Enterobacter* sp.	Jing et al. (2014)
Cd	*Sedum alfredii*	**Sphingomonas* sp.	Chen et al. (2014a)
Cd, Zn	*Salix schwerinii* *Salix viminalis*	**Bacillus pumilus* **Pantoea agglomerans* **Paenibacillus lactis* *Caulobacter* sp. *Leifsonia* sp.	Weyens et al. (2013)
Cd, Zn	*Salix caprea*	*Burkholderia* sp. **Actinobacteria*	Kuffner et al. (2010)
Ni, Pb, Cr	*Cicer arietinum* *Zea mays*	*Pseudomonas* sp. *Ralstonia* sp.	Tank and Saraf (2009) Braud et al. (2009)
Metalloid(s)			
As	*Pteris vittata*	*Delftia* sp. *Bacillus* sp. *Variovorax* sp. *Pseudomonas* sp. *Pseudoxanthomonas* sp.	Lampis et al. (2015)

Table 7.3. (Continued)

Target pollutant(s)	Plant(s) used	Bacterial strain(s)	Reference(s)
Metalloid(s)			
Se	*Stanleya pinnata* *Astragalus bisulcatus*	*Bacillus* sp. *Pantoea* sp. *Pseudomonas* sp. *Staphylococcus* sp. *Paenibacillus* sp.	Sura-de Jong et al. (2015)
As	*Zea mays*	*Pseudomonas koressnsis* *Rhodococcus aetherivorans*	Shagol et al. (2014)
As	---	*Pseudomonas putida*	Chen et al. (2014b)
Sb	---	*Bacillales*	Abin and Hollibaugh (2014)
Se	*Astragalus* sp. *Stanleya* sp.	Endophytic bacteria	Lindblom et al. (2013)
As	*Heliantus annuus*	*Alcaligenes* sp.	Cavalca et al. (2013)
Sb	*Pteris fauriei* *Humata tyermanii* *Pteris ensiformis*	---	Feng et al. (2013)
As	*Pteris vittata*	Arsenic reducing bacteria	Yang et al. (2012)
As	*Populus deltoids*	*Agrobacterium radiobacter*	Wang et al. (2011)
AS, Sb	*Pteris cretica*	---	Feng et al. (2011)
Radionuclide(s)			
U, Sr	*Agrostis capillaris* *Deschampsia flexuosa* *Festuca rubra* *Helianthus annuus*	Microbial consortia	Langella et al. (2014)
Cs, Sr, Pu, Am	Agricultural plans	---	Kozhakhanov et al. (2014)
U, Cs, Sr	*Helianthus annuus*	---	Soudek et al. (2006) Prasad (2007)
U, Ra	*Helianthus annuus*	---	Rodriguez et al. (2006)
U	---	*Halomonas* sp. *Clostridium* sp.	Francis (2006) Francis et al. (2004)

*Endophytes

The establishment of plants in a contaminated soil is a major constraint in successful phytoremediation, which could be overcome by the inoculation of PGP bacteria. The other important factor is selection of suitable crop specie, which has central role in metal decontamination. Keeping in view above constraints researchers have defined basic criteria for hyperaccumulator, it 1) should have fast and short growth cycle, 2) tolerate high concentration of exposed pollutant and 3) accumulate greater amount of exposed pollutant in a harvestable part. The capacity to accumulate metals may change with the exposed metal for example if plants accumulate 100 mg kg^{-1} Cd, ≥1000 mg kg^{-1} Pb/Ni/Cu and ≥10,000 mg kg^{-1} Zn/Mn on dry weight basis are considered as hyperaccumulators (Vangronsveld et al., 2009; Sessitsch et al., 2013). Till now, about 45 plant families (*Fabaceae*, *Brassicaceae*, *Euphorbiaceae*, *Lamiaceae*) and 500 plant species are known to accumulate higher amount of heavy metals in their biomass (Jaffré et al., 2013). Here, we are discussed some case studies about metals, metalloids and radionuclides remediation through plant-bacteria association (Table 7.3, above).

2.2.1. Heavy Metals

Several plant species are being deployed to remediate metal contaminated soil but only some of them are good hyperaccumulator, their efficiency to accumulate metals can be increased with the inoculation of PGP bacteria. The establishment of plant in metal contaminated soil is a first prerequisite for successful phytoremediation. Both plants and bacteria have ability to alleviate metal toxicity, which can be further stimulated by their beneficial association. These bacteria have ability to increase plant growth through production of various characters such as siderophores, exopolysacchrides, biosurfactants, phosphorus solubilization, auxin and organic acid production (Glick et al., 2010; Chen et al., 2014a; Ahmad et al., 2014, 2016). Several bacterial species have been reported which either established plant in contaminated soil and promote metal accumulation or release some organic compounds to bind metals and promote metal stabilization in soil. In a study, Sheng et al. (2008b) reported higher accumulation of Cd 230-292 mg kg^{-1} in *Solanum nigrum* inoculated with PGP bacterial strain LK9 identified as *Pseudomonas* sp. Similarly, *Mycobacterium* sp. ACC14 and *Pseudomonas* sp. ACC9, ACC23 bacterial strains enhanced Cd uptake in *Brassica napus* (Dell'Amico et al., 2008). Inoculation of *Ralstonia eutropha* and *Chryseobacterium humi* to maize plant increased its biomass up to 63% and decreased Cd uptake up to 81% in shoot in Cd contaminated soil, while, bacterial profile in maize rhizosphere altered continuously depending upon Cd levels in soil. They concluded that these bacterial species can be used for short term phytostabilization of Cd along with biomass production for energy purposes (Moreira et al., 2014). In another study, Ma et al. (2015) reported accumulation and translocation of Ni-Zn-Fe in *Brassica juncea* and *Ricinus communis* inoculated by the rhizospheric and endophytic bacteria. Regardless of plant species, both of the bacterial strains increased plant biomass in multimetal serpentine soil, however, exhibited <1 translocation factor in both plant species for all the metals indicating that plant-bacteria combinations are suitable for phytostabilization of metals in serpentine soil. Similarly, Ahmad et al. (2014) and Kuffner et al. (2010) reported stabilization of Cd and Cd/Zn, respectively, by the inoculation of rhizosphere bacteria, however, they also observed hyperaccumulation of metals upon inoculation of rhizospheric (Ahmad et al., 2016) and endophytic (Kuffner et al., 2008, 2010) bacteria.

Inoculation of Cd resistant PGP bacterial strains enhanced root elongation and biomass of *Brassica napus* cultivated in Cd contaminated soil (Sheng and Xia, 2006), nitrogen fixing, phosphorus and potassium solubilizing consortium of PGP bacteria also improved *Brassica juncea* biomass cultivated in Pb-Zn mine tailings (Wu et al., 2006). Metal resistant and PGP bacteria isolated from plants grown in a metal contaminated soil were belonged to *Klebsiella* sp. and *Enterobacter* sp., which increased plant growth and metal (Cd, Zn, Pb) accumulation in *Brassica napus* upon inoculation (Jing et al., 2014). The bacteria having high resistance to metals may survive better in metal contaminated soil. The previous studies showed good correlation with heavy metal resistance and mobilization/immobilization of metals in soil and subsequent uptake in plants (Gadd, 2000, 2004; Kuffner et al., 2008, 2010). Metal resistant *Pseudomonas putida* strain CZ1 increased Cu uptake up to 212% and translocation up to 1.86% in Cu tolerant plant *Elsholtzia splendens* in comparison to un-inoculated control plants, moreover, synchrotron-based X-ray fluorescence microscopy confirmed presence of Cu in root cortex and xylem (Xu et al., 2015). The ability of maize and wheat to accumulate Cd are found to be regulated by the inoculation of PGP bacterial species *Klebsiella* and *Bacillus*, both of the species have ability to proliferate at higher Cd concentration (4.45 mmol L^{-1}) (Ahmad et al., 2014). This study concludes that PGP bacterial inoculation could be deployed to stabilize Cd in rhizosphere. In another study, Ahmad et al. (2016) further investigated the role of root colonization of PGP bacteria and Cd uptake in maize. They found that root colonization of PGP bacteria (CIK-521R i.e., *Enterobacter* sp.) is a key to sustain plants in stress conditions; this association was improving maize growth and biomass while promoting Cd translocation and accumulation in maize tissues. In the same study, another bacterial strain (CIK-518 i.e., *Bacillus* sp.) was promoting maize biomass while suppressing Cd uptake and translocation in maize tissues. The in depth study showed that both of the strains showing different characters like (CIK-518 i.e., *Bacillus* sp.) was good colonizer in Cd contaminated soil that might be responsible for stabilization of Cd in rhizosphere while (CIK-521R i.e., *Enterobacter* sp.) was good colonizer in normal soil and produces auxin that might be responsible in promoting maize biomass leading to Cd accumulation in aerial part of plant. This study highlights that PGP characteristics, root colonization and tolerance to Cd are important tools for inoculated plants to combat in stress conditions.

Lead resisting endophytic *Pseudomonas* sp. and *Microbacterium* sp. increased Pb uptake in *Brassica napus* (Sheng et al., 2008b). In another study, inoculation of *Burkholderia cepacia* to *Sedum alfredii* a metal hyperaccumulating plant significantly enhanced plant growth, Cd/Zn uptake and translocation of these metals from root to shoot (Li et al., 2007), the authors further concluded that secretion of organic acids by the inoculated roots of *Sedum alfredii* was the resistance mechanism because these formed stable complexes with metals and protect plant from the damaging effects of exposed metals. Hyperaccumulation of metals is not always appearing to be best strategy for rapid reduction of metals from soil. Therefore, in recent study, Janssen et al. (2015) emphasized on selection of best performing clones of willow plant and its subsequent augmentation with PGP bacteria for enhanced metal extraction using plant-bacteria association. They found that *Salix viminalis* clone performed better by extracting 74 and 91% Cd and Zn respectively, in comparison to other commercial colnes. They isolated five bacterial strains from the selected clones and upon re-inoculation of *Rahnella* sp. to *Salix viminalis* again exhibited good results by increasing twig biomass in multimetal (Cd-Pb-Zn) contaminated soil. In another study, Weyens et al. (2013) reported that ability of willow clones to accumulate higher concentrations of metals (Cd, Zn) is partly related to bacterial population associated to that clone. They studied two willow clones "Tora which have ability to accumulate twice amount of metals as Belgisch Rood", it is because Tora has greater percentage of metal resistant endophytic bacterial population in comparison to Belgisch rood. These studies imply that inducing heavy metal tolerance in plants through PGP bacterial inoculation is a good strategy to manage metal contaminated soils.

Above studies impart that bacterial association to plants remove heavy metals more efficiently in comparison to sole plantation. Actually bacteria have fascinating properties like ACC-deaminase activity, phosphate solubilization, indole acetic acid and siderophore production which not only improve plant growth but also solubilize heavy metals in rhizosphere, from which plants can take up metals and accumulate in aerial parts. ACC-deaminase producing bacteria are capable to hydrolyze ACC into α ketobutyrate and ammonia (Ullah et al., 2015), which is a source of N for inoculated plant and important nutrient for promoting plant growth in stress conditions. Several studies showed that metal resistant PGP bacteria containing ACC-deaminase protected plant species against metal stress and promote plant growth (Belimov, 2001; Arshad et al., 2007). The presence of ACC-deaminase producing bacteria in the rhizosphere of metal hyperaccumulating *Thlaspi goesingense* confirms its role in alleviating metal toxicity (Idris et al., 2004). Another phytohormone, auxin is important for plant biomass production and many bacterial species such as *Burkholderia, Bacillus, Klebsiella* and *Pseudomonas* are known to produce IAA in metal stressed conditions and govern metal uptake (Kuffner et al., 2008, 2010; Sessitsch et al., 2013; Ahmad et al., 2014; Glick, 2014). PGP bacteria producing IAA were able to improve *Brassica juncea* biomass in the presence of hexavalent Cr (Rajkumar et al., 2006), reduced Zn toxicity in *Trifolium repens* (Vivas et al., 2006) and improved Ni accumulation in *Brassica juncea* (Zaidi et al., 2006). The diazotrophic bacteria for example *Rhizobium leguminosarum* can fix atmospheric nitrogen even in the presence of heavy metals (Nonnoi et al., 2012). *Rhizobium* sp. strain RL9 and *Bradyrhizobium* sp. strain RM8 are tolerant of Zn and Ni and produces substantial amount of IAA (Wani et al., 2007, 2008). The number of PGP microorganisms are reported to produce siderophores (metal chelating agents which bind metals and enhance their uptake to plants through solubilization), many of the PGP bacteria promotes siderophore production in stress conditions (Gadd, 2010; Kuffner et al., 2010; Rajkumar et al., 2010). Inoculation of siderophore producing *Pseudomonas* sp. increased bioavailable fractions of metals (Ni, Pb, Cr) in soil and their subsequent accumulation in plant (Braud et al., 2009; Tank and Saraf, 2009). Idris et al. (2004) identified many siderophore producing rhizosphere and endophytic bacteria from Ni hyperaccumulating *Thlaspi goesingense* and proposed their possible involvement in Ni accumulation.

Furthermore, PGP bacteria have ability to produce organic acids such as oxalic and citric acids, which induces mobilization of heavy metals through complexation process (Ullah et al., 2015). Inoculation of endophytic bacterial strain *Sphingomonas* SaMR12 induced secretions of organic acids (citric, oxalic and succinic), which enhance Cd mobility and bioavailability in soil and increase Cd tolerance in shoot by alleviating Cd induced oxidative stress (Chen et al., 2014a). PGP bacteria also acidify their surroundings by fixing carbon dioxide through oxidation-reduction reactions with iron and sulfur minerals, which cause mobilization of heavy metals due to sulfuric acid production. This process also adds in metal solubilization and phytoremediation (Gadd, 2004). In response to heavy metal stress some bacteria/fungi produce phytochelatin, which are metal loving compounds and mobilize heavy metals such as Cd, Pb and Cu. The synthesis of phytochelatin from glutathione through enzymatic reaction is crucial to protect cell from metal toxicity such as Cd because it forms Cd-phytochelatin complex and shuttled into the vacuole for long term sequestration of this metal within cell or in some cases out of cell (Jasinski et al., 2003; Verbruggen et al., 2009). In a study, Kang et al. (2007) found >7 times higher Cd accumulation by overexpression of PC Synthase. Another cysteine-rich protein i.e., metallothionines form complexes with heavy metals such as Zn, Cd and Cu. PGP bacteria having ability to produce metallothionines are capable to sequester/accumulate greater amount of heavy metals (Gadd, 2010). PGP bacteria can improve plant growth using one or more than one mechanisms describing above, mobilize heavy metals and thus assist in phytoremediation of metals from soil.

2.2.2. Metalloids

The activities of human beings and geological processes are the main sources of metalloid accumulation in the environment. Some of the metalloids are essential for biological system but their accumulation in the environment may cause threat to living organisms. Thanks to nature, that creates some plants and microorganisms to tackle these toxicants.

Arsenic (As) mobilization and immobilization through bacterial inoculation is an important way of bioremediation (Wang and Zhao 2009). The substantial amount of As accumulated in metal hyperaccumulating fern upon inoculation of rhizobacteria whereas no As accumulation was observed upon fungal inoculation (Jankong et al., 2007). Bacteria associated with the roots of *Cirsium arvense* were able to reduce toxicity of As because these bacteria show ACC deaminase, IAA and siderophore production, which promote plant growth in As contaminated soil (Cavalca et al., 2010), another, bacterial strain isolated from root nodules of leguminous plant showed greater biomass of sunflower grown in As contaminated soil (Reichman 2007). Shagol et al. (2014) isolated 21 As-tolerant bacteria, studied their diversity and potential to be used in phytoremediation. They identified six genera *Brevibacterium,*

Paenibacillus, Pseudomonas, Microbacterium, Rhodococcus and *Rahnella*, which have potential to tolerate high concentrations of arsenite [As(III)] and arsenate [As(V)], most of the isolates showed reduction of As(V) whereas only one strain *Pseudomonas koressnsis* JS123 showed oxidation and reduction of As. Some of the strains especially *Rhodococcus aetherivorans* JS2210, *Pseudomonas koreensis* JS2214, and *Pseudomonas* sp. JS238 increased root length even in the presence of As, this might be due to production of IAA, siderophore, ACC-deaminase and solubilization of inorganic phosphate by the inoculated strains. The isolated bacteria were tolerant to higher concentrations of As and positive for various PGP activities made them good option for phytoremediation of As. In addition to various PGP traits, colonization of desired strain is essential to mediate metal uptake. This factor was studied by Cavalca et al. (2013) in which they quantified ACR3(2) type arsenate efflux pump in bacterium *Alcaligenes* sp. Dhal-L using real time PCR technique, inoculated sunflower plant showed 53% higher amount of As and presence of 100 times higher ACR3(2) gene copy in the rhizosphere indicating that colonization is responsible for mobilization and subsequent uptake of As in sunflower. The detoxification of As through microbial application is a viable strategy, however, it is based on expression of *Ars* operon, which reduces As(V) into As(III) by arsenate reductase, once As(III) produced it is effluxed via transmembrane ARC3 or ArsB types pump. Recent studies showed that plant inoculated with As resistant and arsenate to arsenite reducing PGP bacteria increased accumulation of As in *Pteris vittata* (Yang et al., 2012; Lampis et al., 2015), these studies also confirmed As resistant bacteria in the rhizosphere *Pteris vittata*, that exhibited PGP characters such as IAA and siderophore production.

The bacteria residing in the rhizosphere may convert some metals/ metalloids (Hg, Te, Tn, Se, As, Pb, Sb) into volatile form, which considered as detoxification mechanism. Rhizobacteria involves in Se accumulation in plants and its subsequent volatilization into the atmosphere (De Souza et al., 1999), similarly, degradation of methyl mercury and reduction of Hg^{+2} into Hg^0 is the mediated reactions of rhizobacteria assisted plant volatilization (Barkay et al., 2003). The bioavailability and toxicity of many metalloids influence through microbial oxidation, reduction and methylation process. For example, biomethylation of Se and Te, similarly, the reduction of oxyanions of Se (Selenite) and Te (Tellurite) into an elemental form is considered as detoxification mechanism of these metalloids (Gadd, 2010). Sura-de Jong et al. (2015) investigated the endophytic bacterial diversity of Se hyperaccumulators *Stanleya pinnata*/*Astragalus bisulcatus* and non-accumulators *Physaria bellii/Medicago sativa* belonged to Brassicaceae and Fabaceae, they observed that Se hyperaccumulators and non-accumulators shared similar diverse bacterial community, however, bacteria harbor hyperaccumulator were much more resistant to Se, having ability to reduce selenite into red elemental Se and promoting plant biomass. Another study, showed presence of diverse endophytic microbial community (bacteria, fungi) in Se hyperaccumulator and re-inoculation of these microbes to same or new plant species are responsible for accumulation of Se in different parts of plant including root, shoot, seed and nodules (Lindblom et al., 2013), moreover, those microbes having ability to reduce selenite to elemental Se and tolerance to Se may further be processed for production of Se nanoparticles and employed for wastewater treatment (Staicu et al., 2015). Several microorganisms are mediated the methylation of As and form monomethyl arsine, dimethyl arsine and trimethyl arsine. Many bacteria use arsenite as electron donor and oxidized it into arsenate in the presence of arsenite oxidase (Gadd, 2010). *Pseudomonas putida* KT2440 containing arsM gene methylated As and enhanced its removal through bacterial mediated methylation and volatilization from As contaminated soil (Chen et al., 2014b). Microbial mediated oxidation-reduction can change solubility of many metals/metalloids [Hg(II), Fe(III), Mn(IV), Se(VI), Cr(VI) and U(VI)] using as terminal electron acceptor (Gadd 2004; Di Gregorio et al., 2006). There are less studies available on removal of Sb through plant-bacterial association, however, it has recently been reported that some microbes have ability to use antimonate [Sb(V)] as terminal electron acceptor and convert it into antimonite [Sb(III)] during anaerobic respiration (Abin and Hollibaugh, 2014).

2.2.3. Radionuclides

Radionuclides are being released into the environment from the nuclear power plants, waste disposal sites of nuclear power plants, accidental release (U, Pu) and as fission product (Cs, Sr). Similar to other metals, their presences also threaten life on the planet. Fortunately, nature has blessed this planet with such plants and microbes, which have ability to transform, solubilize, and accumulate radionuclides (Francis, 2006), moreover, sulfur and iron oxidizing bacteria increases dissolution of U because these bacteria acidify the medium by producing sulfuric acid and oxidizing agent ferric sulfate. Some bacteria have ability to accumulate U on their surfaces through biosorption mechanism; this is actually takes place owing to affiliation of U with carboxyl, hydroxyl and phosphate functional groups present in cell wall of bacteria. In a study, Francis et al. (2004) reported intercellular accumulation of U in *Halomonas* sp., EXAFS analysis confirmed U binding with that of phosphate group. Another bacterial specie *Clostridium* mediated reduction of U(VI) into U(IV) (Francis, 2006).

Apart from microbes, plants also have potential to accumulate radionuclides as they do in metals and organic pollutants. It is widely accepted that accumulation of pollutant to plant biomass is a two-steps process, at first step radionuclides/metals/organic pollutants taken up by the plant roots through positive affiliation to negative charged surfaces followed by passive transport through cell wall whereas second step involves accumulation of radionuclides into the protoplast through cell membrane utilizing energy (Cecal et al., 2002). The management of radionuclides using plants and grasses is widely accepted technology, however, less work has been reported in radionuclides remediation as compared to metals/metalloids. The rhizofilteraion of U, Cs and Sr; whereas, accumulation of Cs and Sr in sunflower has previously been reported (Dushenkov et al., 1997; Soudek et al., 2006; Prasad, 2007). Soil to plant translocation factor of U and Ra was studied using sunflower, which removed 42 and 24% of radionuclides respectively, most of the Ra translocated to shoot while U accumulated in root (Rodriguez et al., 2006). The agricultural crops showed accumulation of radionuclides "Cs, Sr, Pu, Am" in vegetative part of crops (Kozhakhanov et al., 2014). Moreover, it is assumed that bacterial association with plants may further increase radionuclide accumulation or stabilization in contaminated soil. In this regard, Langella et al. (2014) revealed the ability of grasses (*Agrostis capillaris, Deschampsia flexuosa, Festuca rubra*) and herb (*Helianthus annuus*) inoculated with microbial consortia for phytoextraction and phytostabilization of multimetals at a contaminated site known as the area of uranium mine in Germany and copper mine in Sweden. Results indicated the role of plant species and microbial consortia on accumulation of heavy metals "Cd, Ni, Mn, Zn" and radionuclides "Sr, U". *Helianthus annuus* showed extraction of almost all the metals at German uranium mining area regardless of microbial consortia; however, *Festuca rubra* enhanced the extraction of Sr, whereas reduced U in plants inoculated with microbial consortia.

3. Factors Affecting Plant-Bacteria Partnership

Plant-bacteria partnership is dependent on various factors including the genetic diversity of bacteria itself and host plant, the climatic conditions, soil type and competing community in it. Therefore, it is important to know about the specificity of plant-bacteria relationship and the ways to facilitate that partnership. Bio-augmentation (the addition of bacterial culture to enhance the rate of biodegradation process) is a strategy used to facilitate the desired bacterial species in its functioning. However, the root colonization ability of the bacterial species is also of utmost importance for developing better association and get maximum benefit in remediation process. Moreover, certain biotic (pathogens, protozoa, fungi) and abiotic (pH, temperature, moisture and nutrient availability) factors influence the functioning of plant-bacteria association. How this association influenced by the method of inoculation and root colonization is elucidated here.

3.1. Method of Inoculation

The inoculation methods can influence the survival and colonization capability of an applied microorganism and its bioremediation potential. Method of inoculation directly influences the colonization, persistence, and activity of an applied microorganism in the soil, root zone (rhizosphere) and different compartments of the plant (root/shoot) (Afzal et al., 2010, 2012, 2013). There are four methods of inoculation such as i) seed inoculation: coating of seeds with peat based slurry in sugar solution, ii) soil inoculation: spreading of bacterial liquid culture on the surface of soil with irrigation water, iii) rhizosphere inoculation: introduction of bacteria directly into the rhizosphere by dripping or injection, iv) foliar inoculation: spraying of bacterial culture onto the surface of plant or soil (Bressan and Borges, 2004; Rosenblueth and Martinez-Romero, 2006; Weyens et al., 2009; Afzal et al., 2012, 2013). Various researchers have evaluated these methods of inoculation and elaborated their findings. Ryegrass was planted by Afzal et al. (2013) in hydrocarbon-contaminated soil. They inoculated the plants with *Burkholderia phytofirmans* PsJN through different methods of inoculation and observed soil inoculation as a prominent strategy in improving plant growth, hydrocarbon degradation and phytoremediation. Similarly, the soil inoculation of *Pantoea* sp. in combination with *Pseudomonas* sp. showed highest increases in the rates of hydrocarbon biodegradation and improved ryegrass growth as compared to seed imbibement (Afzal, 2010). The root or seed inoculation of plant beneficial bacteria to metal accumulating plants like spinach and maize, increased the bioavailability of metals to plants and ultimately their uptake (Ali et al., 2013; Ahmad et al., 2016). In another study, Álvarez-López et al. (2015) investigated various method of inoculation including soil, seed, dual soil and soil+seed inoculation using *Rhodococcus erythropolis* and observed their effect on phytoexraction of metals (Cd, Pb, Zn) in tobacco plant. Soil inoculation outperformed the other methods by improving plant biomass upto 45%, however, either less or differential response of inocluation methods was observed on metal uptake. Soil and soil+seed inoculation increased Pb uptake by 29 and 34%, respectively, however, seed inoculation increased Zn uptake by 24%. This study endorsed the previous conclusion that soil inoculation is the best method for getting good plant biomass in contaminated soil, however, soil and/or seed might be useful for phytoremediation depending on metal type. The literature for the inoculation methods is very rare, even though the studies up till now show the soil inoculation method as a best way of inoculation, which has improved plant biomass and rate of amelioration of contaminated soils.

3.2. Root Colonization by Bacteria

The plant-bacteria association is highly specific to each other (Hussain et al., 2014a, b). This is because plants release flavonoids for specific bacteria whereas bacteria secrete nod factor to identify their host plants. That chemical communication (chemotaxis) between the plant and microorganisms leads towards the development of a mutual association (Thimmaraju et al., 2008) where plants provide carbon substrate (root exudates) to the microbes on their part (Bais et al., 2008; Shukla et al., 2011) whereas microbes perform PGP attributes to support plant growth under normal as well as stressful conditions. Population of beneficial bacteria in the rhizosphere of plant or on the surface of plant roots must be higher to facilitate the bioremediation processes and improve the growth of the host plant. As mentioned earlier, the inoculation method influences the colonization potential of bacteria to plants and well established population of bacteria in the rhizosphere have strong relevance in bacterial-assisted phytoremediation of pollutants (Afzal et al., 2013). Shukla et al. (2011) described that the beneficial bacteria must have the capability to degrade the selective soil contaminant and colonization of roots to accelerate the rhizoremediation (remediation of rhizosphere soil by degradation/ transformation of contaminants) process. However, certain bacteria have varied capability to colonize the roots or interior of the plants. The ability *Pseudomonas putida* strain PCL1444 was found 100 fold higher in rhizosphere than the other efficient colonizer *Pseudomonas fluorescens* strain WCS365, this is because former bacterium was efficiently utilized root exudates of

Lolium multiflorum cv. Barmultra and degrade heavily polluted naphthalene soil (Kuiper et al., 2002). Similar kind of relationship was explored by the Germaine et al. (2009) in plants inoculated with *Pseudomonas putida*. They observed that this strain was efficient colonizer of root exterior as well as interior and showed capability to degarde naphthalene up to 40%. Moreover, inoculated plants showed greater seed germination and transpiration rate in comparison to un-inoculated control plants. Ahmad et al. (2016) observed improvement in the growth of maize in Cd contaminated soils due to seed inoculation of *Klebsiella* and *Enterobacter* strains CIK-518 and CIK-521R, respectively. These strains were tolerant to high concentrations of Cd and colonized the plant roots at greater extent. Similarly, about 53% increase in the uptake of arsenic in sunflower was recorded by Cavalca et al. (2013) with the inoculation of *Alcaligenes* sp. strain Dhal-L. The authors observed increased colonization of inoculated bacteria in the rhizosphere soil of sunflower. In another study, molecular analysis revealed microbial community diversity in rhizosphere soil contaminated with Cd (Moreira et al., 2014). They found that microbial community was efficient in short-term phytostabilization of Cd in soil. This part emphasis on root colonization of bacterial strains used in remediation process, above studies showed strong correlation of root colonization and pollutant degradation/stabilization/extraction. Root colonization is a good tool for selection or screening of efficient bacterial strains which assist phytoremediation process.

CONCLUSION

Phytoremediation is an innovative and safe technique for management of soil contaminated with organic and inorganic pollutants, however, the establishment of plants in contaminated soil is a major concern. The present review warrants that PGP bacteria resided in the rhizosphere/ endosphere have successfully been employed to overcome this problem, moreover, rhizosphere have rich source of carbon and nutrients so that bacteria utilize them to get energy and survive better in harsh conditions and indirectly solubilize or stabilize pollutants in the soil. This review highlighted the importance of inoculating PGP bacteria to produce greater plant biomass and to remediate contaminated soil. Plant-bacterial association speeds up the process of decontamination of contaminated soil through phytostabilization, phytoextraction, phytodegradation and phytovolatilization of organic and inorganic pollutants. Plant-bacterial interaction also altered bioavailable fraction of pollutants to plants by inducing production of siderophores, IAA, ACC-deaminase, organic acids and phytochelatins leading to amelioration of stress conditions. Metalloids and radionuclides are mostly transformed either through oxidation-reduction or methylation-volatilization processes by positive association of plant-bacteria, however, this association has also enhanced pollutant uptake in above part of plant supported by fewer studies. There are various factors either biotic or abiotic which may influence bacterial-plant association, but here we discussed two key factors i.e., method of inoculation and root colonization in relation to their impact on pollutant remediation. This chapter reviewed soil inoculation as best method for inoculation in contaminated soil whereas root colonization has detected as indicator of remediation process. Those bacteria having good colonization efficiency are more likely to improve plant biomass and pollutant degradation in comparison to less efficient colonizers. There are several studies indicating the role of plant-bacteria partnership on organic and metal pollutants, however, less studies are available which document plant-bacteria interaction on metalloids and radionuclides remediation, therefore, future research should focus on this less investigated area. Although, number of studies reported in the context of bacterial assisted phytoremediation but our understanding to the mechanisms involved in this process still poor and thereby, it is obvious to conduct future research to gain better insight into the mechanisms involved behind these complex soil-plant-microbe interactions in contaminated soil. Moreover, there is need to exploit bench or pot scale results into a more diverse ecological conditions at field scale to warrant this technology.

References

Abin, C. A., Hollibaugh, J. T. (2014). Dissimilatory antimonate reduction and production of antimony trioxide microcrystals by a novel microorganism. *Environmental Science and Technology 48*, 681-688.

Adiloglu, S., Turgut Saglam Ali Süme, M. (2015). Chrome (Cr) pollution in agricultural areas improvement by phytoremediation method with canola (*Brassica napus* L.) plant growing. *Journal of Essential Oil Bearing Plants 18*, 1180-1186.

Afzal, M. (2010). Plant-microbe interactions for the remediation of hydrocarbon contaminated soil. Dissertation for PhD. *University of Natural Resources and Life Sciences*, Vienna.

Afzal, M., Khan, S., Iqbal, S., Mirza, M. S., Khan, Q. M. (2013). Inoculation method affects colonization and activity of *Burkholderia phytofirmans* PsJN during phytoremediation of diesel-contaminated soil. *International Biodeterioration and Biodegradation Journal 85*, 331-336.

Afzal, M., Yousaf, S., Reichenauer, T. G., Sessitsch, A. (2012). The inoculation method affects colonization and performance of bacterial inoculant strains in the phytoremediation of soil contaminated with diesel oil. *International Journal of Phytoremediation 14*, 35-47.

Ahemad, M., Khan, M. S. (2012). Effects of pesticides on plant growth promoting traits of *Mesorhizobium* strain MRC4. *Journal of Saudi Society of Agricultural Sciences 11*, 63-71.

Ahmad, F., Iqbal, S., Anwar, S., Afzal, M., Islam, E., Mustafa, T., Khan, Q. M. (2012a). Enhanced remediation of chlorpyrifos from soil using ryegrass (*Lollium multiflorum*) and chlorpyrifos-degrading bacterium *Bacillus pumilus* C2A1. *Journal of Hazardous Materials 237*, 110-115.

Ahmad, I., Akhtar, M. J., Asghar, H. N., Ghafoor, U., Shahid, M. (2016). Differential effects of plant growth-promoting rhizobacteria on maize growth and cadmium uptake. *Journal of Plant Growth Regulation 35*, 303-315.

Ahmad, I., Akhtar, M. J., Asghar, H. N., Zahir, Z. A. (2013). Comparative efficacy of growth media in causing cadmium toxicity to wheat at seed germination stage. *International Journal of Agriculture and Biology 15*, 517-522.

Ahmad, I., Akhtar, M. J., Zahir, Z. A., Jamil, A. (2012). Effect of cadmium on seed germination and seedling growth of four wheat (*Triticum aestivum* L.) cultivars. *Pakistan Journal of Botany 44*, 1569-1574.

Ahmad, I., Akhtar, M. J., Zahir, Z. A., Naveed, M., Mitter, B., Sessitsch, A. (2014). Cadmium-tolerant bacteria induce metal stress tolerance in cereals. *Environmental Science and Pollution Research 21*, 11054-11065.

Aitken, M. D., Stringfellow, W. T., Nagel, R. D., Kazunga, C., Chen, S. H. (1998). Characteristics of phenanthrene-degrading bacteria isolated from soils contaminated with polycyclic aromatic hydrocarbons. *Canadian Journal of Microbiology 44*, 743-752.

Al-Awadhi, H., El-Nemr, I., Mahmoud, H., Sorkhoh, N. A., Radwan, S. (2009). Plant-associated bacteria as tools for the phytoremediation of oily nitrogen-poor soils. *International Journal of Phytoremediation 11*, 11-27.

Ali, T., Mahmood, S., Khan, M. Y., Aslam, A., Hussain, M. B., Asghar, H. N., Akhtar, M. J. (2013). Phytoremediation of cadmium contaminated soil by auxin assisted bacterial inoculation. *Asian Journal of Agriculture and Biology 1*, 79-84.

Alrumman, S. A., Standing, D. B., Paton, G. I. (2015). Effects of hydrocarbon contamination on soil microbial community and enzyme activity. *Journal of King Saud University Science 27*, 31-41.

Álvarez, A., Benimeli, C. S., Saez, J. M., Giuliano, A., Amoroso, M. J. (2015). Lindane removal using *Streptomyces* strains and maize plants: a biological system for reducing pesticides in soils. *Plant and Soil 395*, 401-413.

Anderson, S. E., Meade, B. J. (2014). Potential health effects associated with dermal exposure to occupational chemicals. *Environmental Health Insights 8*, 51-62.

Andreolli, M., Lampis, S., Poli, M., Gullner, G., Biró, B., Vallini, G. (2013). Endophytic *Burkholderia fungorum* DBT1 can improve phytoremediation efficiency of polycyclic aromatic hydrocarbons. *Chemosphere 92*, 688-694.

Ankohina, T., Kochetkov, V., Zelenkova, N., Balakshina, V., Boronin, A. (2004). Biodegradation of phenanthrene by Pseudomonas bacteria bearing rhizospheric plasmids in model plant-microbial associations. *Applied Biochemistry and Microbiology 40*, 568-572.

Arshad, M., Saleem, M., Hussain, S. (2007). Perspectives of bacterial ACC deaminase in phytoremediation. *Trends in Biotechnology 25*, 356-362.

Atterby, H., Smith, N., Chaudhry, Q., Stead, D. (2002). Exploiting microbes and plants to clean up pesticide contaminated environments. *Pesticide Outlook 13*, 9-13.

Badri, D. V., Weir, T. L., Lelie, D. V. D., Vivanco, J. M. (2009). Rhizosphere chemical dialogues: plant-microbe interactions. *Current Opinion in Biotechnology 20*, 642-650.

Bais, H. P., Broeckling, C. D., Vivanco, J. M. (2008). Root exudates modulate plant–microbe interactions in the rhizosphere in secondary metabolites in soil ecology. In: Karlovsky, P. (ed.) *Secondary Metabolites in Soil Ecology-Soil Biology.*, Springer Berlin Heidelberg, pp. 241-252.

Barac, T., Taghavi, S., Borremans, B., Provoost, A., Oeyen, L., Colpaert, J. V., van der Lelie, D. (2004). Engineered endophytic bacteria improve phytoremediation of water soluble, volatile, organic pollutants. *Nature Biotechnology*, *22*, 583-588.

Barkay, T., Miller, S. M., Summers, A. O. (2003). Bacterial mercury resistance from atoms to ecosystems. *FEMS Microbiology Reviews 27*, 355-384.

Becerra-Castro, C., Kidd, P. S., Rodríguez-Garrido, B., Monterroso, C., Santos-Ucha, P., Prieto-Fernández, Á. (2013). Phytoremediation of hexachlorocyclohexane (HCH)-contaminated soils using *Cytisus striatus* and bacterial inoculants in soils with distinct organic matter content. *Environment Pollution 178*, 202-210.

Belimov, A. A., Safronova, V. I., Sergeyeva, T. A., Egorova, T. N., Matveyeva, V. A., Tsyganov, V. E. Borisov, A. Y., Tikhonovich, I. A., Kluge, C., Preisfeld, A., Dietz, K. J., Stepanok, V. V. (2001). Characterization of plant growth-prompting rhizobacteria isolated from polluted soils and containing 1-aminocyclopropane-1-carboxylate deaminase. *Canadian Journal of Microbiology 47*, 642-652.

Bouchez, M., Blanchet, D., Bardin, V., Haeseler, F., Vandecasteele, J. P. (1999). Efficiency of defined strains and of soil consortia in the biodegradation of polycyclic aromatic hydrocarbon (PAH): mixtures. *Biodegradation 10*, 429-435.

Braud, A., Jézéquel, K., Bazot, S., Lebeau, T. (2009). Enhanced phytoextraction of an agricultural Cr, Hg- and Pb-contaminated soil by bioaugmentation with siderophore producing bacteria. *Chemosphere 74*, 280-286.

Brazil, G. M., Kenefick, L., Callanan, M., Haro, A., de Lorenzo, V., Dowling, D. N., O'Gara, F. (1995). Construction of a rhizosphere Pseudomonad with potential to degrade polychlorinated biphenyls and detection of *bph* gene expression in the rhizosphere. *Applied and Environmental Microbiology*, *61*, 1946-1952.

Bressan, W., Borges, M. (2004). Delivery methods for introducing endophytic bacteria into maize. *Biocontrol 49*, 315-322.

Castelo-Grande, T., Augusto, P. A., Monteiro, P., Estevez, A. M., Barbosa, D. (2010). Remediation of soils contaminated with pesticides: a review. *International Journal of Environment and Analytical Chemistry 90*, 438-467.

Cavalca, L., Corsini, A., Bachate, S. P., Andreoni, V. (2013). Rhizosphere colonization and arsenic translocation in sunflower (*Helianthus annuus* L.) by arsenate reducing *Alcaligenes* sp. strain. Dhal-L. *World Journal of Microbiology and Biotechnology* doi:10.1007/s11274-013-1359-4.

Cavalca, L., Zanchi, R., Corsini, A., Colombo, M., Romagnoli, C., Canzi, E., Andreoni, V. (2010). Arsenic-resistant bacteria associated with roots of the wild *Cirsium arvense* (L.) plant from an arsenic polluted soil, and screening of potential plant growth-promoting characteristics. *Systems of Applied Microbiology 33*, 154-164.

Cecal, A., Popa, K., Caraus, I., Craciun, I. I. (2002). Uranium and thorium uptake on the hydrophilic plants. In: Merkel, J. B., Planer-Friederich, B., Wolkersdorfer, C. (eds.) *Uranium in the Aquatic Environment III*. Springer, Freiberg.

Chaudhry, Q., Blom-Zandstra, M., Gupta, S., Joner, E. J. (2005). Utilizing the synergy between plants and rhizosphere microorganisms to enhance breakdown of organic pollutants in the environment. *Environmental Science and Pollution Research 12*, 34-48.

Chaudhry, Q., Schroeder, P., Werck-Reichhart, D., Grajek, W., Marecik, R. (2001). Prospects and limitations of phytoremediation for the removal of persistent pesticides in the environment. *Environmental Science and Pollution Research 9*, 4-17.

Chekol, T., Vough, L. R., Chaney, R. L. (2004). Phytoremediation of polychlorinated biphenyl-contaminated soils: the rhizosphere effect. *Environment International 30*, 799-804.

Chen, B., Zhang, Y., Rafiq, M. T., Khan, K. Y., Pan, F., Yang, X., Feng, Y. (2014a). Improvement of cadmium uptake and accumulation in *Sedum alfredii* by endophytic bacteria *Sphingomonas* SaMR12: effects on plant growth and root exudates. *Chemosphere 117*, 367-373.

Chen, J., Sun, G. X., Wang, X. X., de Lorenzo, V., Rosen, B. P., Zhu, Y. G. (2014b). Volatilization of arsenic from polluted soil by *Pseudomonas putida* engineered for expression of the arsM arsenic(III) S-adenosine methyltransferase gene. *Environmental Science and Technology 48*, 10337-10344.

Chen, L., Luo, S., Xiao, X., Guo, H., Chen, J., Wan, Y., Li, B., Xu, T., Xi, Q., Rao, C., Liu, C., Zeng, G. (2010). Application of plant growth-promoting endophytes (PGPE) isolated from *Solanum nigrum* L. for phytoextraction of Cd-polluted soils. *Applied Soil Ecology 46*, 383-389.

Chigbo, C., Batty, L. (2014). Phytoremediation for co-contaminated soils of chromium and benzo [a] pyrene using *Zea mays* L. *Environmental Science and Pollution Research 21*, 3051-3059.

Daane, L. L., Harjono, I., Zylstra, G. J., Haggblom, M. M. (2001). Isolation and characterization of polycyclic aromatic hydrocarbon degrading bacteria associated with the rhizosphere of salt marsh plants. *Applied and Environmental Microbiology 67*, 2683-2691.

Dams, R. I., Paton, G. I., Killham, K. (2007). Rhizoremediation of pentachlorophenol by *Sphingobium chlorophenolicum* ATCC 39723. *Chemosphere 68*, 864-870.

De Souza, M. P., Huang, C. P. A., Chee, N., Terry, N. (1999). Rhizosphere bacteria enhance the accumulation of selenium and mercury in wetland plants. *Planta 209*, 259-263.

Dell' Amico, E., Cavalca, L., Andreoni, V. (2008). Improvement of *Brassica napus* growth under cadmium stress by cadmium-resistant rhizobacteria. *Soil Biology and Biochemistry 40*, 74-84.

Desneux, N., Decourtye, A., Delpuech, J. M. (2007). The sublethal effects of pesticides on beneficial arthropods. *Annual Review of Entomology 52*, 81-106.

Di Gregorio, S., Lampis, S., Malorgio, F., Petruzzelli, G., Pezzarossa, B., Vallini, G. (2006). *Brassica juncea* can improve selenite and selenate abatement in selenium contaminated soils through the aid of its rhizosperic bacterial population. *Plant and Soil 285*, 233-244.

Donnelly, P. K., Hegde, R. S., Fletcher, J. S. (1994). Growth of PCB degrading bacteria on compounds from photosynthetic plants. *Chemosphere 28*, 981-988.

Doty, S. L. (2008). Enhancing phytoremediation through the use of transgenics and endophytes. *New Phytologist 179*, 318-33.

Duruibe, J. O., Ogwuegbu, M. O. C., Egwurugwu, J. N. (2007). Heavy metal pollution and human biotoxic effects. *International Journal of Physical Sciences 2*, 112-118.

Dushenkov, S., Vasudev, D., Kapulnik, Y., Gleba, D., Fleisher, D., Ting, K. C., Ensley, B. (1997). Phytoremediation: A novel approach to an old problem. In: Wise, D. L. (ed.) *Global Environmental Biotechnology*., Elsevier, Amsterdam, pp 563–572.

Farid, M., Ali, S., Shakoor, M. B., Bharwana, S. A., Rizvi, H., Ehsan, S., Hannan, F. (2013). EDTA assisted phytoremediation of cadmium, lead and zinc. *International Journal of Agronomy and Plant Production 4*, 2833-2846.

Fava, F., Di Gioia, D. (1998). Effects of triton X-100 and *Quillaya saponin* on the ex situ bioremediation of a chronically polychlorobiphenyl-contaminated soil. *Applied Microbiology and Biotechnology 50*, 623-630.

Favas, P. J., Pratas, J., Varun, M., D'Souza, R., Paul, M. S. (2014). Phytoremediation of soils contaminated with metals and metalloids at mining areas: potential of native flora. In: Hernández-Soriano, M. C. (ed.) *Environmental Risk Assessment of Soil Contamination*., 485-517.

Feng, R. W., Wang, X. L., Wei, C. Y., Tu, S. X. (2013). The accumulation and subcellular distribution of arsenic and antimony in four fern plants. *International Journal of Phytoremediation* doi.org/10.1080/15226514. 2013.

Feng, R. W., Wei, C. Y., Tu, S. X., Tang, S. R., Wu, F. C. (2011). Simultaneous hyperaccumulation of arsenic and antimony in Cretan brake fern: evidence of plant uptake and subcellular distributions. *Microchemical Journal 97*, 38-43.

Fletcher, J. S., Hegde, R. S. (1995). Release of phenols by perennial plant roots and their potential importance in bioremediation. *Chemosphere 31*, 3009-3016.

Francis, A. J. (2006). Microbial transformations of radionuclides released from nuclear fuel reprocessing plants. International Symposium on Environmental Modeling and Radioecology, Rakkasho, Aomori, Japan Oct. 18-20.

Francis, A. J., Gillow, J. B., Dodge, C. J., Harris, R., Beveridge, T. J., Papenguth, H. W. (2004). Association of uranium with halophilic and nonhalophilic bacteria and archaea. *Radiochemica Acta* 92, 481-488.

French, C. E., Rosser, S. J., Davies, G. J., Nicklin, S., Bruce, N. C. (1999). Biodegradation of explosives by transgenic plants expressing pentaerythritol tetranitrate reductase. *Natural Biotechnology 17*, 491-494.

Gadd, G. M. (2000). Bioremedial potential of microbial mechanisms of metal mobilization and immobilization. *Current Opinion in Biotechnology 11*, 271-279.

Gadd, G. M. (2004). Microbial influence on metal mobility and application for bioremediation. *Geoderma 122*, 109-119.

Gadd, G. M. (2010). Metals: minerals and microbes: geomicrobiology and bioremediation. *Microbiology 156*, 609-643.

Garba, S. T., Osemeahon, A. S., Maina, H. M., Barminas, J. T. (2012). Ethylene diamine tetra acetate (EDTA)-assisted phytoremediation of heavy metal contaminated soil by *Eleusine indica* L. Gearth. *Journal of Environmental Chemistry and Ecotoxicology 4*, 103-109.

Gerhardt, K. E., Huang, X. D., Glick, B. R., Greenberg, B. M. (2009). Phytoremediation and rhizoremediation of organic soil contaminants: potential and challenges. *Plant Science 176*, 20-30.

Germaine, K., Keogh, E., Garcia-Cabellos, G., Borremans, B., van der Lelie, D., Barac, T., Oeyen, L., Vangronsveld, J., Moore, F. P., Moore, E. R., Campbell, C. D., Ryan, D., Dowling, D. N. (2004). Colonization of poplar trees by gfp expressing bacterial endophytes. *FEMS Microbiology Ecology 48*, 109-118.

Germaine, K. J., Keogh, E., Ryan, D., Dowling, D. (2009). Bacterial endophyte-mediated naphthalene phytoprotection and phytoremediation. *FEMS Microbiology Letters 296*, 226-234.

Germaine, K. J., Liu, X., Cabellos, G. G., Hogan, J. P., Ryan, D., Dowling, D. N. (2006). Bacterial endophyte enhanced phytoremediation of the organochlorine herbicide 2,4-dichlorophenoxyacetic acid. *FEMS Microbiology Ecology 57*, 302-310.

Glick, B. R. (2010). Using soil bacteria to facilitate phytoremediation. *Biotechnology Advances 28*, 367-374.

Glick, B. R. (2014). Bacteria with ACC deaminase can promote plant growth and help to feed the world. *Microbiology Research* 169, 30-39.

Guo, H., Yao, J., Cai, M., Qian, Y., Guo, Y., Richnow, H. H., Blake, R. E., Doni, S., Ceccanti, B. (2012). Effects of petroleum contamination on soil microbial numbers, metabolic activity and urease activity. *Chemosphere 87*, 1273-1280.

Gupta, D. K., Walther, C. (2014). *Radionuclide Contamination and Remediation through Plants.*, Zofingen: Springer.

Hannink, N., Rosser, S. J., French, C. E., Basran, A., Murray, J. A. H., Nicklin, S., Bruce, N. C. (2001). Phytodetoxification of TNT by transgenic plants expressing a bacterial nitroreductase. *Nature Biotechnology 19*, 1168-1172.

He, Y., Xu, J. M., Tang, C. X., Wu, Y. P. (2005). Facilitation of pentachlorophenol degradation in the rhizosphere of ryegrass (*Lolium perenne* L.). *Soil Biology and Biochemistry 37*, 2017-2024.

Hegde, R. S., Fletcher, J. S. (1996). Influence of plant growth stage and season on the release of root phenolics by mulberry as related to development of phytoremediation technology. *Chemosphere 32*, 2471-2479.

Hoagland, R. E., Zablotowicz, R. M., Locke, M. A. (1997). An integrated phytoremediation strategy for chloroacetamide herbicides in soil. In: Hoagland, R. E. (ed.) *Phytoremediation of Soil and Water Contaminants*, American Chemical Society, Washington USA, 92-105.

Hodgson, E., Rose, R. L., Ryu, D. Y., Falls, B. L., Levi, P. E. (1995). Pesticide-metabolizing enzymes. *Toxicology Letters* 82/83, 73-81.

Huang, X. D., El-Alawi, Y., Penrose, D. M., Glick, B. R., Greenberg, B. M. (2004). A multi-process phytoremediation system for removal of polyaromatic hydrocarbons from contaminated soils. *Environment Pollution 130*, 465-476.

Hussain, M. B., Zahir, Z. A., Asghar, H. N., Asgher, M. (2014a). Can catalase and EPS producing rhizobia ameliorate drought in wheat. *International Journal of Agriculture and Biology 16*, 3-13.

Hussain, M. B., Zahir, Z. A., Asghar, H. N., Mahmood, S. (2014b). Scrutinizing rhizobia to rescue maize growth under reduced water conditions. *Soil Science Society of America Journal 78*, 538-545.

Hussain, S., Devers-Lamrani, M., El-Azahari, N., Martin-Laurent, F. (2011). Isolation and characterization of an isoproturon mineralizing *Sphingomonas* sp. strain SH from a French agricultural soil. *Biodegradation 22*, 637-650.

Hussain, S., Maqbool, Z., Ali, S., Yasmin, T., Imran, M., Mehmood, F., Abbas, F. (2013). Biodecolorization of Reactive Black-5 by a metal and salt tolerant bacterial strain *Pseudomonas* sp. RA20 isolated from Paharang Drain effluents in Pakistan. *Ecotoxicology and Environmental Safety 98*, 331-338.

Idris, R., Trifinova, R., Puschenreiter, M., Wenzel, W. W., Sessitsch, A. (2004). Bacterial communities associated with flowering plants of the Ni hyperaccumulator *Thlaspi goesingense. Applied and Environmental Microbiology 70*, 2667-2677.

Imran, M., Arshad, M., Khalid, A., Hussain, S., Mumtaz, M. W., Crowley, D. E. (2015). Decolorization of reactive black-5 by Shewanella sp. in the presence of metal ions and salts. *Water and Environment Research 87*, 579-586.

Isoda, H., Talorete, T. P., Han, J., Oka, S., Abe, Y., Inamori, Y. (2005). Effects of organophosphorous pesticides used in china on various mammalian cells. *Environmental Science 12*, 9-19.

Jacobsen, C. S. (1997). Plant protection and rhizosphere colonization of barley by seed inoculated herbicide degrading *Burkholderia* (Pseudomonas) *cepacia* DBO1(pRO101) in 2,4-D contaminated soil. *Plant and Soil 189*, 139-144.

Jaffré, T., Pillon, Y., Thomine, S., Merlot, S. (2013). The metal hyperaccumulators from New Caledonia can broaden our understanding of nickel accumulation in plants. *Frontiers in Plant Science 4*, 1-7.

Jankong, P., Visoottiviseth, P., Khokiattiwong, S. (2007). Enhanced phytoremediation of arsenic contaminated land. *Chemosphere 68*, 1906-1912.

Janssen, J., Weyens, N., Croes, S., Beckers, B., Meiresonne, L., Van Peteghem, P., Carleer, R., Vangronsveld, J. (2015). Phytoremediation of metal contaminated soil using willow: exploiting plant-associated bacteria to improve biomass production and metal uptake. *International Journal of Phytoremediation 17*, 1123-1136.

Jasinski, M., Ducos, E., Martinoia, E., Boutry, M. (2003). The ATP-binding cassette transporters: structures, function, gene family comparison between rice and Arabidopsis. *Plant Physiology 131*, 1169-1177.

Jelusic, M., Lestan, D. (2014). Effect of EDTA washing of metal polluted garden soils. Part I: toxicity hazards and impact on soil properties. *Science of the Total Environment 475*, 132-141.

Jing, Y. X., Yan, J. L., He, H. D., Yang, D. J., Xiao, L., Zhong, T., Yuan, M., Cai, X. D., Li, S. B. (2014). Characterization of bacteria in the rhizosphere soils of polygonum pubescens and their potential in promoting growth and Cd Pb, Zn uptake by *Brassica napus*. *International Journal of Phytoremediation 16*, 321-333.

Johnson, D. L., Maguire, K. L., Anderson, D. R., McGrath, S. P. (2004). Enhanced dissipation of chrysene in planted soil; the impact of a rhizobial inoculum. *Soil Biology and Biochemistry 36*, 33-38.

Joner, E. J., Leyval, C. (2003). Rhizosphere gradients of polycyclic aromatic hydrocarbon (PAH) dissipation in two industrial soils, and the impact of arbuscular mycorrhiza. *Environmental Science and Technology 37*, 2371-2375.

Kang, S. H., Singh, S., Kim, J. Y., Lee, W., Mulchandani, A., Chen, W. (2007). Bacteria metabolically engineered for enhanced phytochelatin production and cadmium accumulation. *Applied and Environmental Microbiology 73*, 6317-6320.

Khan, Z., Doty, S. (2011). Endophyte assisted phytoremediation. *Current Topics in Plant Biology 12*, 97-104.

Khan, Z., Roman, D., Kintz, T., delas Alas, M., Yap, R., Doty, S. (2014). Degradation, phytoprotection and phytoremediation of phenanthrene by endophyte *Pseudomonas putida*, PD1. *Environmental Science and Technology 48*, 12221-12228.

Kim, J., Vipulanandan, C. (2006). Removal of lead from contaminated water and clay soil using a biosurfactant. *Journal of Environmental Engineering 132*, 777-786.

Kingsley, M. T., Fredrickson, J. K., Metting, F. B., Seidler, R. J. (1994). Environmental restoration using plant-microbe bioaugmentation. In: Hinchee, R. E., Leeson, A., Semprini, L., Ong, S. K. (eds.) *Bioremediation of Chlorinated and Polyaromatic Hydrocarbon Compounds*. Lewis Publishers, Boca Raton, FL, U.S.A. 287-292.

Kopittke, P. M., Blamey, F. P. C., Asher, C. J., Menzies, N. W. (2010). Trace metal phytotoxicity in solution culture: a review. *Journal of Experimental Botany*, *61*, 945-954.

Kozhakhanov, T. E., Lukashenko, S. N., Larionova, N. V. (2014). Accumulation of artificial radionuclides in agricultural plants in the area used for surface nuclear tests. *Journal of Environmental Radiation 137*, 217-226.

Kuffner, M., De Maria, S., Puschenreiter, M., Fallmann, K., Wieshammer, G., Gorfer, M., Strauss, J., Rivelli, A. R., Sessitsch, A. (2010). Culturable bacteria from Zn- and Cd-accumulating *Salix caprea* with differential effects on plant growth and heavy metal availability. *Journal of Applied Microbiology 108*, 1471-1484.

Kuffner, M., Puschenreiter, M., Wieshammer, G., Gorfer, M., Sessitsch, A. (2008). Rhizosphere bacteria affect growth and metal uptake of heavy metal accumulating willows. *Plant and Soil 304*, 35-44.

Kuiper, I., Kravchenko, L. V., Bloemberg, G. V., Lugtenberg, B. J. J. (2002). *Pseudomonas putida* strain PCL 1444, selected for efficient root colonization and naphthalene degradation, effectively utilizes root exudates components. *Molecular Plant Microbe Interaction 15*, 734-741.

Kuiper, I., Lagendijk, E. L., Bloemberg, G. V., Lugtenberg, B. J. J. (2004). Rhizoremediation: a beneficial plant-microbe interaction. *Molecular Plant Microbe Interaction 17*, 6-15.

Kumar, B., Gaur, R., Goel, G., Mishra, M., Singh, S. K., Prakash, D., Sharma, C. S. (2012). Residues of pesticides and herbicides in soils from agriculture areas of Delhi region, India. *Electronic Journal of Environmental Agricultural and Food Chemistry 11*, 328-338.

Kuyukina, M. S., Ivshina, I. B., Makarov, S. O., Litvinenko, L. V., Cunningham, C. J., Philp, J. C. (2005). Effect of biosurfactants on crude oil desorption and mobilization in a soil system. *Environment International 31*, 155-161.

Lampis, S., Santi, C., Ciurli, A., Andreolli, M., Vallini, G. (2015). Promotion of arsenic phytoextraction efficiency in the fern *Pteris vittata* by the inoculation of As-resistant bacteria: a soil bioremediation perspective. *Frontiers in Plant Science* doi:10.3389/fpls.2015.00080.

Langella, F., Grawunder, A., Stark, R., Weist, A., Merten, D., Haferburg, G., Büchel, G., Kothe, E. (2014). Microbially assisted phytoremediation approaches for two multi-element contaminated sites. *Environmental Science and Pollution Research 21*, 6845-6858.

Leigh, M. B., Fletcher, J. S., Fu, X., Schmitz, F. J. (2002). Root turnover: An important source of microbial substrates in rhizosphere remediation of recalcitrant contaminants. *Environmental Science and Technology 36*, 1579-1583.

Leigh, M. B., Prouzova, P., Mackova, M., Macek, T., Nagle, D. O., Fletcher, J. S. (2006). Polychlorinated biphenyl (PCB)-degrading bacteria associated with trees in a PCB-contaminated site. *Applied and Environmental Microbiology 72*, 2331-2342.

Li, H. Y., Wei, D. Q., Shen, M., Zhou, Z. P. (2012). Endophytes and their role in phytoremediation. *Fungal Diversity 54*, 11-18.

Li, W. C., Ye, Z. H., Wong, M. H. (2007). Effects of bacteria on enhanced metal uptake of the Cd/Zn hyperaccumulating plant, *Sedum alfredii*. *Journal of Experimental Botany 58*, 4173-4182.

Lin, X., Li, X., Li, F., Zhang, L., Zhou, Q. (2008). Evaluation of plant-microorganism synergy for the remediation of diesel fuel contaminated soil. *Bulletin of Environmental Contamination and Toxicology 81*, 19-24.

Lindblom, S. D., Valdez-Barillas, J. R., Fakra, S. C., Marcus, M. A., Wangeline, A. L., Pilon-Smits, E. A. H. (2013). Influence of microbial associations on selenium localization and speciation in roots of Astragalus and Stanleya hyperaccumulators. *Environmental and Experimental Botany 88*, 33-42.

Liste, H., Alexander, M. (2000). Plant-promoted pyrene degradation in soil. *Chemosphere 40*, 7-10.

Longley, M., Sotherton, N. W. (1997). Factors determining the effects of pesticides upon butterflies inhabiting arable farmland. *Agriculture Ecosystems and Environment 61*, 1-12.

Luo, C., Wang, S., Wang, Y., Yang, R., Zhang, G., Shen, Z. (2015). Effects of EDDS and plant-growth-promoting bacteria on plant uptake of trace metals and PCBs from e-waste–contaminated soil. *Journal of Hazardous Material*s *286*, 379-385.

Ma, B., He, Y., Chen, H. H., Xu, J. M., Rengel, Z. (2010). Dissipation of polycyclic aromatic hydrocarbons (PAHs) in the rhizosphere: synthesis through meta-analysis. *Environment Pollution 158*, 855-861.

Ma, Y., Rajkumar, M., Rocha, I., Oliveira, R. S., Freitas, H. (2015). Serpentine bacteria influence metal translocation and bioconcentration of *Brassica juncea* and *Ricinus communis* grown in multi-metal polluted soils. *Frontiers in Plant Science* doi:10.3389/fpls.2014.00757.

Makris, K. C., Sarkar, D., Datta, R. (2009). Coupling indigenous biostimulation and phytoremediation for the restoration of 2,4,6-trinitrotoluene-contaminated sites. *Journal of Environmental Monitoring 12*, 399-403.

Manickam, N., Bajaj, A., Saini, H. S., Shanker, R. (2012). Surfactant mediated enhanced biodegradation of hexachlorocyclohexane (HCH) isomers by *Sphingomonas* sp. NM05. *Biodegradation 23*, 673-682.

Mannisto, M. K., Tiirola, M. A., Puhakka, J. A. (2001). Degradation of 2,3,4,6-tetraclorophenol at low temperature and low dioxygen concentrations by phylogenetically different groundwater and bioreactor bacteria. *Biodegradation 12*, 291-301.

Megharaj, M., Ramakrishnan, B., Venkateswarlu, K., Sethunathan, N., Naidu, R. (2011). Bioremediation approaches for organic pollutants: A critical perspective. *Environment International 37*, 1362-1375.

Meng, L., Qiao, M., Arp, H. P. H. (2011). Phytoremediation efficiency of a PAH-contaminated industrial soil using ryegrass, white clover, and celery as mono-and mixed cultures. *Journal of Soils and Sediment 11*, 482-490.

Miller, E. K., Boydston, J. A., Dyer, W. E. (1998). Mechanisms of pentachlorophenol phytoremediation in soil. Pacific Northwest Regional Meeting of the ASAE, Fairmont Hot Springs, Montana, USA, 10-12 Sep. 1998, pp 7.

Moreira, H., Marques, A. P. G. C., Franco, A. R., Rangel, A. O. S. S., Castro, P. M. L. (2014). Phytomanagement of Cd-contaminated soils using maize (*Zea mays* L.) assisted by plant growth-promoting rhizobacteria. *Environmental Science and Pollution Research 21*, 9742-9753.

Narasimhan, K., Basheer, C., Bajic, V. B., Swarup, S. (2003). Enhancement of plant-microbe interactions using a rhizosphere metabolomics-driven approach and its application in the removal of polychlorinated biphenyls. *Plant Physiology 132*, 146-153.

Naz, N., Young, H. K., Ahmed, N., Gadd, G. M. (2005). Cadmium accumulation and DNA homology with metal resistance genes in sulfate-reducing bacteria. *Applied and Environmental Microbiology 71*, 4610-4618.

Nie, M., Zhang, X., Wang, J., Jiang, L., Yang, J., Quan, Z., Cui, X., Fang, C., Li, B. (2009). Rhizosphere effects on soil bacterial abundance and diversity in the Yellow River Deltaic ecosystem as influenced by petroleum contamination and soil salinization. *Soil Biology and Biochemistry 41*, 2535-2542.

Nonnoi, F., Chinnaswamy, A., García de la Torre, V. S., Coba de la Peña, T., Lucas, M. M., Pueyo, J. J. (2012). Metal tolerance of rhizobial strains isolated from nodules of herbaceous legumes *Medicago* spp. and *Trifolium* spp. growing in mercury contaminated soils. *Applied Soil Ecology 61*, 49-59.

Passatore, L., Rossetti, S., Juwarkar, A. A., Massacci, A. (2014). Phytoremediation and bioremediation of polychlorinated biphenyls (PCBs): state of knowledge and research perspectives. *Journal of Hazardous Materials 278*, 189-202.

Phillips, L. A., Germida, J. J., Farrell, R. E., Greer, C. W. (2008). Hydrocarbon degradation potential and activity of endophytic bacteria associated with prairie plants. *Soil Biology and Biochemistry 40*, 3054-3064.

Prasad, M. N. V. (2007). Sunflower (*Helinathus annuus* L.) - a potential crop for environmental industry. *Helia 30*, 167-174.

Rajkumar, M., Ae, N., Prasad, M. N. V., Freitas, H. (2010). Potential of siderophore producing bacteria for improving heavy metal phytoextraction. *Trends in Biotechnology 28*, 142-149.

Rajkumar, M., Nagendran, R., Lee, K. J., Lee, W. H., Kim, S. Z. (2006). Influence of plant growth promoting bacteria and Cr^{6+} on the growth of Indian mustard. *Chemosphere 62*, 741-748.

Reichman, S. M. (2007). The potential of the legume-rhizobium symbiosis for the remediation of arsenic contaminated sites. *Soil Biology and Biochemistry*, *39*, 2587-2593.

Rodriguez, P. B., Tome, F. V., Fernandez, M. P., Lozano, J. C. (2006). Linearity assumption in soil-to-plant transfer factors of natural uranium and radium in *Helianthus annuus* L. *Science of Total Environment 361*, 1-7.

Rosenblueth, M. N., Martinez-Romero, E. (2006). Bacterial endophytes and their interactions with hosts. *Molecular Plant-Microbe Interaction 19*, 827-837.

Ryan, P. R., Delhaize, E., Jones, D. L. (2001). Function and mechanism of organic anion exudation from plant roots. *Annual Review of Plant Physiology and Plant Molecular Biology 52*, 527-560.

Ryan, R. P., Germaine, K., Franks, A., Ryan, D. J., Dowling, D. N. (2008). Bacterial endophytes: recent developments and applications. *FEMS Microbiology Letter 278*, 1-9.

Rylott, E. L., Jackson, R. G., Edwards, J., Womack, G. L., Seth-Smith, H. M., Rathbone, D. A., Strand, S. E., Bruce, N. C. (2006). An explosive-degrading cytochrome P450 activity and its targeted application for the phytoremediation of RDX. *Nature Biotechnology 24*, 216-219.

Saiki, Y., Habe, H., Yuuki, T., Ikeda, M., Yoshida, T., Nojiri, H., Omori, T. (2003). Rhizoremediation of dioxin-like compounds by a recombinant *Rhizobium tropici* strain expressing carbazole 1,9a-dioxygenase constitutively. *Bioscience Biotechnology and Biochemistry 67*, 1144-1148.

Samardjieva, K. A., Gonçalves, R. F., Valentão, P., Andrade, P. B., Pissarra, J., Pereira, S., Tavares, F. (2015). Zinc accumulation and tolerance in *Solanum nigrum* are plant growth dependent. *International Journal of Phytoremediation 17*, 272-279.

Sessitsch, A., Kuffner, M., Kidd, P., Vangronsveld, J., Wenzel, W. W., Fallmann, K., Puschenreiter, M. (2013). The role of plant-associated bacteria in the mobilization and phytoextraction of trace elements in contaminated soils. *Soil Biology and Biochemistry 60*, 182-194.

Shagol, C. C., Krishnamoorthy, R., Kim, K., Sundaram, S., Sa, T. (2014). Arsenic-tolerant plant-growth-promoting bacteria isolated from arsenic-polluted soils in South Korea. *Environmental Science and Pollution Research 21*, 9356-9365.

Shann, J. R. (1995). The role of plants and plant/microbial systems in the reduction of exposure. *Environmental Health Perspective 103*, 13-15.

Shaw, L. J., Burns, R. G. (2004). Enhanced mineralisation of [U-14C] 2,4-dichlorophenoxyacetic acid in the soil from the rhizosphere of *Trifolium pratense*. *Applied and Environmental Microbiology 70*, 4766-4774.

Sheng, X., Chen, X., He, L. (2008a). Characteristics of an endophytic pyrene-degrading bacterium of *Enterobacter* sp. 12J1 from *Allium macrostemon* Bunge. *International Biodeterioration and Biodegradation 62*, 88-95.

Sheng, X. F., He, L. Y., Zhou, L., Shen, Y. Y. (2009). Characterization of *Microbacterium* sp. F10a and its role in polycyclic aromatic hydrocarbon removal in low-temperature soil. *Canadian Journal of Microbiology 55*, 529-535.

Sheng, X. F., Xia, J. J. (2006). Improvement of rape (*Brassica napus*) plant growth and cadmium uptake by cadmium resistant bacteria. *Chemosphere 64*, 1036-1042.

Sheng, X. F., Xia, J. J., Jiang, C. Y., He, L. Y., Qian, M. (2008b). Characterization of heavy metal-resistant endophytic bacteria from rape (*Brassica napus*) roots and their potential in promoting the growth and lead accumulation of rape. *Environment Pollution 156*, 1164-1170.

Shukla, K. P., Sharma, S., Singh, N. K., Singh, V., Tiwari, K., Singh, S. (2011). Nature and role of root exudates: Efficacy in bioremediation. *African Journal of Biotechnology 10*, 9717-9724.

Siciliano, S. D., Fortin, N., Mihoc, A., Wisse, G., Labelle, S., Beaumier, D., Ouellette, D., Roy, R., Whyte, L. G., Banks, M. K., Schwab, P., Lee, K., Greer, C. W. (2001). Selection of specific endophytic bacterial genotypes by plants in response to soil contamination. *Applied and Environmental Microbiology 6*, 2469-2475.

Siciliano, S. D., Greer, C. W. (2000). Plant-bacterial combinations to phytoremediate soil contaminated with high concentrations of 2,4,6-trinitrotoluene. *Journal of Environmental Quality 29*, 311-316.

Siddiqui, Z. S., Ahmed, S. (2006). Combined effects of pesticide on growth and nutritive composition of soybean plants. *Pakistan Journal of Botany 38*, 721-733.

Singh, N. (2003). Enhanced degradation of hexachlorocyclohexane isomers in rhizosphere soil of Kochia sp. *Bulletin of Environmental Contamination and Toxicology 70*, 775-782.

Smith, M. J., Flowers, T. H., Duncan, H. J., Alder, J. (2006). Effects of polycyclic aromatic hydrocarbons on germination and subsequent growth of grasses and legumes in freshly contaminated soil and soil with aged PAHs residues. *Environment Pollution 141*, 519-525.

Soudek, P., Tykva, R., Vaňková, R., Vaněk, T. (2006). Accumulation of radioiodine from aqueous-solution by sunflower (*Helianthus annuus* L.) at the laboratory scale. *Environmental and Experimental Botany 57*, 220-225.

Staicu, L. C., van Hullebusch, E. D., Lens, P. N. L., Pilon-Smits, E. A. H., Oturan, M. A. (2015). Electrocoagulation of colloidal biogenic selenium. *Environmental Science and Pollution Research International 22*, 3127-3137.

Stomp, A. M., Han, K. H., Wilbert, S., Gordon, M. P., Cunningham, S. D. (1994). Genetic strategies for enhancing phytoremediation. *Annals of the New York Academy of Sciences 721*, 481-492.

Sun, H., Xu, J., Yang, S., Liu, G., Dai, S. (2004). Plant uptake of aldicarb from contaminated soil and its enhanced degradation in the rhizosphere. *Chemosphere 54*, 569-574.

Sun, M., Fu, D., Teng, Y., Shen, Y., Luo, Y., Li, Z., Christie, P. (2011). In situ phytoremediation of PAH-contaminated soil by intercropping alfalfa (*Medicago sativa* L.) with tall fescue (*Festuca arundinacea* Schreb.) and associated soil microbial activity. *Journal of Soils and Sediment 11*, 980-989.

Sura-de Jong, M., Reynolds, R. J. B., Richterova, K., Musilova, L., Staicu, L. C., Chocholata, I., Cappa, J. J., Taghavi, S., van der Lelie, D., Frantik, T., Dolinova, I., Strejcek, M., Cochran, A. T., Lovecka, P., Pilon-Smits, E. A. H. (2015). Selenium hyperaccumulators harbor a diverse endophytic bacterial community characterized by high selenium resistance and plant growth promoting properties. *Frontiers in Plant Science* doi:10.3389/fpls.2015.00113.

Tank, N., Saraf, M. (2009). Enhancement of plant growth and decontamination of nickel-spiked soil using PGPR. *Journal of Basic Microbiology 49*, 195-204.

Teng, Y., Li, X., Chen, T., Zhang, M., Wang, X., Li, Z., Luo, Y. (2015). Isolation of the PCB-degrading bacteria *Mesorhizobium sp.* ZY1 and its combined remediation with *Astragalus sinicus* L. for contaminated soil. *International Journal of Phytoremediation* doi:10.1080/15226514.2015.1073667.

Teng, Y., Luo, Y., Sun, X., Tu, C., Xu, L., Liu, W., Christie, P. (2010). Influence of arbuscular mycorrhiza and rhizobium on phytoremediation by alfalfa of an agricultural soil contaminated with weathered PCBs: a field study. *International Journal of Phytoremediation 12*, 516-533.

Teng, Y., Shen, Y., Luo, Y., Sun, X., Sun, M., Fu, D., Li, Z., Christie, P. (2011). Influence of *Rhizobium meliloti* on phytoremediation of polycyclic aromatic hydrocarbons by alfalfa in an aged contaminated soil. *Journal of Hazardous Material*s doi:10.1016/j.jhazmat.2010.11.126.

Testiati, E., Parinet, J., Massiani, C., Laffont-Schwob, I., Rabier, J., Pfeifer, H. R., Prudent, P. (2013). Trace metal and metalloid contamination levels in soils and in two native plant species of a former industrial site: evaluation of the phytostabilization potential. *Journal of Hazardous Material*s *248*, 131-141.

Thimmaraju, R., Czymmek, K. J., Pare, P. W., Bais, H. P. (2008). Root-secreted malic acid recruits beneficial soil bacteria. *Plant Physiology 148*, 1547-1556.

Travis, E. R., Hannink, N. K., Van der Gast, C. J., Thompson, I. P., Rosser, S. J., Bruce, N. C. (2007). Impact of transgenic tobacco on trinitrotoluene (TNT) contaminated soil community. *Environmental Science and Technology 41*, 5854-5861.

Uhlik, O., Jecna, K., Macknova, M., Vlcek, C., Hroudova, M., Demnerova, K., Paces, V., Macek, T. (2009). Biphenyl metabolizing bacteria in the rhizosphere of horseradish and bulk soil contaminated by polychlorinated biphenyls as revealed by stable isotope probing. *Applied and Environmental Microbiology 75*, 6471-6477.

Ullah, A., Mushtaq, H., Ali, H., Munis, M. F. H., Javed, M. T., Chaudhary, H. J. (2015). Diazotrophs-assisted phytoremediation of heavy metals: a novel approach. *Environmental Science and Pollution Research 22*, 2505-2514.

Van Aken, B., Peres, C. M., Doty, S. L., Yoon, J. M., Schnoor, J. L. (2004). *Methylobacterium populi* sp. nov., a novel aerobic, pink-pigmented, facultatively methylotrophic, methane-utilizing bacterium isolated from poplar trees (*Populus deltoides* x nigra DN34). *International Journal of Systematic and Evolutionary Microbiology 54*, 1191-1196.

Vane, C. H., Kim, A. W., Beriro, D. J., Cave, M. R., Knights, K., Moss-Hayes, V., Nathanail, P. C. (2014). Polycyclic aromatic hydrocarbons (PAH) and polychlorinated biphenyls (PCB) in urban soils of Greater London, UK. *Applied Geochemistry 51*, 303-314.

Vangronsveld, J., Herzig, R., Weyens, N., Boulet, J., Adriaensen, K., Ruttens, A., Thewys, T., Vassilev, A., Meers, E., Nehnelajova, E., van der Lelie, D., Mench, M. (2009). Phytoremediation of contaminated soils and groundwater: lessons from the field. *Environmental Science and Pollution Research 16*, 765-794.

Verbruggen, N., Hermans, C., Schat, H. (2009). Mechanisms to cope with arsenic or cadmium excess in plants. *Current Opinion in Plant Biology 12*, 364-372.

Vivas, A., Biró, B., Ruíz-Lozano, J. M., Barea, J. M., Azcón, R. (2006). Two bacterial strains isolated from a Zn-polluted soil enhance plant growth and mycorrhizal efficiency under Zn toxicity. *Chemosphere 62*, 1523-1533.

Wang, Q., Xiong, D., Zhao, P., Yu, X., Tu, B., Wang, G. (2011). Effect of applying an arsenic-resistant and plant growth-promoting rhizobacterium to enhance soil arsenic phytoremediation by Populus deltoides LH05-17. *Journal of Applied Microbiology 111*, 1065-1074.

Wang, S., Zhao, X. (2009). On the potential of biological treatment for arsenic contaminated soils and groundwater. *Journal of Environmental Management 90*, 2367-2376.

Wani, P. A., Khan, S., Zaidi, A. (2007). Effect of metal tolerant plant growth promoting *Bradyrhizobium* sp (vigna) on growth, symbiosis, seed yield and metal uptake by greengram plants. *Chemosphere 70*, 36-45.

Wani, P. A., Khan, S., Zaidi, A. (2008). Effect of metal-tolerant plant growth promoting rhizobium on the performance of pea grown in metal-amended soil. *Archives of Environmental Contamination and Toxicology 55*, 33-42.

Weyens, N., Beckers, B., Schellingen, K., Ceulemans, R., Croes, S., Janssen, J., Haenen, S., Witters, N., Vangronsveld, J. (2013). Plant-associated bacteria and their role in the success or failure of metal phytoextraction projects: first observations of a field-related experiment. *Microbial Biotechnology 6*, 288-299.

Weyens, N., Truyens, S., Dupae, J., Newman, L., Taghavi, S., Lelie, D., Carleer, R., Vangronsveld, J. (2010). Potential of the TCE degrading endophyte Pseudomonas putida W619-TCE to improve plant growth and reduce TCE phytotoxicity and evapotranspiration in poplar cuttings. *Environment Pollution 158*, 2915-2919.

Weyens, N., van der Lelie, D., Taghavi, S., Newman, L., Vangronsveld, J. (2009). Exploiting plant-microbe partnerships to improve biomass production and remediation. *Trends in Biotechnology 27*, 591-598.

White, J. C. (2001). Plant-facilitated mobilization and translocation of weathered 2,2-bis(p-chlorophenyl):-1,1-dichloroethylene (p,p'-DDE) from an agricultural soil. *Environmental Toxicology and Chemistry 20*, 2047-2052.

Wu, S. C., Cheung, K. C., Luo, Y. M., Wong, M. H. (2006). Effects of inoculation of plant growth-promoting rhizobacteria on metal uptake by *Brassica juncea*. *Environment Pollution 140*, 124-135.

Xu, C., Chen, X., Duan, D., Peng, C., Le, T., Shi, J. (2015). Effect of heavy-metal-resistant bacteria on enhanced metal uptake and translocation of the Cu-tolerant plant, *Elsholtzia splendens*. *Environmental Science and Pollution Research International 22*, 5070-5081.

Xun, F., Xie, B., Liu, S., Guo, C. (2015). Effect of plant growth-promoting bacteria (PGPR) and arbuscular mycorrhizal fungi (AMF) inoculation on oats in saline-alkali soil contaminated by petroleum to enhance phytoremediation. *Environmental Science and Pollution Research 22*, 598-608.

Yang, C., Zhou, Q., Wei, S., Hu, Y., Bao, Y. (2011). Chemical-assisted phytoremediation of CD-PAHs contaminated soils using *Solanum nigrum* L. *International Journal of Phytoremediation 13*, 818-833.

Yang, Q., Tu, S., Wang, G., Liao, X., Yan, X. (2012). Effectiveness of applying arsenate reducing bacteria to enhance arsenic removal from polluted soils by *Pteris Vittata* L. *International Journal of Phytoremediation 14*, 89-99.

Yenn, R., Borah, M., Boruah, H. D., Roy, A. S., Baruah, R., Saikia, N., Tamuli, A. K. (2014). Phytoremediation of abandoned crude oil contaminated drill sites of Assam with the aid of a hydrocarbon-degrading bacterial formulation. *International Journal of Phytoremediation 16*, 909-925.

Yu, X. Z., Wu, S. C., Wu, F. Y., Wong, M. H. (2011). Enhanced dissipation of PAHs from soil using mycorrhizal ryegrass and PAH-degrading bacteria. *Journal of Hazardous Material 186*, 1206-1217.

Yu, Y. L., Chen, Y. X., Luo, Y. M., Pan, X. D., He, Y. F., Wong, M. H. (2003). Rapid degradation of butachlor in wheat rhizosphere soil. *Chemosphere 50*, 771-774.

Zaidi, S., Usmani, S., Singh, B. R., Musarrat, J. (2006). Significance of *Bacillus subtilis* strain SJ-101 as a bioinoculant for concurrent plant growth promotion and nickel accumulation in *Brassica juncea. Chemosphere 64*, 991-997.

Zhang, J., Yin, R., Lin, X., Liu, W., Chen, R., Li, X. (2010). Interactive effect of biosurfactant and microorganism to enhance phytoremediation for removal of aged polycyclic aromatic hydrocarbons from contaminated soils. *Journal of Health Science 56*, 257-266.

Zhang, Y., He, L., Chen, Z., Zhang, W., Wang, Q., Qian, M., Sheng, X. (2011). Characterization of lead-resistant and ACC deaminase-producing endophytic bacteria and their potential in promoting lead accumulation of rape. *Journal of Hazardous Materials 186*, 1720-1725.

Zhu, Y. G., Shaw, G. (2000). Soil contamination with radionuclides and potential remediation. *Chemosphere 41*, 121-128.

In: Chemical Pollution Control … ISBN: 978-1-53611-034-0
Editor: Naser A. Anjum

Chapter 8

HAIRY ROOT *IN VITRO* SYSTEMS: A SUITABLE BIOLOGICAL MATRIX FOR PLANT BASED REMEDIATION OF ENVIRONMENTAL POLLUTANTS

Shakti Mehrotra[1,*] and Vikas Srivastava[2]
[1]Plant Biotechnology Division, Central Institute of Medicinal and Aromatic Plants, PO – CIMAP, Picnic Spot Road, Lucknow, India
[2]Department of Plant Sciences, Central University of Jammu, Rahya-Suchani (Bagla), Dist. Samba, Jammu (J&K), India

ABSTRACT

In recent years, the ability of hairy root cultures (HRCs) to simulate enzyme based degradation, tolerance and hyperaccumulation of pollutant molecules has been scientifically focused. The versatility and close resemblance with native plants have made HRCs a preferred system for studying various *in planta* physiological and biochemical events during pollutant stress. Additionally, use of HRCs also facilitated the study of catabolic pathways and identification of related enzymes involved in pollutant metabolism. This chapter presents a comprehensive account on various aspects of remediation of environmental pollutants utilizing

[*]Corresponding author: Shakti Mehrotra, Plant Biotechnology Division, Central Institute of Medicinal & Aromatic Plants, PO – CIMAP, Picnic Spot Road, Lucknow, India; E-mail: shaktimehrotra22@gmail.com.

HRCs of different plant species. In addition, the enzymatic degradation of phenol and its derivatives, polyaromatic hydrocarbons, polychlorinated biphenyl ethers, azodyes, antibiotics etc. have also been focused. Besides, development of up-scale culture system for large scale and incessant production of effective enzymes, genetic engineering for expression of candidate genes for the enhancement of remedial potential and microbe assisted hairy root remediation system has also been discussed as potential prospects for exploring HRCs based phytoremediation.

Keywords: hairy roots, hyperaccumulation, phytoremediation, tolerance, phytodegradation

ABBREVIATIONS

PTC	Plant Tissue culture
HRCs	Hairy root cultures
HR	Hairy roots
PHAs	Polyaromatic hydrocarbons
PCBs	Polychlorinated biphenyl ethers
HM	Heavy metals
DNT	2,4-dinitrotoluene
TNT	2,4,6- trinitrotoluene
SOD	Super oxide dismutase

1. PHYTOREMEDIATION THROUGH PLANT TISSUE CULTURE: INTRODUCTION TO A MORE FEASIBLE STRATEGY

Biotechnological interventions to agriculture and environmental science have greatly transformed the conventional concepts and practices of crop production, improvement and protection. Continuous efforts are being taken to improve the in-built production potential of plants as they are natural producers of food, medicine, fiber and building materials. Besides, biotechnological mediation also offers the exploration of some unique features of plants for their improvement and protection. Natural ability of plants to deal with environmental pollutants is one such unique feature (Cobbett et al., 2000; Dhankher et al., 2009; Doran, 2009). To be more precise, some plants utilize hazardous organic and inorganic pollutants in their metabolism and either remove them completely or render them into less harmful form. This kind of remedial degradation or conversion of environmental pollutants through plants is studied under the subject area of phytoremediation (Cobbett et al., 2000; Meagher et al., 2000; Arthur et al., 2005; Kramer, 2005; Marmiroli et al., 2006; Lee et al., 2013). In recent years, the ability of plants to remove, tolerate or even degrade pollutant molecules into harmless forms has been scientifically focused. Various strategies under phytoremediation utilize the ability of plants to clean-up soil, air and water from a range of pollutants with the help of their inherent or acquired metabolism (Garbisu, 2002; Macek et al., 2004; Brazier-Hicks et al., 2007). In contrast to conventional strategies of environment clean-up, the plant based remediation offers utilization of natural phenomenon that too on less expense and without any threat to the natural populations (Cunningham et al., 1996; Brazier-Hicks et al., 2007; Wu et al., 2010). Moreover, the plant-based remediation offers a renewable system that can also be utilized *in-situ* (Suresh and Ravishshanker, 2004; Doty, 2007; Schroder et al., 2008; Gerhardt et al., 2009). Although, the microbial remediation is considered as a fast and very efficient remedial system yet their maintenance at polluted sites needs additional efforts. For example, performance of microbial system for longer periods relies on adequate nutritional supply (Eapen et al., 2007; Macek et al., 2000; 2008). Moreover, it often involves manipulation of environmental parameters to allow uninterrupted microbial growth. The other conventional remedial systems involving physical (irradiation, neutralization and incineration of contaminated medium) and chemical (vapourization, oxidation and extraction of toxic compounds from contaminated medium) methods require tedious efforts, high cost and intense labor. These challenges and limitations in conventional methods have diverted the global concern towards phytoremediation methods that utilize physiological activities of plants.

In recent past, substantial efforts are being made to develop practically more feasible strategies for plant-based remediation. This includes "plant tissue culture technology" where undifferentiated (cell suspensions and callus), differentiated (whole plant, shoot and root) and transgenic cultures are frequently utilized for this purpose (Doran, 2009; Govindwar and Kagalkar, 2010). Plant tissue culture (PTC) technology offers significant advantages in investigating inherent potential of plant cells and their capability to deal with toxic molecules with lesser instances of microbial interference. However, treating *in-vitro* cultures as a representative of whole plant needs little attention with reference to the bioavailability of pollutants, their uptake and accumulation efficiency that are likely to be different in two systems. In spite of this difference, the results of the PTC based remediation studies are extremely helpful in predicting the physiological behavior of plants in the presence of toxic pollutants. In this reference, hairy root cultures (HRCs) that are developed as a consequence of deliberate bacterial infection to different parts of the host plant can be used as a preferred biological matrix to explore phytoremediation reactions (Doran, 2011; Georgive et al., 2012; Agostini et al., 2013; Zhou et al., 2013). The versatility and close resemblance of HRCs with the native plants makes this system more appropriate for studying different *in planta* physiological and biochemical events. This chapter focuses on the use of HRCs to investigate plant based processes and mechanism for remediation of hazardous environmental pollutants. The chapter also points out specific advantages and limitations of HRCs based phytoremediation research.

2. Hairy Root Cultures (HRCs): Potential Matrix for Phytoremediation

Hairy roots are the fibrous and thread like structures developed from the plant tissues infected with soil dwelling Gram-negative bacteria *Agrobacterium rhizogenes*. During the infection, the bacterial cells transfers a particular DNA segment (T- DNA) from their root inducing (Ri) plasmids to the host cell, which randomly incorporates into host genome (Gelvin, 2003; Veena and Taylor, 2007; Georgiev et al., 2007; 2010; Chandra and Chandra, 2011; Chandra, 2012). Successful gene transfer involves adequate synchronization between an array of molecular activities at bacteria and plant level. Integration into plant genome and simultaneous expression of bacterial genes presents on T-DNA attribute to the consequent altered morphology, physiology and biochemistry of host (Christey and Braun, 2005). One of the visible symptoms of this gene transfer phenomenon is the emergence of hair like protrusions or hairy roots (HR) at infection sites. In laboratory, intentional infection of different strains of *A. rhizogenes* to various plant species may result in the development of hairy root *in vitro* systems (Tepfer, 1989; Ono and Tian, 2011; Mehrotra et al., 2015). Being differentiated and organized, hairy roots closely mirror the physiological and biochemical activities of the native plant and by this means offer authenticity to a large extent, in investigating plant behavior under controlled conditions. The hairy roots are genetically stable and show several specific properties like steady metabolic potential, fast growth and easy maintenance in hormone free defined growth medium. Besides, low cost implementation and ease of handling allows their long-term maintenance and development of large-scale bioreactor systems (Kim et al., 2002; Srivastava and Srivastava 2007; Mehrotra et al., 2015). Similar to the intact plant roots, the *in vitro* hairy roots possess more area of contact with the culture medium. Further, being isolated cultures, the hairy roots also avoid the interference of translocation of compounds to the other tissues. The absence of soil matrix also rules out the interactions with rhizosphere microbial communities. Moreover, HRCs offer an easy *in vitro* system for heterologous gene transfer, characterization and expression (Guillon et al., 2006a,b; Mehrotra et al., 2010; Zhou et al., 2011; Georgiev et al., 2012). In this instance, myriad reports are available demonstrating genetic engineering in HRCs and enhancement of their inherent enzymatic potential for secondary metabolite production (Georgiev et al., 2007; Mehrotra et al., 2015; Srivastava et al., 2016). Additionally, reports are also available where enzymatic pathways in HRCs have been modulated for several value added applications like recombinant protein production and biotransformation etc. (Georgiev et al., 2012; Srivastava, 2015, 2016). To sum up, in the context of plant based remediation of environmental pollutants, such striking features of HRCs make them superior and advantageous over microbial/whole plant based conversion and degradation of toxic compounds. A comparative account on properties of different phytoremediation system is given in Table 8.1.

Table 8.1. Comparison of hairy root *In-Vitro* systems with microbial and whole plant systems of remediation of toxic compounds

Hairy root *in-vitro* systems	Microbial systems	Whole plants
Rapid growth in microbe free defined medium	Rapid growth and low doubling time	Slow growth with absolute microbial interactions during uptake and absorption of nutrients
Ease in maintenance for virtually infinite growth and recycled for several generations	Indefinite growth with continuous nutrient supply	Definite life span. Individual plants are needed to be grown and re-established for individual experiment
Sustained growth with genetic and biochemical stability in medium variables and give highly reproducible results	Growth dynamics largely depends upon nutritional supply. Uniform growth conditions generate reproducible results	Variation in physical/chemical conditions may lead to varying and non-reproducible results

Table 8.1. (Continued)

Hairy root *in-vitro* systems	Microbial systems	Whole plants
Being isolated cultures no possible interference of translocation of compounds to other tissues, soil matrix, and also interactions with rhizosphere microbial communities	Simple cellular uptake of pollutant molecules and degradation through simplified metabolic activities	Uptake and translocation is influenced and interrupted by barriers like microbes, cellular matrix, physical and chemical properties of growth medium
Amenable to gene transfer and over expression for enhanced rate of remediation even in isolated culture forms.	Easy system for gene transfer and over expression to enhance catabolic activities against candidate molecule	Enhanced rate of remediation can be achieved only after the establishment of transgenic plants
High degree of physiological and biochemical resemblance with isolated normal root cultures or intact plant system.	Resemble with intact plants in certain metabolic pathways, enzymes and transformation reactions.	--------------
Ideal for investigating cell/or tissue-specific metabolic activities. Phytoremediation mechanisms restricted to roots can only be simulated	Ideal for comparative assessment of only those pathways and reactions that are similar in plants	Different Phytoremediation mechanism involving aerial parts (volatilization etc.) can be explored. Not adequate for studying cell/tissue specific metabolic activity
System easy to harvest for isolation of intermediates or bi-products of bioconversion reactions. In some cultures products leaches out in the medium that can be separated without destroying the culture	System easy to harvest for intermediates or bi-products of bioconversion reactions. Products in some cultures leaches out in the medium that can be separated without destroying the culture	Harvesting sometimes lead to complete eradication of plants
Excellent possibility of development in large scale bioreactor systems for transgenic and non-transgenic hairy roots	Development in large scale remedial system is easy for normal and transgenic microbes	Large-scale cultivation of candidate plants requires geo-physical, climatic and seasonal barriers. Additionally, transgenic cultivation may require authoritarian supervision
Possess large surface area for contact and absorption of contaminants from growth medium	------	Absolute soil-microbe-plant matrix functions for bioavailability and absorption of candidate molecule
Performance at natural contaminated sites still has to face several bottle necks	Uninterrupted performance at natural sites require continuous inoculation and nutrient supply	Wide applications in remediation at original contaminated sites.

3. Mechanism of Remediation: In-Planta Fate of Pollutant Molecules

A generalized classification of chemical pollutants to which plants and other life forms come across is based on their chemical structure. Broadly they are grouped as organic and inorganic pollutants that are released into the environment as a result of some natural events and various human activities. Introduction and surplus presence of these toxic chemicals cause imbalance in the environment and consequent soil, air and water pollution. As these compounds have structural differences, plants possess differential mechanism for their uptake and metabolism (Clemens et al., 2006; Abhilash et al., 2009; Zhou et al., 2013; Agostini et al., 2013). Major organic pollutants include oil and petroleum refined products, polyaromatic hydrocarbons (PHAs), polychlorinated biphenyl ethers (PCBs), pesticides, herbicides, alcohols, phenols, detergents, azodyes etc. Among inorganic contaminants the heavy metals, fertilizers, perchlorates are the major types that make soil, water and air toxic.

Different mechanisms that are acquired by the plants for uptake, translocation and metabolism of organic and inorganic pollutant molecules majorly include phytoextraction which involves uptake of pollutant molecules through roots followed by their translocation and subsequent storage in cells (hyperaccumulation) and metabolic degradation into simple, less toxic forms (phytodegradation) (Doty, 2008). Further, some plants also volatilize the hazardous molecules into the environment with the help of transpiration pull following their uptake from polluted soil or water. For example, trichloroethylene (TCE) is one of such volatile organic compounds (VOCs) of which uptake is influenced by the rate of transpiration. TCE is further released from the plants in its volatile form through the mechanism known as phytovolatilization (Xingmao and Burken, 2003) (Figure 8.1).

Earlier, Sandermann (1992 and 1994) proposed phase wise categorization of all kinds of detoxification mechanisms acquired by the plants. In a three phase operation, in phase I, the pollutant molecules undergo phytotransformation reactions where it gets transformed into less toxic forms through enzymatic degradation. This is followed by conjugation in phase II where transformants conjugates with the cellular compounds like saccharides, proteins, lignins etc. The final and phase III involves sequestration or compartmentation of conjugates into various parts of cell such as vacuole and cell walls (Georgiev et al., 2012; Agostini et al., 2013).

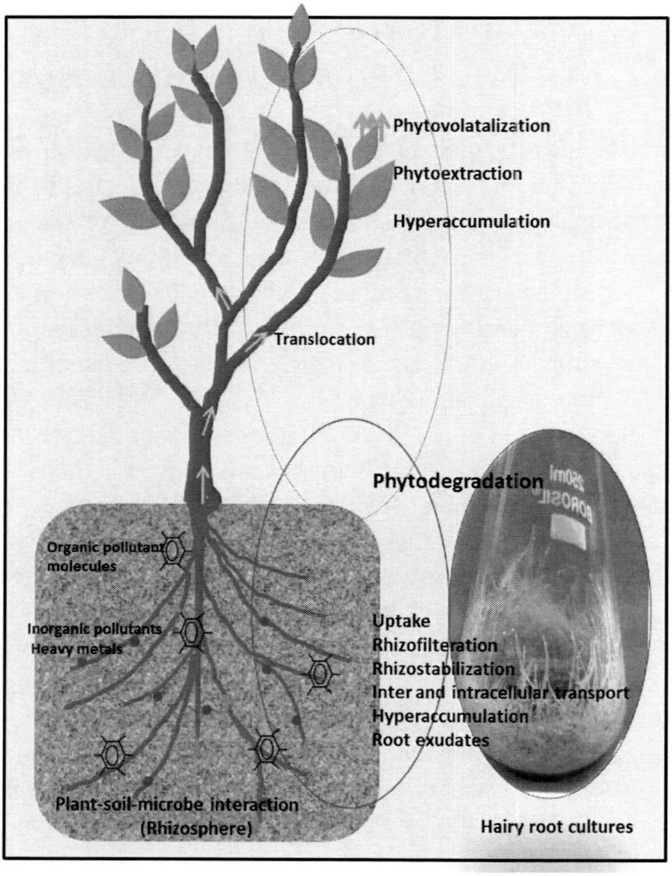

Figure 8.1. Mechanism of uptake and metabolism of pollutant molecules in plants; Various plant root based remedial activities can be simulated and investigated utilizing *in-vitro* HRCs.

Roots have specially developed surfaces for the maximum contact and absorption of molecules from the rhizosphere. These surfaces are distinguished by the presence of specific chemical receptors (Tangahu et al., 2011). Pollutants bind with the surface receptors as effectively as nutrient molecules and accumulate on root surface and subsequently absorbed in the root tissue. This kind of activity becomes helpful during rhizofilteration and phytostabilization (reducing the mobility of heavy metal ions in soil/medium) of metal contaminated soils and water (Alkorta et al., 2010). Also, the plant roots often secrete exudates that enhance the bioavailability of specific metal ions. Exudates from plant roots contain some oxidoreductive enzymes and metal chelators, which form metal-chelates and other complexes with the metal ions that are otherwise firmly bound with soil particles and complexes thus, facilitates the process of rhizofilteration (Gleba et al., 1999; Meagher, 2000). Roots exudation depends upon several internal and external factors like plant species, metal/nutrient stress, rhizospheric microbial activities and chemical properties of soil etc. For example, in an iron-deficient soil, plant roots of Poaceae members often secrete deoxymugenineic acid (siderophores), which is highly selective for Fe^{3+}. Iron in the form of Fe^{3+}–phytosiderophore complex get accumulated in rhizosphere and subsequently absorbed by the roots (Romheld, 1991; Yehuda et al., 1996; Morrissey et al., 2009).

Although selective, the uptake of inorganic pollutants through root cells is as efficient as the uptake of other nutrients and largely depends upon the redox potential. Redox potential (Eh) and pH of growth medium are the two master factors that influence the mobilization and bioavailability of metal particularly heavy metal ions (Reddy and Patrick, 1977; Husson et al., 2013). Though in traces, various inorganic pollutants are also important for plants as other nutrient molecules. Thus, plants have similar mechanism for the uptake and translocation of inorganic nutrient as well as pollutant molecules. Various families of transmembrane proteins for example, Zn transporters (ZIP) proteins play an important role in uptake and translocation of various inorganic metal ions. Mediation of ZIP and other proteins in uptake and translocation of iron in hairy root cultures of *Medicago truncatula* has recently been proposed (Tejada-Jiménez, 2015). In contrast, the organic pollutant molecules act as xenobiotics and thus, plant system do not possess a proper transport for these molecules. The uptake and metabolism of such compounds largely depend upon diffusion force, physical/chemical properties of the compound and medium (soil or water) pH, temperature, rhizosphere microflora etc.

3.1. Enzymatic Degradation of Various Organic Pollutants: Hairy Root Cultures Based Case Studies

In-planta conversion of organic contaminants into simple and less/non-toxic forms involves cellular enzymatic paraphernalia. Several studies reveal the involvement of plant endogenous peroxidases, catalases, laccases and H_2O_2 in uptake and degradation of toxic compounds (Boominathan and Doran, 2003; Singh et al., 2006; Ghodake et al., 2009; Talano et al., 2010). Mimicing the intact plant behavior, cultured roots and hairy roots react in a similar fashion if harmful molecules are externally supplied into the growth medium. The enzymatic system of hairy root tissues actively participates in degradation of toxic compounds. Hairy root cultures of *Cichorium intybus* and *Brassica juncea* were initiated to investigate uptake and degradation of DDT (Suresh et al., 2005). The TLC analysis of root tissue extract revealed the presence of 1,1-dichloro- 2,2-bis(*p*-chlorophenyl)ethylene (DDE) and 1,1-dichloro-2,2-bis(p-chlorophenyl)ethane (DDD), the common transformation products of DDT. These transformants (DDE and DDD) are also formed as a result of microbial degradation of DDT under natural conditions (Hecht et al., 2004; Mwangi et al., 2010). Detection of these compounds in the HRCs of test plants suggested the involvement of cellular enzymes of root tissues in DDT degradation.

Phenols and associated compounds like chlorophenols, catechol, nitrophenols, bisphenolsetc are the common environmental pollutants. These compounds exist in the environment due to the activities of chemical, petroleum, pesticides and pharmaceutical industries. Infiltration of phenolic substances in ecosystem leads to the histopathological changes, mutagenecity and carcinogenicity in its living components. To avoid the limitations and challenges of conventional remediation methods of phenols and derivatives, in recent advances, the plant tissue cultures are being used to screen the capabilities of various plants for remediation of these compounds. In this context, owing to their properties, the HRCs are observed as efficient system to study the cellular accumulation, degradation and plant based removal of these toxic compounds (Gonzalez et al., 2012a, b; Ibanez et al., 2011; Sosa-Alderate, 2009, 2012a). These studies may help in identifying common and/or species specific metabolic pathways for phenols and *in-planta* fate of toxic pollutants even after their transformation. The upcoming text demonstrates some relevant examples of HRC based studies where uptake and phytoremediation of phenol and its derivative compounds have been explored.

In a study, HRCs of *Solanum aviculare, Daucus carota and Ipomea batata* have been established to investigate uptake and transformation of phenols and chlorophenols through the involvement of inherent peroxidases (Santos de Araujo et al., 2006). Out of three, hairy roots of *S. aviculare* exhibited best response in terms of sustained growth (in the medium containing 2000 μmol l^{-1} phenol) along with high degree of enzymatic removal of phenolic compounds from the aqueous medium (98.6%). In addition to this, majority of the metabolites detected from transformation of phenolic compounds labeled with ^{14}C were polar conjugates and get accumulated into the cell walls or membranes. Furthermore, while studying comparative enzyme kinetics, it was observed that peroxidases from these three different plants exhibited differential affinity towards different phenolic substrates like phenol, catechol, 2,4-dichloro and 2,4,6-trichlorophenol. The study demonstrated not only variation in substrate preference at individual plant level but also the peroxidase isozymes based variation and metabolic potential for different phenolic substrates (Coniglio et al., 2008). Thus, HRCs also offers a significant matrix in understanding the substrate affinity, specificity and catalytic activity of various isoforms of enzymes during the metabolism of pollutants. Such studies can be helpful in screening of plant species that have high degree of inherent potential of removal of phenols and related toxic compounds. Such species with high remedial potential can also be explored for the identification of possible pathways and candidate genes involved in remedial reactions. Further, the activity of peroxidase and its isoforms during phenol removal from the growth medium through HRCs of different plants was also studied (Coniglio et al., 2008; González et al., 2008; Sosa Alderete et al., 2012b). In this context, the HRCs of *Lycopersicon* that naturally contains high peroxidase activity justify attention. In the enzyme kinetic study of extracts of *Lycopersicon* hairy roots, the basic peroxidases were found to be the main enzymes that actively participated in phenol consumption from aqueous medium (González et al., 2006). Similar results were obtained in HRCs of *Nicotiana tabaccum* where basic peroxidase was found to have more affinity towards its phenolic substrate (Sosa Alderete et al., 2012a). Contrary to this, in *Brassica napus* HRCs, acidic peroxidase exhibited greater affinity and increased degrading activity towards phenolic compounds as compared to basic and neutral peroxidases (Agostini et al., 2003; Coniglio et al., 2008). Thus, the ability of plant species to degrade pollutant molecule may supposedly depend upon the biochemical properties of metabolizing enzyme present in its tissues. Therefore, these are the issues that have to be focused while selecting a potential plant based system for degradation of phenolic compounds. HRCs of *Daucus carota* and *Lycopersicon* also exhibited increased peroxidase activity upon exogenous supplementation of phenolic compounds in the growth medium (Santa de Arauzo et al., 2006; Talano et al., 2003). Involvement of cellular peroxidases in removal of 2,4-dichlorophenol was also observed in *Nicotiana* HRCs (Talano et al., 2010). In a recent study, hairy roots of *Atropa belladonna* were utilized for phenol removal from waste-waters (Mazaheri et al., 2015). In the presence of optimized concentrations of H_2O_2, the *A. belladonna* hairy roots were able to remove 98% of phenol-based toxins from the aqueous medium. Such studies confirm the significant role of cellular peroxidases during the catabolism of phenolic compounds. Existence of plant specific variations in enzymes for substrate specificity as well as variations in substrate affinity and catalytic behavior of various isoforms of enzymes may lead to the difference in consequent products formed after the HRCs based degradation of phenols and similar compounds. This indicates towards the existence of differential mechanisms in which phytotransformation of xenobiotic compounds takes place in different plant species.

Explosive leftovers including nitro-substituted compounds like 2,4-dinitroroluene (DNT), 2,4,6- trinitrotoluene (TNT), hexahydro-1,3,5-trinitro-1,3,5-triazine (RDX) and glycerol trinitrate (GNT) etc. are also the major contributor of environmental pollution. Plants efficiently take up these residual explosives and partially transform them through metabolic activities. However, for explosive pollutants, simulation of plant based remedial process utilizing *in-vitro* cultures is less explored yet, HRCs have been used in few studies of plant based degradation of DNT and TNT. Uptake, transformation and formation of conjugation products of these explosives and other related compounds as well as their effects on enzymatic activity in hairy root cultures of *Catharanthus roseus* and *Armoracia rusticana* has been reported (Hughes et al., 1997; Bhadra et al., 1999; Nepovím et al., 2004). The study suggested severe adverse effects of these xenobiotic compounds on enzymatic activities of hairy root cells. Upon treatment with TNT, axenic cultures of *C. roseus* hairy roots exhibited the presence of aminated nitrotoluenes and other soluble conjugates as transformed products (Bhadra et al., 1999). Since HRCs possess similar intrinsic physiology to that of intact plants, these results can be visualized while investigating whole plant based remediation of TNT and other explosive molecules. Nevertheless, the need of present hour is to encourage efforts to achieve practical solutions to translate this remediation phenomenon from laboratory to the actual polluted sites.

The process of transport, storage and degradation of complex xenobiotics and the formation of less toxic products are far more complex to investigate in whole plants. In contrast, being an isolated *in-vitro* system, the hairy roots are easy to handle and thus, appear more suitable to study such processes (Schröder et al., 2007). Enzymatic degradation of polychlorinated biphenyl ethers (PCBs) through hairy root cultures of *Atropa belladonna* and *Solanum nigrum* was evident through the identification of metabolites of complex PCBs in root tissue (Mackova et al., 1997a, b; Morita et al., 2001; Kucerova et al., 2000; Rezek et al., 2007). In a comparative study of different *S. nigrum in-vitro* cultures, the HRCs exhibited greater potential to transform PCBs than undifferentiated cultures. The detection of mono and di-hydroxy-chlorobiphenyls as metabolites in HR extracts confirmed the transformation of PCBs into comparatively these less toxic products (Kucerova et al., 2000). To further extension, Rezek et al., (2012) demonstrated the formation of methoxy-PCBs and hydroxyl-methoxy-PCBs as novel compounds along with di-hydroxy-chlorobiphenyls as a result of PCBs transformation. Methoxy-PCBs are the metabolites accumulated in cultured tissues of some plant species as a result of detoxification of polychlorinated biphenyl ethers compounds (Rezek et al., 2008). This has paved the way to extend such studies to the identification of possible catabolic activities and products of PCBs degradation in HRCs. Further, HRCs can be considered as ideal system for these studies as detection of pathways and products would indicate the existence of similar in the intact native plants.

Another important group of organic pollutants comprised of azo dyes. Azo dye is one of the major classes of detrimental environmental pollutants largely present in effluents of textile and leather industries. *In-vitro* systems including HRCs offers a suitable milieu to investigate enzymatic degradation mechanism and intermediate/final transformed products during remediation of azo dyes. HRCs of *Brassica juncea* and *Tagetus patula* were found to have high potential for removal of textile dyes like methyl orange and Reactive red-198 (Patil et al., 2009; Telke et al., 2011). Involvement of enzymes in the metabolism of azo dyes is recently characterized through the HRCs of *Physalis minima* (Jha et al., 2015). In this study, temporal activation of different antioxidant enzymes like SOD, peroxidases and catalase was observed in hairy roots exposed to the dye Reactive Black-8 (RB8). This kind of enzyme kinetics in HRCs during azo dye stress suggested the coordinated and synchronized performance of cellular antioxidant enzymatic machinery to rescue the tissues from the oxidative damage triggered by the azo dyes. Involvement of intracellular laccases and other enzymes in hairy roots and microbes based removal of dyes is also reported (Ghodake et al., 2009; Patil et al., 2009; Kagalkar et al., 2009, 2010). As HRCs reveal a common spectrum of major metabolic reactions and metabolites to that of intact plants, such studies can be extended and visualized to the whole plant level. Such information may provide help in understanding the role of various enzymes in plant based dye decolorization and further identification of responsible candidate genes. The expression of such candidate genes in transgenic systems may enhance their remedial potential and thus, can be helpful in elucidating molecular mechanism behind xenobiotic metabolism, hyperaccumulation and tolerance. In a recent study, hairy roots of a halophyte plant *Sesuvium portulacastrum*were screened for remediation of 10 structurally different toxic textile dyes including Reactive green 19A-HE4BD (Lokhande et al., 2015). This plant possesses the natural ability of phytoremediation of heavy metals and degradation of textile dyes. HRCs of *S. portulacastrum* exhibited degradation ranged from 55 to 98% for different dyes and this differential degradation was attributed to the structural difference and molecular weight of dyes. Maximum degradation was observed with Reactive green 19A-HE4BD (98%), which was confirmed by HPLC and FTIR analysis. Study also proposed that hairy root based dye decolorization and degradation is a time dependent process, which is directly proportional to the concentration of dye.

Further, the constant exposure of antibiotic sources to the environment and insufficient regulation on their disposal has created major environmental concern (Martinez et al., 2009). Constant exposure to antibiotics may lead to multi antibiotic resistance in living components of ecosystem. Tetracycline (TC) is one of the major antibiotics used for various human and veternery health purposes. Under varying environmental conditions tetracycline readily transforms into its isomers epi-TC and anhydro-TC, which possess even higher toxicity to life forms. A number of plant species like sunflower, vetiver grass, water lettuce, tomato and mustard etc. has natural ability to degrade antibiotics (Gujarathi et al., 2005). HRCs of *Helianthus annuus* (sunflower) were found efficient to catalyze the degradation of tetracycline (TC) and oxytetracycline (OTC) from aqueous media indicating their natural ability to remediate these two antibiotics (Gujarathi et al., 2005). Increased level of endogenous reactive oxygen species (ROS) in HRCs of *Helianthus annuus* were found when grown in the medium supplemented with tetracycline. This was supposedly associated with the remediation of tetracycline. Further, *in-planta* fate and metabolism of paracetamol was also studied with the help of hairy root cultures of *Armoraria rusticana.* The results showed considerable potential of hairy roots for detoxification of paracetamol and subsequent formation of a paracetamol-glucoside conjugate (Huber et al., 2009).

3.2. Utilization of HRCs in Determining Fate of Transformants

Following enzymatic degradation, the simple forms of organic compounds tend to produce conjugates with cellular biomolecules. Thus, phytotransformation/phytodegradation of toxic molecules facilitates their cellular compartmentation in conjugated complex forms. Phytotransformation products are less toxic and more soluble thus easily polymerise as oligo-, tri-, poly- or form complexes with other cellular metabolites like amino acids, lignin, organic acids etc. Like metal ions and other inorganic pollutant molecules, majority of soluble complexes resulted from phytotransformation get accumulated in vacuoles with the help of ATP-binding cassette (ABC) transporters. However, the transformed products that conjugate with complex polysaccharides produce insoluble complexes and subsequently transport in the form of bound residues and accumulate on cell wall.

In addition to the utility of HRCs in investigating enzymatic degradation of toxic compounds, these *in-vitro* systems have proven their worth to get an insight of product formation, related mechanism and their *in-planta* effect on other cellular activities (Talano et al., 2006; 2010). Sosa-Alderate et al., (2011, 2012) demonstrated oxidative stress responses and lipid peroxidation in wild type, transgenic HRs and non-transgenic HRs of *Nicotiana tabaccum* after phenol treatment. The increased phospholipase activity and phospholipid metabolism in these systems was proposed to be associated with phenol treatment. Till date, little information is available regarding conjugation of phytotransformants with molecular repository of cell. However, *in-planta* existence of different conjugation-detoxification reactions and enzymes may lead to the clue for exact mechanism behind this phase II of xenobiotic detoxification mechanism of plants.

3.2.1. Hyperaccumulation of Heavy Metals

The heavy metal ions that are absorbed from root surface through diffusion are transported with the help of polychelatins and metaliothioneins in the presence of enzyme polychelatin synthase (Cobbett and Goldsbrough 2002). The chelatins are the low molecular weight proteins that have high metal binding properties. They facilitate their transport into vacuole through readily forming chelate-metal complexes. HRCs of *Solanum nigrum* are reported to accumulate 98% of the total exogenously supplemented Zn (Subroto et al., 2007). Some of the plant species are highly tolerant to the pollutants and they can easily propagate in contaminated environment that too without showing any morphological or behavioral change. The pollutant molecules accumulate in the different parts of these plant species. The complexity of heavy metal tolerance involves a synchronized set of molecular regulation and plants tend to neutralize heavy metal stress through overexpression of various stress related proteins. Nevertheless, the plant response to heavy metals may vary from plant to plant and determined by type of metal, physiology of plant, plant age, physical and chemical characteristics of soil etc. (Manara, 2012; Abolghassem et al., 2015). Following uptake and translocation, inter and intracellular immobilization and further compartmentation of hazardous compounds in the cell is the result of different synchronized molecular activities. Though there is a lacuna in comprehensive information, the immobilization through chelation in cytoplasm and vacuole compartmentation has largely been reported (Miller et al., 2008; Sun et al., 2009). During vacuolar compartmentation the heavy metals accumulate either in the form of soluble conjugates or insoluble precipitates and thus, remain separated from cytoplasm and normal metabolic activities. Several species are recognized that hyperaccumulate toxic metal ions with the help of vacuolar compartmentalization. Hairy root cultures of a number of such hyperaccumulator plant species have been developed to investigate the physiology of hyperaccumulation and tolerance/growth performance of plants. Hairy root cultures of *Solanum nigrum* and *Thlaspi caerulescens* have largely been explored for the metal hyperaccumulation and growth performance under severe Zn and Cd toxicity, respectively (Subroto et al., 2007; Nedelkoska and Doran, 2000). Furthermore, the HRCs of *Alyssum bertolonii* for Ni tolerance and hyperaccumulation (Boominathan and Doran, 2003) have been developed and explored. The HRCs of *Brassica juncea* and *Chenopodium amaranticolor* have been established for the removal of exogenously supplemented uranium in the liquid growth medium (Eapen et al., 2003). Hairy roots of *T. caerulescens*, a known Cd hyperaccumulator species, exhibited higher catalase activity both in the presence (>500 μM) and absence of Cd in the medium as compared to the hairy roots of *Nicotiana*, which is a non-hyperaccumulator species (Boominathan and Doran, 2003).

Conclusion, Current Incentives and Future Perspectives

HRCs can be considered as model *in-vitro* system to investigate phytoremediation as they are found to be helpful in understanding plant behavior under xenobiotic stress. HRCs mimic the intrinsic potential of native plants in remediation of organic and inorganic pollutant molecules at different levels. Various root based phenomenon that facilitate phytoremediation such as plant-soil-microbe interaction, role of root exudates, accumulation of pollutant molecules in rhizosphere, phytostabalization, rhizodegradation, rhizofilteration and uptake can be simulated *in-vitro* with the help of hairy root cultures of various plant species. HRCs of a number of tolerant and hyperaccumulator plant species are being initiated and developed to study the mechanism of xenobiotic metabolism, tolerance and hyperaccumulation of pollutant molecules (Figure 8.2).

Exploration of enzymatic degradation of xenobiotics in phase I of xenobiotic metabolism in HRCs of model plants can provide new opportunities to research and understand *in-planta* fate of pollutant molecules. Besides, development of HRCs of heavy metal hyperaccumator plant species can provide better understanding of the mechanism of hyperaccumulation. Molecular insights of xenobiotic management, hyperaccumulation and tolerance by plants can pave the way for genetic modifications in important plant species in order to enhance their inbuilt potential to deal with environmental pollutants. The major challenge that needs attention in using HRCs as remedial system is the implementation of effective leads obtained from these *in-vitro* systems to *in-situ* treatment. Development of hairy root mediated transgenic plants with superior remediation ability and their field cultivation can be a future perspective in this regard. Secondly, the development of microbe assisted phyto and rhizoremedial systems can also be the point of future concern. Further, identification of catabolic pathways, related enzymes and their exploitation through genetic engineering could be a possible way for the improvement of plants with phytoremediation abilities. Inclusion of information gathered in 'omic repositories' of plants of lower taxa, bacteria and fungi would certainly provide additional advantages to this direction.

Hairy root culture mediated phytoremediation is a less-explored area and restricted to the laboratory only. However, in laboratory, a number of studies have been performed to develop transgenic hairy root lines expressing the candidate genes involved in metabolic degradation of pollutant molecules. This not only helps in better understanding of physiology of remediation but also in developing transgenic plants with desired traits such as efficient ability of removal and tolerance of toxic compounds (Abhilash et al., 2009). In this context, transgenic and non-transgenic tomato hairy root cultures have been developed to investigate the removal of phenol supplemented in growth medium (Oller et al., 2005; Gonzalez et al., 2006, 2008). Stable over-expression of peroxidase gene (*tpx 1*) in transgenic tomato hairy root clones reported increased removal of phenolic compounds from the growth medium. Similarly, over-expression of peroxidase genes *tpx1* and *tpx2* in transgenic HRCs of *Nicotiana tabaccum* was found to be associated with greater phenol removal efficiency of roots (Sosa-Alderete et al., 2011). Establishment of *Atropa belladonna* transgenic hairy root clones over expressing mammalian P-4502E1, a mammalian liver enzyme known to metabolize a wide range of toxic pollutants, represent another relevant example in this regard (Banerjee et al., 2002). These transgenic hairy root lines exhibit efficient metabolic activity against trichloroethylene (TCE), a common environmental pollutant. The degradation of TCE was evident from the identification of chloral and trichloroethanol, the common TCE metabolites. The fundamental idea of genetic engineering in hairy roots for enhancing remedial potential lies in the expression of candidate genes that can bestow efficient uptake, translocation and degradation of xenobiotics. The global 'omic repositories' prepared for biodegradation phenomenon at molecular level in lower plants like algae and fungi, microbes and even in animals can be used to retrieve information about effective gene sequences. Additionally, development of transgenic plants with superior ability of xenobiotic remediation through *Agrobacterium rhizogenes* mediated HRCs and transfer of these transgenics to the contaminated sites may give feasible solutions for *in-situ* remediation (Abhilash et al., 2009).

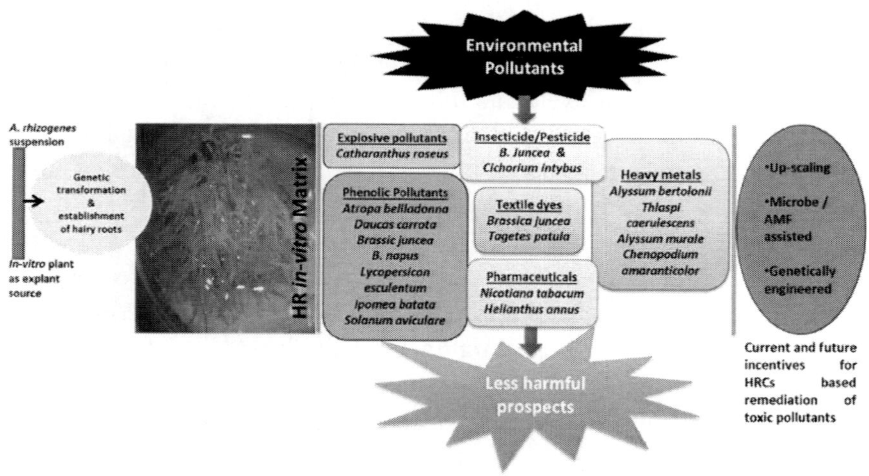

Figure 8.2. Schematic representation of the development of HRCs of different plant species (selected examples) utilized for remediation studies of organic and inorganic pollutants, and current and future incentives of HRCs based phytoremediation.

Microbe assisted phytoremediation: One of the major aspects in plant-based remediation of toxic compounds is phyto-microbial interaction (Abhilash et al., 2012). It is evident that root physiology is greatly influenced by the rhizospheric microflora. These microfloras include symbiotic and free living bacterial and fungal communities present in and/or around rhizosphere. Plant associated microbes like mycorrhiza and plant growth promoting bacteria etc. release several organic and inorganic compounds like organic acids, enzymes and siderophores etc. These compounds have acidifying, oxido-reductive and chelating abilities thus, have been proposed to be directly involved in several rhizosphere based biological activities. These microbial secretions play an important role in metal accumulation in rhizosphere and also facilitate their uptake by plants. Applications of plant-associated microbes have paved the way for promising research in phytoremediation (Nie et al., 2011; Rajkumar et al., 2012). Hairy roots have been successfully explored to study nutrient uptake and their metabolism, signal transduction and other molecular phenomenon behind physiology and biochemistry of plant behavior during root-microbial interactions. However, utilization of HRCs to investigate root-microbe interaction behind phytoremediation is in infancy. Utilizing HRCs of some plant species, few attempts have also been made to simulate phyto-microbial association to understand the mechanism behind microbe-assisted enhancement of remedial potential of plants. In this context, *Brassica napus* HRCs were grown in association of two rhizobacteria, *Burkholderia kururiensis* KP 23 and *Agrobacterium rhizogenes* LBA 9402, which have natural capability to tolerate high concentrations of phenols and dichlorophenols. The presence of both rhizospheric microbes in the medium having *B. napus* hairy roots result enhanced phenol degradation as compared to the *B. napus* hairy roots growing in the medium without microbes (Gonzalez et al., 2013). Further, the HRCs of *B. napus* were also hydroponically grown in association of bacterium from the genus *Pantoea*, the genus of plant growth promoting bacterial species. The bacterium scavenges on phenol for energy and also has the ability to convert chromium IV to Cr III. In this rhizoremedial system, the hairy roots exhibited enhanced phenol removal and Cr accumulation efficiency when compared to hairy roots not inoculated with bacterium. Reports are available that describe that plant roots in association with arbuscular mychorrhizal fungi (AMF) can also be beneficial for phytoremediation. AMFs are a group of obligate biotrophs that develop a symbiotic relation with host plant by facilitating soil, water and mineral uptake in exchange of photosynthetic products from the host. Besides, AMFs are beneficial to host in several other ways including the enhancement of heavy metal tolerance and improving phytoremediation potential of the host plant. Hairy roots have also been explored to study the physiological and molecular mechanism behind the establishment of this plant-AMF association (Sriniwasan et al., 2014). In hairy roots of *Vigna vexillata* and *Daucus carota*, various stages of mycorrhizal infection followed by vesicles and arbuscule formation were investigated through co-culturing of *Glomus microcarpum* and *G. intraradices*, respectively (Chandran, 2010). To understand the mineral and heavy metal uptake through plant root-AMF interaction, few reports are available that describe the co-culturing of hairy roots with AMF. Co-culturing of *Nicotiana tabaccum* HRCs with *Glomus intraradices* prevented the roots from phenol induced oxidative stress (Ibáñez et al., 2011). In this study, over expression of peroxidase coding *tpx1* was also proposed to be associated with root safety mechanism during oxidative stress. Further, *Dacaus carota* hairy roots co-cultivated with *Glomus intraradices* exhibited alleviated Cd-induced growth inhibition (Janoušková and Vosátka, 2005). These studies suggested that co-cultivation of AMF in transgenic hairy roots can be a powerful tool to study its biology. This further offers an easy approach to investigate potential and limits of the monoxenic system in studying the interaction between plant-AMF and heavy metals. Plant-AMF association induces an array of changes in root physiology and nutrient availability, which may determine the outcome of a plant based remediation attempt. However, the enhancement of heavy metal toxicity through plant-AMF association depends upon the type and concentration of heavy metal, soil conditions, fungal symbiotic plant partner and its physiological state. Further, plant-fungal symbiotic association stimulates expression of various plant genes that are supposedly involved in heavy metal detoxification and/or tolerance by plants (Karthikeyan and Kulakow, 2003; Nie et al., 2011). Thus, at present, some serious scientific attention is required to decode the plant-microbe association in favor of enhancing the natural phytoremediation potential of plants. For this purpose, HRCs owing to their uniqueness and resemblance to intact native plant, offer an excellent system to study species specific and microbe specific association and its impact on remediation potential of the root tissues.

Up-scaling of HRCs for incessant and safe treatment of xenobiotic compounds at large scale is another major aspect. One of the major advantages of HRCs based production systems is their possible up-scaling in large bioreactors. Several reports are available for large-scale hairy root cultivation for high biomass production (Kim et al., 2002, Srivastava and Srivastava, et al., 2007; Mishra and Ranjan, 2008; Mehrotra et al., 2015a, b). This root biomass can be further utilized for harvesting of important secondary metabolites of pharmaceutical interest. However, for phytoremediation purposes, the upscale culture of HRCs is still not very well attended. Considering the enzyme-based degradation of xenobiotic compounds at industrial level, the major issue to be addressed is the development of low cost enzyme sources. As enzymes in purified form cannot solve the problem due to the unstable nature thus, utilization of enzymatic potential while it is trapped in living tissue is far more convenient. From this point of view, the HRCs can play a significant role as these cultures are easy to handle at large scale. In this context, tobacco HRCs were used to produce inherent peroxidses on pilot scale for the removal of phenolic substances from industrial effluents (Sosa-Alderate et al., 2012). Another relevant example demonstrated the scale up of *Brassica napus* HRCs with external supplementation of H_2O_2 in the medium resulted efficient 2,4-dichlorophenol removal in a discontinuous stirred tank reactor (Angelini et al., 2010). These kinds of studies may help to develop a strategy for safe and incessant treatment of effluents containing chlorophenols at larger levels.

REFERENCES

Abhilash, P. C., Jamil, S., Singh, N. (2009). Transgenic plants for enhanced biodegradation and phytoremediation of organic xenobiotics. *Biotechnology Advances* 27, 474–488.

Abhilash, P. C., Powell, J. R., Singh, H. B., Singh, B. K. (2012). Plant–microbe interactions: novel applications for exploitation in multipurpose remediation technologies. *Trends in Biotechnology* 30, 416–420.

Abolghassem, E., Yulong, D., Farzad, M., Yinfeng, X. (2015). Heavy metal stress and some mechanisms of plant defense response. *The Scientific World Journal* Article ID 756120, 18 pages doi.org/10.1155/2015/756120.

Agostini, E., Coniglio, M. S., Milrad, S., Tigier, H., Giulietti, A. (2003). Phytoremediation of 2,4-dichlorophenol by *Brassica napus* hairy roots cultures. *Biotechnology and Applied Biochemistry* 37, 139–144.

Agostini, E., Talano, M. A., González, S. P., Oller, A. L. W., Medina, I. M. (2013). Application of hairy roots for phytoremediation: what makes them an interesting tool for this purpose? *Applied Microbiology Biotechnology* 97, 1017–1030.

Alkorta, I., Becerril, J. M., Garbisu, C. (2010). Phytostabilization of metal contaminated soils. Rev Environ Health. 25(2):135-46.

Angelini, V. A., Orejas, J., Medina, M. I., Agostini, E. (2011). Scale up of 2,4-dichlorophenol removal from aqueous solutions using *Brassica napus* hairy roots. *Journal of Hazardous Materials* 85, 269–274.

Arthur, E. L., Rice, P. J., Rice, P. J., Anderson, T. A., Baladi, S. M., Henderson, K. L. D., Coats, J. R. (2005). Phytoremediation - An overview. *Critical Review in Plant Science* 24, 109–122.

Banerjee, S., Singh, S., Rahman, L. U. (2012). Biotransformation studies using hairy root cultures - a review. *Biotechnology Advances* 30, 461–468.

Bhadra, R., Wayment, D. G., Hughes, J. B., Shanks, J. V. (1999). Confirmation of conjugation processes during TNT metabolism by axenic plant roots. *Enviromentl Science and Technology* 33, 446–452.

Boominathan, R., Doran, P. M. (2003). Organic acid complexation, heavy metal distribution and the effect of ATPase inhibition in hairy roots of hyperaccumulator plant species. *Journal of Biotechnology* 101, 131– 146.

Brazier-Hicks, M., Edwards, L. A., Edwards, R. (2007). Selection of plants for roles in phytoremediation: The importance of glucosylation. *Plant Biotechnology Journal* 5, 627–635.

Chandra, S. (2012). Natural plant genetic engineer *Agrobacterium rhizogenes*: role of T-DNA in plant secondary metabolism. *Biotechnology Letters* 34, 407–415.

Chandra, S., Chandra, R. (2011). Engineering secondary metabolite production in hairy roots. *Phytochemistry Review* 10, 371–395.

Chandran, R. P., Potty, V. P. (2010). *In vitro* Co-culture of VAM fungi *Glomus microcarpum* in Ri T-DNA transformed hairy roots of *Vigna vexillata. International Journal of Agricultuarl and Enviromental Biotechnology* 3(3), 333-338.

Christey, M. C., Braun, R. H. (2005). Production of hairy root cultures and transgenic plants by *Agrobacterium rhizogenes*- rhizogenes-mediated transformation. *Methods in Molecular Biology* 286, 47–60.

Clemens, S. (2006). Toxic metal accumulation, responses to exposure and mechanisms of tolerance in plants. *Biochimie* 88, 1707–171919.

Cobbett, C. S. (2000). Phytochelatins and their roles in heavy metal detoxification. *Plant Physiology* 123, 825–832.

Cobbett, C., Goldsbrough, P. (2002). Phytochelatins and metallothioneins: roles in heavy metal detoxification and homeostasis. *Annual Review of Plant Biology* 53, 159-82.

Coniglio, M. S., Busto, V. D., González, P. S., Medina, M. I., Milrad, S., Agostini, E. (2008). Application of *Brassica napus* hairy root cultures for phenol removal from aqueous solutions. Chemosphere 72:1035–1042.

Cunningham, S. D., Ow, D. W. (1996). Promises and prospects of phytoremediation. *Plant Physiology* 110, 715–719.

Dhankher, O. P., Pilon-Smits, E. A. H., Meagher, R. B., Doty, S. (2012). Biotechnological approaches for phytoremediation. In: Altman A, Hasegawa M (eds) Plant biotechnology and agriculture. Academic Press, San Diego, CA, USA, pp. 309-328.

Doran, P. M. (2009). Application of plant tissue cultures in Phytoremediation research: incentives and limitations. *Biotechnology and Bioengineering* 103, 60–76.

Doran, P. M. (2011). Hairy root studies in phytoremediation and phytomining. In: Golubev IA (ed) Handbook of phytoremediation. Nova Science, New York, pp. 591–612.

Doty, S. L., James, C. A., Moore, A. L., Vajzovic, A., Singleton, G. L., Ma, C., Khan, Z., Xin, G., Kang, J. W. et al., (2007). Enhanced phytoremediation of volatile environmental pollutants with transgenic trees. *Proceedings of the National Academy of Sciences USA* 104(43), 16816-16821.

Doty, S. L. (2008). Enhancing phytoremediation through the use of transgenics and endophytes. *New Phytologist* 179, 318–333.

Eapen, S., Singh, S., D'Souza, S. F. (2007). Advances in development of transgenic plants for remediation of xenobiotic pollutants. *Biotechnology Advances* 25, 442-451.

Eapen, S., Suseelan, K. N., Tivarekar, S., Kotwal, S. A., Mitra, R. (2003). Potential for rhizofiltration of uranium using hairy root cultures of *Brassica juncea* and *Chenopodium amaranticolor*. *Environmental Research* 91,127–133.

Garbisu, C., Hernandez-Allica, J., Barrutia, O., Alkorta, I., Becerril, J. M. (2002). Phytoremediation: A technology using green plants to remove contaminants from polluted areas. Rev Environ Health 17, 75–90.

Gelvin, S. B. (2003). Agrobacterium-mediated plant transformation: the biology behind the "gene-jockeying" tool. *Microbiology and Molecular Biology Review* 67, 16–37.

Georgiev, M. I., Agostini, E., Ludwig-Muller, J., Xu, J. (2012). Genetically transformed roots: from plant disease to biotechnological resource. *Trends in Biotechnology*30(10), 528-37.

Georgiev, M. I., Jutta, L. M., Bley, T. (2010). Hairy root culture: copying nature in new bioprocesses. *Medicinal Plant Biotechnology* 156-175.

Georgiev, M. I., Pavlov, A. I., Bley, T. (2007). Hairy root type plant *in vitro* systems as sources of bioactive substances. *Applied Microbiology Biotechnology* 74, 1175–1185.

Gerhardt, K. E., Huang, X. D., Bernard, R., Bruce, M. (2009). Phytoremediation and rhizoremediation of organic soil contaminants: Potential and challenges. *Plant Science* 176, 20–30.

Ghodake, G. S., Telke, A. A., Jadhav, J. P., Govindwar, S. P. (2009). Potential of *Brassica juncea* in order to treat textile effluent contaminated sites. *International Journal of Phytoremediation* 11, 297–312.

Gleba, D., Borisjuk, M. V., Borisjuk, L. G., Kneer, R., Poulev, A., Skarzhinskaya, M., Dushenkov, S., Logendra, S., Gleba, Y. Y., Raskin, I. (1999). Use of plant roots for phytoremediation and molecular farming. *Proceedings of Natural Academy of Science* USA 96, 5973-5977.

González, P. S., Agostini, E., Milrad, S. R. (2008). Comparison of the removal of 2,4-dichlorophenol and phenol from polluted water, by peroxidases from tomato hairy roots, and protective effect of polyethylene glycol. *Chemosphere* 70, 982–989.

González, P. S., Capozucca, C., Tigier, H., Milrad, S., Agostini, E. (2006). Phytoremediation of phenol from wastewater, by peroxidases of tomato hairy root cultures. *Enzyme Microbial Technology* 39, 647–653.

González, P. S., Maglione, G. A., Giordana, M., Paisio, C. E., Talano, M. A., Agostini, E. (2012). Evaluation of phenol detoxification by *Brassica napus* hairy roots, using *Allium cepa* test. *Environmental Science and Pollution Research* 19, 482–491.

González, P. S., Ontañon, O. M., Armendariz, A. L., Talano, M. A., Paisio, C. E., Agostini, E. (2013). *Brassica napus* hairy roots and rhizobacteria for phenolic compounds removal. *Environmental Science and Pollution Research* 20, 1310-1317.

Govindwar, S. P., Kagalkar, A. N. (2010). The use of hydroponics and plant tissue culture technologies for dye degradation. In: Phytoremediation technologies for the removal of textile dyes/; an overview and future prospects. Nova Science Publishers, New York, pp. 27-29.

Guillon, S., Jocelyne, T. G., Pati, P. K., Marc, R., Pascal, G. (2006a). Harnessing the potential of hairy roots: dawn of a new era. *Trends in Biotechnology* 24, 403–409.

Guillon, S., Trémouillaux-Guiller, J., Pati, P. K., Rideau, M., Gantet, P. (2006b). Hairy root research: recent scenario and exciting prospects. *Current Opinion in Plant Biology* 9, 341–346.

Gujarathi, N. P., Haney, B. J., Park, H. J., Wickrama Singhe, S. R., Linden, J. C. (2005). Hairy roots of *Helianthus annuus*: a model system to study phytoremediation of tetracycline and oxytetracycline. *Biotechnology Progress* 21, 775–780.

Hecht, N., Guenzi, W., Beard, W. (2004). Anaerobic conversion of DDT to DDD and aerobic stability of DDT in soil. *Society of America proceedings*, *Soil Science* 32, 522-527.

Huber, C., Bartha, B., Harpaintner, R., Schröder, P. (2009). Metabolism of acetaminophen (paracetamol) in plants-two independent pathways result in the formation of a glutathione and a glucose-conjugate. *Environtal Science and Pollution Research* 16, 206–213.

Hughes, J. B., Shanks, J., Vanderford, M., Lauritzen, J., Bhadra, R. (1997). Transformation of TNT by aquatic plants and plant tissue cultures. *Environtal Science and Technology* 31, 266–271.

Husson, O. (2013). Redox potential (Eh) and pH as drivers of soil/plant/microorganism systems: a transdisciplinary overview pointing to integrative opportunities for agronomy. *Plant and Soil* 362, 389–417.

Ibáñez, S. G., Medina, M. I., Agostini, E. (2011). Phenol tolerance, changes of antioxidative enzymes and cellular damage in transgenic tobacco hairy roots colonized by arbuscular mycorrhizal fungi. *Chemosphere* 83, 700–705.

Janoušková, M., Vosátka, M. (2005). Response to cadmium of *Daucus carota* hairy roots dual cultures with *Glomus intraradices* or *Gigaspora margarita*. *Mycorrhiza* 15, 217–224.

Jha, P., Modi, N., Jobby, R., Desai, N. (2015). Differential expression of antioxidant enzymes during degradation of azo dye Reactive Black 8 in hairy roots of *Physalis minima* L. *International Journal of Phytoremediation*17, 305-312.

Kagalkar, A. N., Jagatap, U. B., Jadhav, J. P., Bapat, V. A., Govindwar, S. P. (2009). Biotechnological strategies for phytoremediation of the sulphonated azo dye Direct Red 5B using *Blumea malcolmii* Hook. *Bioresource Technology* 100, 4104–4110.

Kagalkar, A. N., Jagatap, U. B., Jadhav, J. P., Bapat, V. A., Govindwar, S. P., Bapat, V. A. (2010). Studies on phytoremediation potentiality of *Typhonium flagelliforme* for the degradation of Brilliant Blue R. *Planta* 232, 271–285.

Karthikeyan, R., Kulakow, P. A. (2003). Soil plant microbe interactions in phytoremediation. *Advanced Biochemical Engineering and Biotechnology* 78, 51-74.

Kim, Y. J., Wyslouzil, B. E., Weathers, P. J. (2002). Secondary metabolism of hairy root cultures in bioreactors. *In Vitro Cell Developmental Biology Plant* 38, 1–10.

Krämer, U. (2005). Phytoremediation: novel approaches to cleaning up polluted soils. *Current Opinion in Biotechnology* 16, 133-141.

Kučerová, P., Macková, M., Chromá, L., Burkhard, J., Triska, J., Demnerová, K., Macek, T. (2000). Metabolism of polychlorinated biphenyls by *Solanum nigrum* hairy root clone SNC-9O and analysis of transformation products. *Plant and Soil* 225, 109–115.

Lee, J. H. (2013). An overview of phytoremediation as a potentially promising technology for environmental pollution control. *Biotechnology and Bioprocess Engineering* 18, 431-439.

Lokhande, V. H., Kudale, S., Nikalje, G., Desai, N., Suprasanna, P. (2015). Hairy root induction and phytoremediation of textile dye, Reactive green 19A-HE4BD, in a halophyte, *Sesuvium portulacastrum* (L.) L. *Biotechnol Reports* 8, 56-63.

Macek, T., Francová, K., Kochánková, L., Lovecká, P., Ryslavá, E., Rezek, J., Surá, M., Triska, J., Demnerová, K., Macková, M. (2004). Phytoremediation: Biological Cleaning of a Polluted Environment, *Reviews on Environmental Health* 19, 63-82.

Macek, T., Mackova, M., Kas, J. (2000). Exploitation of plants for the removal of organics in environmental remediation. *Biotechnology Advance* 8, 23–34.

Mackova, M., Macek, T., Kucerova, P., Burkhard, J., Pazlarova, J., Demnerova, K. (1997a). Degradation of polychlorinated biphenyls by hairy root culture *of Solanum nigrum*. *Biotechnology Letters* 19, 787-790.

Macková, M., Macek, T., Ocenaskova, J., Burkhard, J., Demnerová, K., Pazlarová, J. (1997b). Biodegradation of polychlorinated biphenyls by plant cells. *International Biodeterioration and Biodegradation* 39, 317–325.

Manara, A. (2012). Plant responses to heavy metal toxicity. In: Furini A (ed) Plants and Heavy Metals, Briefs in Biometals, Springer, pp, 27-53.

Marmiroli, N., Marmiroli, M., Maestri, E. (2006). Phytoremediation and phytotechnologies: a review for the present and the future. In: Twardowska, I. et al., (eds) Soil and Water Pollution Monitoring, Protection and emediation.Springer Netherlands, pp, 3–23.

Mazaheri, H., &Piri, K. (2015). Removal of phenol by *A. belladonna* L. hairy root. *International Journal of Phytoremediation* 17, 1212-1219.

Meagher, R. B. (2000). Phytoremediation of toxic elemental and organic pollutants. *Current Opinion in Plant Biology* 3, 153–162.

Mehrotra, S., Rahman, L. U., Kukreja, A. K. (2010). An extensive case study of hairy-root cultures for enhanced secondary-metabolite production through metabolic-pathway engineering. *Biotechnology and Applied Biochemistry* 56, 161–172.

Mehrotra, S., Srivastava, V., Rahman, L. U., Kukreja, A. K. (2015). Hairy root biotechnology--indicative timeline to understand missing links and future outlook. *Protoplasma*. 252, 1189-1201.

Mehrotra, S., Goel, M. K., Srivastava, V., Rahman, L. U. (2015). Hairy root biotechnology of *Rauwolfia serpentina*: a potent approach for the production of pharmaceutically important terpenoidindole alkaloids. *Biotechnology Letters* 37, 253-263.

Miller, G., Gregorio, B., Maria, B., Jennifer, N., Oscar, H. (2008). Assessment of the Efficacy of Chelate-Assisted Phytoextraction of Lead by Coffeeweed (*Sesbania exaltata* Raf.). *International Journal of Environmental Research* 5, 428–435.

Mishra, B. N., Ranjan, R. (2008). Growth of hairy-root cultures in various bioreactors for the production of secondary metabolites. *Biotechnology and Applied Biochemistry* 49, 1–10.

Morrissey, J., Guerinot, M. L. (2009). Iron uptake and transport in plants: the good, the bad, and the lonome. *Chemical Reviews*109, 4553–4567.

Mwangi, K., Hamadi, I. B., Anne, W. M., Ciira, K. K., Muniru, K. T. (2010). Degradation of dichlorodiphenyltrichloroethane (DDT) by bacterial isolates from cultivated and uncultivated Soil. *African Journal of Microbiology Research* 4, 185-196.

Nedelkoska, T. V., Doran, P. M. (2000). Hyperaccumulation of cadmium by hairy roots of *Thlaspi caerulescens*. *Biotechnology and Bioengineering* 67,607–615.

Nepovim, A., Podlipná, R., Soudek, P., Schröder, P., Vaněk, T. (2004). Effects of heavy metals and nitroaromatic compounds on horseradish glutathione S-transferase and peroxidase. *Chemosphere* 57, 1007–1015.

Nie, M., Yijing, W., Jiayi, Y., Ming, X., Lifen, J., Ji, Y., Changming, F., Jiakuan, C., Bo, L. (2011). Understanding plant-microbe interactions for phytoremediation of petroleum-polluted soil. *PLOS One* doi:10.1371/journal.pone.0017961

Oller, A. L. W., Agostini, E., Talano, M. A., Capozucca, C., Milrad, S. R., Tigier, H. A., Medina, M. I. (2005). Overexpression of a basic peroxidase in transgenic tomato (*Lycopersicon esculentum* Mill. cv. Pera) hairy roots increases phytoremediation of phenol. *Plant Science* 169, 1102–1111.

Ono, N. N., Li, T. (2011). The multiplicity of hairy root cultures: prolific possibilities. *Plant Science* 180, 439–446.

Patil, P., Desai, N., Govindwar, S., Jadhav, J. P., Bapat, V. (2009). Degradation analysis of Reactive Red 198 by hairy roots of *Tagetes patula* L. (Marigold). *Planta* 230, 725–735.

Rajkumar, M., Sandhya, S., Prasad, M. N., Freitas, H. (2012). Perspectives of plant-associated microbes in heavy metal phytoremediation. *Biotechnology Advance* 30, 1562-1574.

Reddy, C. N., Patrick, W. H. (1977). Effect of redox potential and ph on the uptake of cadmium and lead by rice plants. *Journal of Environmental Quality* 6, 259-262.

Rezek, J., Macek, T., Macková, M., Triska, J. (2007). Plant metabolites of polychlorinated biphenyls in hairy root culture of black nightshade *Solanum nigrum* SNC-9O. *Chemosphere* 69, 1221–1227.

Rezek, J., Macek, T., Mackova, M., Triska, J., Ruzickova, K. (2008). Hydroxy-PCBs, methoxy-PCBs and hydroxy–methoxy-PCBs: metabolites of polychlorinated biphenyls formed in vitro by tobacco cells. *Environmental Science and Technology* 42, 5746–5751.

Sandermann, H. (1992). Plant metabolism of xenobiotics. *Trends in Biochemical Science* 17, 82–84.

Sandermann, H. (1994). Higher plant metabolism of xenobiotics: the "green liver" concept. *Pharmacogenetics* 4, 225–241.

Santos de Araujo, B., Dec, J., Bollag, J. M., Pletsch, M. (2006). Uptake and transformation of phenol and chlorophenols by hairy root cultures of *Daucus carota, Ipomoea batatas* and *Solanum aviculare*. *Chemosphere* 63, 642–651.

Santos de Araujo, B., Omena de Oliveira, J., Salgueiro Machado, S., Plestch, M. (2004). Comparative studies of the peroxidases from hairy roots of *Daucus carota, Ipomea batatas and Solanum aviculare*. *Plant Science* 167, 1151–1157.

Schröder, P., Scheer, C. E., Diekmann, F., Stampfl, A. (2007). How plants cope with foreign compounds: translocation of xenobiotic glutathione conjugates in roots of barley (*Hordeum vulgare*). *Environmental Science and Pollution Research* 14, 114–122.

Singh, S., Melo, J. S., Eapen, S., D'Souza, S. F. (2006). Phenol removal using *Brassica juncea* hairy roots: role of inherent peroxidase and H_2O_2. *Journal of Biotechnology* 123, 43–49.

Sosa-Alderete, L. G., Agostini, E., Medina, M. I. (2011). Antioxidant response of tobacco (*Nicotiana tabacum*) hairy roots after phenol treatment. *Plant Physiology and Biochemistry* 49, 1020–1028.

Sosa-Alderete, L. G., Ibáñez, S. G., Agostini, E., Medina, M. I. (2012a). Phytoremediation of phenol at pilot scale by tobacco hairy roots. *International Journal of Environmental Science* 3, 398–407.

Sosa-Alderete, L. G., Racagni, G., Agostini, E., Medina, M. I. (2012b). Phospholipid turnover and phospholipase D activity in tobacco hairy roots exposed to phenol. *Environmental and Experimental Botany* 77, 141–145.

Sosa-Alderete, L. G., Talano, M. A., Ibañez, S. G., Purro, S., Agostini, E., Milrad, S., Medina, M. I. (2009). Establishment of transgenic tobacco hairy roots expressing basic peroxidases and its application for phenol removal. *Journal of Biotechnology* 139, 273–79.

Srinivasan, M., Kumar, K., Kumutha, K., Marimuthu, P. (2014). Establishing monoxenic culture of arbuscular mycorrhizal fungus *Glomus intraradices* through root organ culture. *Journal of Applied Natural Science* 6, 290-293.

Srivastava, S., Srivastava, A. K. (2007). Hairy root culture for mass production of high-value secondary metabolites. *Critical Review in Biotechnology* 27, 29–43.

Srivastava, V. (2015). Studies on hairy roots as bioreactors for normal and elicited biotransformation of exogenous molecules. Ph.D thesis submitted to Department of Botany, Lucknow University, Lucknow, India.

Srivastava, V., Mehrotra S., Mishra S. (2016). Biotransformation through hairy roots: perspectives, outcomes, and major Challenges. In: Jha S (ed) Transgenesis and Secondary Metabolism, Reference Series in Phytochemistry, Springer International Publishing ©.Switzerland, pp, 1-24.

Subroto, M. A., Priambodo, S., Indrasti, N. S. (2007). Accumulation of zinc by hairy root cultures of *Solanum nigrum*. *Biotechnology* 6,344–348.

Sun, Y. B., Zhou, Q. X., Jing, A., Wei-tao, L., Rui, L. (2009). Chelator-enhanced phytoextraction of heavy metals from contaminated soil irrigated by industrial wastewater with the hyperaccumulator plant (*Sedum alfredii* Hance). *Geoderma* 150,106–112.

Suresh, B., Ravishankar, G. A. (2004). Phytoremediation--a novel and promising approach for environmental clean-up. *Critical Reviews in Biotechnology* 24, 97-124.

Suresh, B., Sherkhane, P. D., Kale, S., Eapen, S., Ravishankar, G. A. (2005). Uptake and degradation of DDT by hairy root cultures of *Cichorium intybus* and *Brassica juncea*. *Chemosphere* 61, 1288–1292.

Talano, M. A., Agostini, E., Medina, M. I., Reinoso, H., Tordable, M. C., Tigier, H. A., de Forchetti, S. M. (2006). Changes in ligno-suberization of cell walls of tomato hairy roots produced by salt treatment: the relationship with the release of a basic peroxidase. *Journal of Plant Physiology* 163, 740–749.

Talano, M. A., Frontera, S., González, P., Medina, M. I., Agostini, E. (2010). Removal of 2,4-diclorophenol from aqueous solutions using tobacco hairy root cultures. *Journal of Hazardous Materials* 176, 784–91.

Tangahu, B. V., Sheikh Abdullah, S. R., Hassan, B., Mushrifah, I., Nurina, A., Muhammad, M. (2011). A review on heavy metals (as, pb, and hg) uptake by plants through phytoremediation. *International Journal of Chemical Engineering* Article ID 939161, 31 pages http://dx.doi.org/10.1155/2011/939161.

Tejada-Jimenez, M., Rosario, C. R., Igor, K., Mercedes Lucas, M., Michael, U., Juan, I., Gonzalez, G. M. (2015). *Medicago truncatula* natural resistance-associated macrophage protein1 is required for iron uptake by rhizobia-infected nodule cells.*Plant Physiology* 168, 258–272.

Telke, A. A., Kagalkar, A. N., Jagtap, U. B., Desai, N. S., Bapat, V. A., Govindwar, S. P. (2011). Biochemical characterization of laccase from hairy root culture of *Brassica juncea* L. and role of redox mediators to enhance its potential for the decolorization of textile dyes. *Planta* 234, 137–1149.

Tepfer, D. (1989). Ri T-DNA from *Agrobacterium rhizogenes*: a source of genes having applications in rhizosphere biology and plant development, ecology and evolution. In: Kosuge T, Nester E (eds), Plant–microbe interactions. McGraw-Hill, New York, pp 294–342.

Römheld, V. (1991). The role of phytosiderophores in acquisition of iron and other micronutrients in graminaceous species: An ecological approach. *Plant and Soil*130, 127-134.

Veena, V., Taylor, C. G. (2007). Agrobacterium rhizogenes: recent developments and promising applications. *In Vitro Cell Developmental Biology - Plant* 43, 383–403.

Wu, G., Kang, H. B., Zhang, X. Y., Shao, H. B., Chu, L. Y., Ruan, C. J. (2010). A critical review on the bioremoval of hazardous heavy metals from contaminated soils: issues, progress, eco-environmental concerns and opportunities. *Journal of Hazardous Materials* 174, 1–8.

Xingmao, M., Burken, J. G. (2003). TCE diffusion to the atmosphere in phytoremediation applications. *Environmental Science and Technology* 37, 2534-2539.

Yehuda, Z., Moshe, S., Volker, R., Horst, M., Yitzhak, H., Yona, C. (1996). The role of ligand exchange in the uptake of iron from microbial siderophores by Cramineous plants 1r2.*Plant Physiology* 11, 1273-1280.

Zhou, M. L., Zhu, X. M., Shao, J. R., Tang, J. R., Wu, Y. M. (2011). Production and metabolic engineering of bioactive substances in plant hairy root culture. *Applied Microbiology and Biotechnology* 90, 1229–1239.

Zhou, M. L., Tang, Y. X., Wu, Y. M. (2013). Plant hairy roots for remediation of aqueous pollutants. *Plant Molecular Biology Reporter* 31, 1–8.

In: Chemical Pollution Control …
Editor: Naser A. Anjum
ISBN: 978-1-53611-034-0

Chapter 9

CHEMICAL POLLUTION CONTROL WITH MICROORGANISMS: APPRAISAL AND PERSPECTIVES

Naser A. Anjum*
CESAM-Centre for Environmental and Marine Studies,
Department of Chemistry, University of Aveiro, Aveiro, Portugal

1. APPRAISAL

Air, water and soil can be impacted by a myriad of chemical pollutants. Notably, the health of animal and human significantly depends on the quality of air, water and soil. Therefore, the level of chemical pollutants in these compartments must be kept at minimum. The volume entitled '*Chemical Pollution Control with Microorganisms*' made an effort to discuss various aspecpts related with the theme.

Exhaustive lists of major chemical pollutants removed by microorganisms including bacteria and fungi, and the microbial processes involved in remediation of contaminants were presented (*Chapter 2*). Metals, nutrients and radionuclides are the major inorganic contaminants removed by microbes. However, the types of species, adsorbate, and the physical conditions were

* Corresponding author: Naser A. Anjum, CESAM-Centre for Environmental and Marine Studies, Department of Chemistry, University of Aveiro, 3810-193 Aveiro, Portugal; E-mail: anjum@ua.pt

argued as major factors controlling the rate of degradation or transformation of contaminants. In particular, aerobic conditions favor the effective degradation of most of the organic contaminants (such as pesticides, hydrocarbons and aromatic compounds). Non-living and living biomass of bacteria and fungi have great potential to remove heavy metals, where metals can either be immobilized or mobilized by microbes. Immobilization of heavy metals can reduce them to a lower redox state and make them less bioactive. Leaching, chelation, methylation and redox transformation are involved in the mobilization process. The bacterial transformation of metals involves different reactions including methylation, precipitation, redox reactions, and dissimilatory sulfate reduction pathway. Fungal cell walls can bind metals through chitins, glucans, mannans, proteins, polysaccharides, and lipids. Bacteria and fungi have the potential to remove radionuclides from the environments, where the uptake of radionuclides is modulated by the cell surface, structure and composition of bacterial species. The radionuclides can be transformed via reductive precipitation mediated by enzymes followed by precipitation within the cell; chelation and immobilization by compounds such as organic acids, siderophores and extracellular metabolites; and change of ionic state. Nitrogen fixing and phosphate solubilizing bacteria can remove nitrogen and phosphorus, respectively. Fungi-mediated removal of nitrate involves nitrification and denitrification pathways. Microorganisms have the capacity to degrade organic compounds such as hydrocarbons, pesticides, solvents, explosives and dioxins. Biodegradation of toxic organic compounds can also be possible plant associated bacteria namely endophytes and rhizophytes. Microbes also possess the ability to completely mineralize aliphatic, aromatic, and heterocyclic hydrocarbons. Having proteolytic or keratinolytic activity bacteria and fungi can remove many volatile organic compounds. Moreover, anaerobic and aerobic bacteria can perform biotransformation of polychlorinated biphenyls.

Outstanding role of Glomeromycota fungi and the associated microflora in the biodegradation of oily sludge in artificially contaminated soil was discussed (*Chapter 3*). Evidence was presented about the rhizodegradation of petrolized sludge in the rhizosphere of *Brachiaria brizanta.* Microorganisms have immense inherent capability to control the trace metal levels in aquatic system. Trace metal pollution in aquatic system was overviewed, and the role and underlying mechanisms of varied microorganisms-mediated control (sorption, degradation, transformation and metabolism) of major trace metals in aquatic system was discussed (*Chapter 4*). Biochemical insights into the rhizospheric zone and the microbial community therein, were discussed and implicated for the remediation of chemical pollutants (*Chapter 5*). Rhizospheric microbial community can modulate the growth, nutrition, development, their susceptibility to disease, and resistance to heavy metals, and their potential to degrade xenobiotics. Critial evaluation of the pollutant′s remediation role and underlying mechanisms of ‘plant–bacteria partnerships’ was performed (*Chapters 6 and 7*). Rhizospheric dewelling PGPRs play significant role in providing a base for contaminants in soil and plant to interact during the process of phytoremediation. Notably, during the phytoremediation process, the plant growth promoting rhizobacterias can release chelating and bioleaching agents in the rizospheric region, and cause acidification, enzymatic reduction and redox changes enhancing the mobility and availability of chemical pollutants. Considering the versatility and close resemblence of hairy root cultures to the intact plants was argued as a potential matrix for the remediation of pollutants of various types (*Chapter 8*). Facilitation of the studies on the catabolic pathways and identification of related enzymes involved in pollutant metabolism could be possible via hairy root cultures.

2. PERSPECTIVES

The cleanup of polluted environmental compartments with the plant-microbial interaction has been emerged as a recent outshoot of bioremediation technology. A lost has been done in this regard; however, a number of unattended or least explored aspects require immediate attention. Knowledge of heavy metals resistance and their generic deviations is essential for designing an effective bioremediation strategy. Genetic engineering can be used to integrate genes into bacteria in order to enhance heavy metals resistance and accumulation. However, very little literature exists on the expression of heavy metal resistant genes. Development of transgenic bacterial system effectively expressing heavy metals resistant gene for bioremediation can accelerate microbial-assisted remediation of chemical pollutants. The role of microorganisms in the minimization of oil spills in soil and water has been least explored. Hence, the studies on the role and underlying mechanisms of arbuscular mycorrhizae in phytoremediation of oil spills or other hydrocarbon debris in tropical polluted soils will be rewarding. New insights on the plant-microbial interaction-assisted minimization of chemical pollutants can be benefited with the optimization of the process with the plants potentially fast growing, better metal uptake ability and high biomass along with their suitability for a range of microbial diversity in their rhizosphere. Translation of the outcomes from the laboratory/simulated studies, into field studies can demonstrate the usefulness of the approach. Studies on the selection and improvement in the expression of catabolic genes in the rhizosphere need to be intensified. Plant growth promoting bacteria residing in the rhizosphere/endosphere have enough potential to protect plants against the impact of chemical pollutants, and also solubilize or stabilize pollutants in the soil system. Extensive reports are available on the role of plant-bacteria partnership on organic and metal pollutants. However, information is scanty on the plant-bacteria interaction on metalloids and radionuclides remediation. Additionally, our understanding on the mechanisms involved in the process of bacteria assisted phytoremediation still poor. Moreover, it would be imperative to exploit the bench or pot scale results into more diverse ecological conditions at field scale. Hairy root cultures have been explored to study the physiological and molecular mechanism behind the establishment of this plant-AMF association, and its role in the uptake of minerals and heavy metals. Development of strategies for safe and incessant treatment of effluents containing chlorophenols at larger levels can also be possible with hairy root cultures. However, for phytoremediation purposes, the upscale culture of hairy root cultures is still not very well attented. Harvesting of important secondary metabolites of pharmaceutical interest can be another use of the hairy root cultures.

About the Editor

Dr. Naser A. Anjum is associated with CESAM-Centre for Environmental and Marine Studies, University of Aveiro, Portugal. Dr. Anjum´s research interests include: (a) Insights into the biology of crop/plant-environment interaction, (b) Strategies for boosting crop-stress defense, and (c) Sustainable tools for the minimization of emerging environmental problems/effects. Dr. Anjum has to his credit more than 70 SCI peer-review publications (Total citations: 900, H-index = 18 at SCOPUS; Google H-index: 22 and i-10 index: 48; Citations: 1674) (https://www.scopus.com/authid/detail.uri?authorId=23097123400) and several edited volumes including: (a) Ascorbate-Glutathione Pathway and Stress Tolerance in Plants (2010), Springer, The Netherlands, (b) Plant Adaptation to Environmental Change: Significance of Amino Acids and Their Derivatives (2014), CABI, UK, (c) Phytotechnologies: Remediation of Environmental Contaminants (2012), CRC Press, USA, and (d) The Plant Family Brassicaceae: Contribution Towards Phytoremediation (2012), Springer, The Netherlands. Dr. Anjum is serving as an Associate/Assistant/Managing Editor in reputed, specialized and multidisciplinary journals including Acta Physiologiae Plantarum, Plant Growth Regulation, Brazilian Journal of Botany, Journal of Metallomics and Nanotechnology, and Frontiers in Marine Science. He is also a life member of Indian Society for Plant Physiology (ISPP), and National Environmental Science Academy (NESA), New Delhi.

INDEX

D

E

F

G

H

I

J

K

L

M

N

O

P

Q

R

S

T

U

V

W

X

Y

Z